Mathematical Thinking

Problem-Solving and Proofs

John P. D'Angelo
Douglas B. West
University of Illinois — Urbana

PRENTICE HALL
Upper Saddle River, NJ 07458

Library of Congress Cataloging-in-Publication Data

D'Angelo, John Philip, and West, Douglas Brent
    Mathematical Thinking: Problem-Solving and Proofs
    John P. D'Angelo and Douglas B. West
       p.   cm.
    Includes index.
    ISBN 0-13-263393-0
    1. Mathematical Thinking. I. Title
    QA166.D   1997
    511'.5--dc20   95-24773
                                                    CIP

Acquisitions Editor: George Lobell
Editorial Assistant: Gale Epps
Editorial Director: Tim Bozik
Editor-in-Chief: Jerome Grant
Assistant Vice President of Production and Manufacturing: David W. Riccardi
Editorial/Production Supervision: Robert C. Walters
Managing Editor: Linda Mihatov Behrens
Executive Managing Editor: Kathleen Schiaparelli
Manufacturing Buyer: Alan Fischer
Manufacturing Manager: Trudy Pisciotti
Creative Director: Paula Maylahn
Art Director: Jayne Conte
Cover Designer: Bruce Kenselaar
Cover Photo: Terrence Moore/KB

Printed in the United States of America

10 9 8 7 6 5 4 3 2

ISBN 0-13-263393-0

PRENTICE-HALL INTERNATIONAL (UK) LIMITED, *London*
PRENTICE-HALL OF AUSTRALIA PTY. LIMITED, *Sydney*
PRENTICE-HALL CANADA INC., *Toronto*
PRENTICE-HALL HISPANOAMERICANA S.A., *Mexico*
PRENTICE-HALL OF INDIA PRIVATE LIMITED, *New Delhi*
PRENTICE-HALL OF JAPAN INC., *Tokyo*
SIMON & SCHUSTER ASIA PTE. LTD., *Singapore*
EDITORA PRENTICE-HALL DO BRASIL, LTDA. , *Rio de Janeiro*

To all who enjoy mathematical puzzles,
and to our loved ones,
who tolerate our enjoyment of them.

Contents

Preface for the Student

This book aims to be both enjoyable and demanding. We present interesting problems and develop the basic undergraduate mathematics needed to solve them. Below we list 36 such problems. We solve most of these in this book, while at the same time developing enough theory to prepare you for upper-division math courses.

This course will differ from other math courses you have taken, because it emphasizes writing and language skills. We do not ask that you memorize formulas, but rather that you learn to express yourself clearly and accurately. You will learn to solve mathematical puzzles as well as to write proofs of theorems from elementary algebra, discrete mathematics, and calculus. This will broaden your knowledge and improve the clarity of your thinking.

How can you improve your writing? Good writing requires practice. Rereading and revising solutions can improve your presentation. You must say what you mean and mean what you say. Mathematics offers a tremendous opportunity for this, because clear decisions can be made about whether sentences contain faulty reasoning. Mathematics uses formulas to express complicated thoughts, and you will learn how to combine well chosen notation with clear explanation in sentences. This will enable you to communicate ideas concisely and accurately.

We invite you to consider some intriguing problems. Solutions to most appear in the text, and we include the others as exercises.

1. Given several piles of pennies, we create a new collection by removing one coin from each old pile to make one new pile. Each original pile shrinks by one; 1,1,2,5 becomes 1,4,4, for example. Which lists of sizes (order is unimportant) are unchanged under this operation?

2. Which natural numbers are sums of consecutive smaller natural numbers? For example, $30 = 9 + 10 + 11$ and $31 = 15 + 16$, but 32 has no such representation.

3. Including squares of all sizes (one-by-one through eight-by-eight), an ordinary eight-by-eight checkerboard has 204 squares. How many squares of all sizes arise using an n-by-n checkerboard? How many triangles of all sizes arise using a triangular grid with sides of length n?

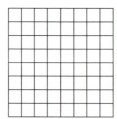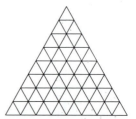

4. At a party with five married couples, no person shakes hands with his or her spouse. Of the nine people other than the host, no two shake hands with the same number of people. With how many people does the hostess shake hands?

5. Is it possible to fill the large region below with non-overlapping copies of the small L-shape? Rotations and translations are allowed.

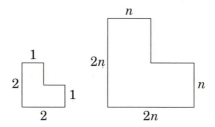

6. We can tell whether two collections of weights have the same total weight by placing them on a balance scale. How many known weights are needed to balance each integer weight from 1 to 121? How should these weights be chosen? (Known weights can be placed on either side or omitted.)

7. If each resident of New York City has 100 coins in a jar, is it possible that no two residents have the same number of coins of each type (pennies, nickels, dimes, quarters, half-dollars)?

8. How can we find the greatest common divisor of two large numbers without factoring them?

9. Why are there infinitely many prime numbers? Why are there arbitrarily long stretches of consecutive non-prime positive integers?

10. Consider a dart board having two regions, one worth a points and the other worth b points, where a and b are positive integers having no common factors. What is the largest point total that cannot be obtained by throwing darts at the board?

11. A math professor cashes a check for x dollars and y cents, but the teller inadvertently pays y dollars and x cents. After the professor buys a newspaper for k cents, the remaining money is twice as much as the original value of the check. If $k = 50$, what was the value of the check? If $k = 75$, why is this situation impossible?

12. Must there be a Friday the 13th in every year?

13. When two digits in the base 10 representation of an integer are interchanged, the difference between the old number and the new number is divisible by nine. Why?

14. A positive integer is *palindromic* if reversing the digits of its base 10 representation doesn't change the number. Why is every palindromic integer with an even number of digits divisible by 11? What happens in other bases?

15. How can one describe all the integer solutions to $42x + 63y = z$, or to $x^2 + y^2 = z^2$?

16. Suppose L is a prime number. For which positive integers K can we express the rational number K/L as the sum of the reciprocals of two positive integers?

17. Are there more rational numbers than integers? Are there more real numbers than rational numbers? What does "more" mean for infinite sets?

18. Can player A have a higher batting average than player B in day games and in night games but a lower batting average than player B over all games?

Player	Day	Night	Overall
A	.333	.250	.286
B	.300	.200	.290

19. Suppose A and B gamble as follows: On each play, each player shows 1 or 2 fingers, and one pays the other x dollars, where x is the total number of fingers showing. If x is odd, then A pays B; if x is even, then B pays A. Who has the advantage, and how can that player exploit it?

20. Given a positive integer k, how can we obtain a formula for the sum $1^k + 2^k + \cdots + n^k$?

21. Suppose candidates A and B in an election receive a and b votes, respectively. If the votes are counted in a random order, what is the probability that candidate A never trails?

22. Can the numbers $0, \ldots, 100$ be written in some order so that no 11 positions contain numbers that successively increase or successively decrease? (An increasing or decreasing set need not occupy consecutive positions or use consecutive numbers.)

23. Suppose each dot in an n by n grid of dots is colored black or white. How large must n be to guarantee the existence of a rectangle whose corners have the same color?

24. How many positive integers less than 1,000,000 have no common factors with 1,000,000?

25. Suppose n students take an exam, and the exam papers are handed back at random for peer grading. What is the probability that no student gets his or her own paper back? What happens to this probability as n goes to infinity?

26. A computer plotter is to draw a figure on a page. How can one determine the minimum number of times the pen must be lifted while drawing the figure?

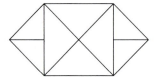

27. Suppose there are n girls and n boys at a party, and each girl likes some of the boys. Under what conditions is it possible to pair the girls with boys so that each girl is paired with a boy that she likes?

28. Suppose n points lie on a circle. How many regions are created by drawing all chords joining these points, assuming that no three chords have a common intersection?

29. A Platonic solid has congruent regular polygons as faces and has the same number of faces meeting at each vertex. Why are the tetrahedron, cube, octahedron, dodecahedron, and icosahedron the only ones?

30. Suppose n spaces are available for parking along the side of a street. We can fill the spaces using Rabbits, which take one space, and/or Cadillacs, which take two spaces. In how many ways can we fill the spaces? In other words, how many lists of 1's and 2's sum to n?

31. Repeatedly pushing the "x^2" button on a calculator generates a sequence tending to 0 if the initial positive value is less than 1 and tending to ∞ if it is greater than 1. What happens with other quadratic functions?

32. What numbers have more than one decimal representation?

33. Suppose that the points in a tennis game are independent and that the server wins each point with probability p. What is the probability that the server wins the game?

34. How is $\lim_{n \to \infty}(1 + x/n)^n$ relevant to compound interest?

35. One type of baseball player hits singles with probability p and otherwise strikes out. Another type hits home runs with probability $p/4$ and otherwise strikes out. Assume that a single advances each runner by two bases. Compare a team composed of the home-run hitters with a team composed of the singles hitters. Which team generates more runs per inning?

36. Suppose two jewel thieves steal a circular necklace with $2m$ gold beads and $2n$ silver beads arranged in some unknown order. Why is it that, for any arrangement of the beads, there is a way to cut the necklace along some diameter so that each thief gets half the beads of each color? Why is it that a heated circular wire always contains two diametrically opposite points where the temperature is the same? How are these questions related?

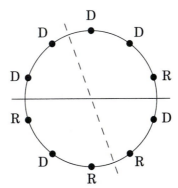

Preface for the Instructor

This book arose from discussions about the undergraduate mathematics curriculum. We asked several questions. Why do students find it difficult to write proofs? What is the role of discrete mathematics? How can the curriculum better integrate diverse topics? Perhaps most important, why don't students enjoy and appreciate mathematics as much as we might hope?

Upper-division courses in mathematics expose serious gaps in the preparation of students; the difficulties are particularly evident in elementary real analysis courses. Such courses present two obstacles to students. First, the concepts of elementary analysis are subtle; it took mathematicians centuries to understand limits. Second, proofs require both attention to exposition and a different intellectual attitude from computation. The combination of the two difficulties has defeated many students. Basic courses in linear or abstract algebra evince similar difficulties and can be overly formal. Due to their specialized focus, upper-class courses often fail to address adequately the need for careful exposition. If students first learn techniques of proof and habits of careful exposition, then they will better appreciate more advanced mathematics when they encounter it.

The excitement of mathematics springs from engaging problems. Students have natural mathematical curiosity about problems such as those listed in the Preface for the Student. They then care about the techniques used to solve them; hence we use these problems as a focus of development. We hope that students and instructors will enjoy this approach as much as we have.

A course introducing techniques of proof should not specialize in one area of mathematics; later courses offer ample opportunities for specialization. This book considers diverse problems and demonstrates relationships among several areas of mathematics. One of the authors studies complex analysis in several variables, the other studies discrete mathematics. We explored the interactions between discrete and continuous mathematics to create a course on problem-solving and proofs.

Content

We present elementary aspects of algebra and number theory, combinatorics, and analysis. We develop such diverse topics as prime factorization, modular arithmetic, Pythagorean triples, techniques of counting, basic graph theory, recurrence relations, sequences and series, the basic theorems of calculus, continuous nowhere differentiable functions, and the fundamental theorem of algebra. We integrate these topics into a coherent whole, choosing material that illustrates techniques of proof and interactions among the topics.

Part I (Elementary Concepts) begins by deriving the quadratic formula and using it to motivate the axioms for the real numbers, which we agree to assume. We discuss sets, logical statements, and functions, paying careful attention to the use of language. The highlight of Part I is the application of induction to several engaging problems.

Part II (Properties of Numbers) studies the number systems $\mathbb{N}$, $\mathbb{Z}$, and $\mathbb{Q}$. We explore q-ary expansions, cardinality, binomial coefficients, the Euclidean algorithm, and prime factorization. Equivalence relations provide the foundation for our development of modular arithmetic and the rational numbers. Features include the Schroeder-Bernstein Theorem, Fermat's Little Theorem, the Chinese Remainder Theorem, criteria for irrationality, Pythagorean triples, Simpson's Paradox, and a bit of probability.

Part III (Discrete Mathematics) explores combinatorial arguments. We consider elementary enumeration, the pigeonhole principle, the inclusion-exclusion principle, graphs, and recurrence relations. Highlights include Bertrand's Ballot Problem (Catalan numbers), more on probability, the Euler totient function, Hall's Theorem on systems of distinct representatives, Platonic solids, and the Fibonacci numbers. Combinatorial problems lead us to recurrence relations and sequences. We develop various techniques to solve recurrences. Familiarity with sequences facilitates the transition to continuous mathematics.

Part IV (Continuous Mathematics) begins with the Least Upper Bound Property for the real numbers. We prove the Bolzano-Weierstrass Theorem and use it to prove that Cauchy sequences converge. We then develop the theory of calculus: sequences, series, continuity, differentiation, uniform convergence, and the Riemann integral. We define the natural logarithm via integration and the exponential function via infinite series, and we prove their inverse relationship. We define trigonometric functions via infinite series, using results on interchange of limiting operations to verify their properties. We include Cantor's proof that $\mathbb{R}$ is uncountable, convexity, and van der Waarden's example of a continuous and nowhere differentiable function. We omit many applications covered adequately in calculus courses, such as Taylor polynomials, analytic geometry, Kepler's laws, polar coordinates, and many of the physical interpretations of derivatives and integrals. We close by developing the properties of complex numbers and proving the Fundamental Theorem of Algebra.

In Appendix A we develop the properties of arithmetic and construct the real number system using Cauchy sequences. There we begin with $\mathbb{N}$ and subsequently construct $\mathbb{Z}$, $\mathbb{Q}$, and $\mathbb{R}$. We include a portion of the construction of $\mathbb{Q}$ in Chapter 8 in order to illustrate the fundamental role of equivalence relations.

Pedagogy

Certain pedagogical issues require careful attention. In order to benefit from this course, students must have a sense of intellectual progress. An axiomatic treatment of the real numbers seems painfully slow and frustrates students. They have learned algebraic computational techniques throughout their schooling, and we want to build on this foundation. This dictates our starting point. In Chapter 1 we list the axioms for the real numbers and their elementary algebraic consequences, and we accept them for computation and reasoning. We defer the construction of the real numbers and verification of the field axioms to Appendix A, for later appreciation. We generally do not assign the exercises in Chapter 1 that request verification of algebraic properties from the axioms.

Chapters 2 and 3 discuss elementary aspects of quantifiers, mathematical language, and functions. This material provides the language for all subsequent discussion. Formal discussion of mathematical language is problematic; students master techniques of proofs through examples of usage, not via memorization of terminology from formal logic. Thus we do not stress the formal manipulation of logical symbols. After the discussion in Chapter 2 that emphasizes the *use* of logic, familiarity with logical concepts is conveyed by repeated usage throughout the book. Chapter 2 can be treated lightly in class; students can refer to it when they need help with manipulating logical statements.

The collection of exercises is a strong feature of this book. Many are fun, some are routine, and some are difficult. Exercises designated by "(–)" or "(+)" are routine or difficult, respectively; those designated by "(!)" are especially interesting or instructive. Having used these designations, we order the exercises for each chapter roughly in parallel to the presentation of material in the text, rather than in order of difficulty. Most exercises emphasize thinking and writing rather than computation. The understanding and communication of mathematics through problem-solving should be the driving force of the course.

The Preface to the Student lists many engaging problems. Some of these begin chapters as motivating "Problems"; others are left to the exercises. Solutions of such problems in the text are designated as "Solutions". Items designated as "Examples" are generally easier than those designated as "Solutions" or "Applications". "Examples" serve primarily to illustrate concepts, whereas "Solutions" or "Applications" employ the concepts being developed and involve additional reasoning. Items designated as "Remarks" contain important commentary for which the designations of "Definition" and "Lemma" are not appropriate; students should not ignore the "Remarks". Many of the exercises in the text carry hints; these represent what we feel will be helpful to most students. We also provide an appendix with more elementary hints for many problems; these are meant to help students get started.

This book does not assume calculus and hence in principle can be used in a course taught to freshmen or to high school students. It does require motivation and commitment from the students, since problems can no longer be solved by plugging numbers into a template. The book should be appropriate for students who have studied calculus computationally and wonder why the computations work. It is ideal for beginning majors in mathematics and computer science. Readers outside mathematics who enjoy careful thinking and are curious about mathematics will also profit by it. High school teachers of mathematics may appreciate the interaction between problem-solving and theory.

Design of Courses

We developed this book through numerous courses, beginning with a version we team-taught in 1991 at the University of Illinois. Various one-semester courses can be constructed from this material. A one-semester course on discrete mathematics that emphasizes proofs will cover Parts I-III, omitting most of Chapter 8 (rational numbers) and the more algebraic material from Chapters 6 and 7. A one-semester course in elementary analysis covers Chapters 4 and 5, part of Chapter 8, and Part IV. The full text is suitable for a one-year course culminating in the Fundamental Theorem of Algebra. Using Part I and a selection of material from Parts II-IV, we have taught one-semester courses introducing students to proofs and to a balanced overview of mathematics.

The book offers considerable flexibility in the design of a balanced course. We cover almost all of Chapters 1-6, including the Schroeder-Bernstein Theorem. In Chapter 7, we skip the section on groups. From Chapter 8, the construction of rational numbers and the existence of irrational numbers are indispensable, but the proof of the former can be treated lightly. The material on Pythagorean triples and on probability in Chapters 8 and 9 enriches the course if time permits. In Chapter 9, the material up to the Summation Identity (and the ballot problem) should be covered to illuminate combinatorial reasoning. Although optional, the treatment of multinomial coefficients helps unify the course. In Chapter 10, the pigeonhole principle illustrates the elegance of concise arguments; inclusion-exclusion via the totient function further enhances cohesiveness. Chapter 11 contains enough material for the instructor to sample according to taste and time. In Chapter 12, most instructors will want to discuss first-order and second-order recurrences and perhaps the Catalan recurrence. Chapters 13 through 15 should be covered thoroughly. Coverage in Chapter 16 depends on the instructor's interests and the students' abilities; reaching the Mean Value Theorem probably requires skipping Chapter 11. The rest can be left as supplemental reading for the interested student. The one-semester balanced course will not reach Chapters 17 and 18.

Acknowledgments

The first author taught a version of the course to freshmen at Princeton in fall 1993; he acknowledges the support of both Princeton University and the Institute for Advanced Study for that semester. He also acknowledges the support of the Mathematical Sciences Research Institute for spring 1994, when extensive electronic correspondence between the authors considerably improved the text. The second author typeset the book and illustrations using "groff", a product of the Free Software Foundation. He acknowledges the support of the Center for Advanced Study for spring 1994.

We acknowledge useful comments by Art Benjamin, Dick Bishop, Kaddour Boukaabar, Peter Braunfeld, Tom Brown, Steve Chiappari, Everett Dade, Harold Diamond, Paul Drelles, Sue Goodman, Dan Grayson, Harvey Greenwald, Deanna Haunsperger, Felix Lazebnik, Steve Post, Sara Robinson, Craig Tovey, Steve Ullom, Josh Yulish, and other readers. We are saddened by the untimely death of our colleague N. Tenney Peck, who taught from early versions of this book and made especially valuable comments. The Mathematics Department of the University of Illinois gave us the opportunity to develop the course that inspired this book; we thank our students for struggling with preliminary versions of it. Our editor George Lobell provided the guidance and prodding needed to bring the book to its final form.

The first author thanks Annette and little John for their love and patience. The second author thanks Blake and Ching for their love, encouragement, and understanding during the completion of this book.

-John P. D'Angelo, jpda@math.uiuc.edu
-Douglas B. West, west@math.uiuc.edu
Urbana, 1996

PART I

ELEMENTARY CONCEPTS

Chapter 1

Sets and Numbers

The ancient Babylonians considered the problem of finding two numbers when given their sum and their product. They expressed the solution in words, not in formulas. We begin this book by deriving the quadratic formula and using it to solve this ancient problem. We will then analyze what properties of numbers we needed and use this to motivate the axioms that underlie the real number system.

THE QUADRATIC FORMULA

Given two numbers s and p, the Babylonians wanted to find x and y such that $x + y = s$ and $xy = p$. To do so, we set $y = s - x$ and substitute to obtain $x(s - x) = p$, which we rewrite as $x^2 - sx + p = 0$. Every solution x to the problem of the Babylonians must satisfy this quadratic equation.

Solving this equation is equivalent to solving the general quadratic equation. We don't change the solutions if we multiply the equation by a nonzero constant a to obtain $ax^2 - asx + ap = 0$, and then we can name $b = -as$ and $c = ap$ to obtain $ax^2 + bx + c = 0$. The quadratic formula of elementary mathematics expresses the solution for x in terms of a, b, and c. By "completing the square", we have

$$0 = a(x^2 + \frac{b}{a}x) + c = a(x^2 + \frac{b}{a}x + \frac{b^2}{4a^2}) - \frac{b^2}{4a} + c$$

$$= a(x + \frac{b}{2a})^2 + c - \frac{b^2}{4a}.$$

The last expression has only one x, with a succession of operations applied to it. Undoing the operations (in reverse order), we have

$$\pm\sqrt{(0 + \frac{b^2}{4a} - c)\frac{1}{a}} - \frac{b}{2a} = x$$

2

Putting the left side over a common denominator, we obtain the familiar **quadratic formula**:

$$x = \frac{-b \pm \sqrt{b^2 - 4ac}}{2a}$$

This formula describes all the solutions to the general quadratic equation. When $b^2 - 4ac > 0$, it yields two values. When $b^2 - 4ac = 0$, the two values are equal. When $b^2 - 4ac < 0$, there is no solution in real numbers. Rewriting the solution formula in terms of s and p yields the expressions

$$\frac{s + \sqrt{s^2 - 4p}}{2}, \quad \frac{s - \sqrt{s^2 - 4p}}{2}$$

to solve the problem of the Babylonians. When $s^2 - 4p < 0$, there is no solution.

What properties of numbers did we use in this derivation? First, we used basic rules about the behavior of addition and multiplication. Whenever we add several numbers, the result does not depend on the order in which we write them down or in what order we perform pairwise additions. Multiplication has the same behavior. We also needed the more subtle "distributive law": $u(v + w) = uv + uw$.

We used properties of subtraction and division and their relationship to addition and multiplication. Every number u has an "additive inverse" $-u$, and subtracting u has the same effect as adding $-u$. We can "add 0" (in the form $-u + u$, for example) without causing a change. Similarly, every nonzero number has a multiplicative inverse, but we cannot divide by 0. The properties of inverses allow us to cancel like terms or common factors from both sides of an equality.

We also took a square root. Because the product of two nonzero numbers with the same sign is positive, square roots exist only for nonnegative numbers. Furthermore, if $u^2 = v$, then also $(-u)^2 = v$. For these reasons we write $\pm$ on the square root sign in the quadratic formula and state that there is no solution in real numbers if $b^2 - 4ac < 0$.

The Babylonians could not have understood our solution, because their number system did not include negative numbers! In the real number system, we accept the formula $(-b \pm \sqrt{b^2 - 4ac})/(2a)$ as the solution. This leaves the problem of expressing the square root of $b^2 - 4ac$ in an acceptable form. Expressing the square root of a number as simple as 2 or 3 in decimal form already requires infinite non-repeating decimal expansions. Understanding decimal expansions and square roots requires a careful study of infinite processes and limits.

NUMBERS AND PROBLEM-SOLVING

Our discussion of the Babylonian problem reflects the structure of this book. Mathematics begins with counting, comparison, and the solution of equations. With the natural numbers, consisting of $1, 2, 3, \cdots$ in their usual order, we can discuss counting and solve problems involving finite collections of objects. Solving equations leads us to introduce negative numbers to describe differences, rational numbers to describe ratios, and real numbers to describe limits. Solving other problems leads to geometrical concepts and objects more complicated than numbers, but the properties of elementary number systems remain fundamental.

The collection of real numbers ($\mathbb{R}$) is the usual setting for scientific work. It contains the rational numbers ($\mathbb{Q}$), the integers ($\mathbb{Z}$), and the natural numbers ($\mathbb{N}$). Precise definition of the real numbers is difficult; ancient Greek mathematicians believed that all numbers were rational. We often think of a real number as a point on a "number line". This notion explains addition and subtraction and distance, but it makes multiplication and division hard to interpret, and it implicitly makes some geometric assumptions. Another approach describes a real number as a decimal expansion. A number is rational when its decimal expansion terminates or eventually repeats; a calculator can show us only rational numbers.

If we accept the natural numbers, then we can construct the other numbers from them, define the arithmetic operations, and carefully prove the familiar arithmetic properties. This uses techniques that are fundamental to all mathematical reasoning, including those developed in this book, but it is rather formal and somewhat dull. We believe that students first encountering proofs are more interested in solving new problems than in justifying arithmetic. Therefore, we leave the construction of the real numbers and the verification of their properties to Appendix A, for later appreciation.

"Verifying the properties of the real numbers" means establishing basic properties called "axioms" from which all other properties follow. We list these at the end of this chapter for reference. Our discusion of the quadratic formula hinted at axioms, which come in several types. The first type is arithmetic, governing addition, subtraction, multiplication, and division; Definition 1.15 lists the arithmetic or "field" axioms for real numbers. The second type concerns positivity and the related notion of inequality; Definition 1.16 lists the axioms of this type. The four subsequent propositions summarize elementary consequences of the arithmetic and positivity axioms.

We take these axioms and their elementary consequences for granted and use them without comment to manipulate numbers and equations. For example, we need not name the distributive law explicitly every time we use it. We also assume that every positive number has a unique positive square root, which we prove in Part IV. This rests on an additional axiom, which we will describe shortly.

The purpose of proof is to explain; we will assume that the reader understands arithmetic operations and will concentrate on reasoning and logical structure. The proof of a theorem must be logically correct; it need not record the manner in which it was discovered. When we want to prove an inequality, for example, we may discover the proof by manipulating the inequality to reduce it to a statement known to be true. When the manipulations are reversible, we can write the proof of the desired inequality by starting from the known true statement.

We illustrate this by proving that the product of two numbers is at most the square of their average. The *arithmetic mean* (or "average") of x and y is $(x + y)/2$. The *geometric mean* of nonnegative numbers x and y is $\sqrt{xy}$. Algebraic manipulations of the desired inequality $xy \leq (\frac{x+y}{2})^2$ reduce it to the known true statement $0 \leq (x - y)^2$, so we start from this when writing the proof.

1.1. Proposition. (AGM Inequality). If x and y are real numbers, then $xy \leq (\frac{x+y}{2})^2$. If s and t are nonnegative real numbers, then $\sqrt{st} \leq (s + t)/2$.

Proof. Since the square of a real number is nonnegative, we have $0 \leq (x - y)^2$. Expanding $(x - y)^2$ yields $0 \leq x^2 - 2xy + y^2$. Adding $4xy$ to both sides makes the right side again a perfect square and yields

$$4xy \leq x^2 + 2xy + y^2 = (x + y)^2.$$

The first claim follows from this by dividing both sides by 4.

Given nonnegative numbers s and t, we now know that $st \leq (\frac{s+t}{2})^2$. The second statement of the proposition follows by taking the positive square root of both sides. This does not change the direction of the inequality (Exercise 1.14). ∎

The rational numbers also satisfy all the properties we have discussed. We make one additional assumption about the set of real numbers that distinguishes it from the set of rational numbers. If S is a collection of real numbers, then the real number β is an *upper bound* for S if $x \leq \beta$ for every element x of S. Similarly, α is a *lower bound* for S if $x \geq \alpha$ for every x in S. The additional assumption we make about $\mathbb{R}$ is the "Least Upper Bound Property" or "Completeness Axiom": every nonempty subset of $\mathbb{R}$ that has an upper bound has a smallest upper bound. This axiom allows us to take limits, which explains infinite decimals and enables us to do calculus. It also implies the existence of square roots of positive numbers.

Later in this book we will prove that $\sqrt{2}$ cannot be expressed as the ratio of two integers. The set of rational numbers whose square is less than 2 has no least rational upper bound; completeness does not hold for $\mathbb{Q}$. We postpone until Part IV all further discussion of completeness. Before that we use only arithmetic and order properties (except for the existence of roots). This enables us to develop techniques useful for both rational and real numbers before considering completeness. A structure satisfying the properties we have assumed is called a "complete ordered field". In Appendix A, we construct the real numbers using only the natural numbers and their properties, and we explain why the real numbers form the only complete ordered field.

Having established our "ground rules", we return to problem-solving. What does it mean to determine the solutions to a mathematical problem? Once we describe candidates for the solutions, we must do two things. We must verify that each candidate solves the problem, and we must prove that there are no other solutions. This proves that our candidates are all the solutions. Our next example illustrates this process. In this book, we use **Solution** and **Application** to designate examples that apply additional reasoning, reserving **Example** as a label for more straightforward illustrations of concepts.

1.2. Solution. *The Penny Problem.* Given various piles of pennies, we create a new configuration by removing one coin from each old pile to make one new pile. Each original pile shrinks by one, so each pile of size one disappears: 1,1,2,5 becomes 1,4,4, for example. We don't distinguish between different orderings of the same list of numbers, so we may consider our configurations to be lists of positive integers in nondecreasing order. What lists are unchanged under the operation we have defined?

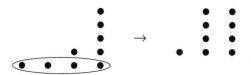

Consider a list of the form $1, 2, \ldots, n$, where n is a positive integer. For each i from 2 to n, the pile of size i becomes a pile of size $i - 1$. The pile of size 1 disappears, and the n piles each contribute one coin to form a new pile of size n. The result is the original list $1, 2, \ldots, n$, so such lists are unchanged.

This does not complete the solution! We must also prove that *no other list* is unchanged; equivalently, that we have described every unchanged list. Suppose that a is an unchanged list, and let n be the largest number in a; we prove that a consists of exactly one copy of each positive integer up to n. Let b be the list obtained from a by applying our operation. Although b is the same as a, we use another letter to emphasize the application of the operation.

Every number in a decreases by 1 to become an entry in b. All old piles now have size less than n. Since b must also have a pile of size n, a must have exactly n piles. Furthermore, b has exactly one pile of size n. If $1 < i \leq n$, then the number of i's in a equals the number of $i-1$'s in b, which equals the number of $i-1$'s in a because a is the same as b. Therefore, a has the same number of i's for each integer i from n down to 1. Since it has exactly one n, it has one copy of each size and hence has the desired form. ∎

SETS

We have used the word "set" in an informal way. We begin our formal development by introducing notation and terminology about sets.

1.3. Definition. A *set* is a collection of distinct objects called its *elements* or *members*. When x is an element of A, we write $x \in A$ and say that x *belongs to* A. When x is not an element of A, we write $x \notin A$. Sets A and B are *equal*, written $A = B$, if they have the same elements.

If every element of A belongs to B, then A is a *subset* of B, and B *contains* A; we write this as $A \subseteq B$. A set A is *strictly contained* in B, written $A \subset B$, if $A \subseteq B$ but $A \neq B$. If A is not contained in B, we write $A \nsubseteq B$.

The *empty set*, written $\varnothing$, has no elements; it is a subset of every set. A *proper subset* of a set A is a subset of A that is not A itself. The *power set* of a set A is the set of all subsets of A.

When we list the elements of a set explicitly, we put braces around the list; "$A = \{-1, 1\}$" specifies the set A consisting of the elements -1 and 1. Writing the elements in a different order does not change a set. Saying that a set consists of "distinct objects" means that if $S = \{x, y, z\}$ and it happens that $x = y$, then $S = \{x, y, z\} = \{x, z\}$. For convenience, we write $x, y \in S$ to mean that both x and y are elements of S.

1.4. Example. Let S be the set {Kansas, Kentucky}. Let T be the set of states in the United States whose names begin with "K". The sets S and T are equal. The set S has four subsets, which are $\varnothing$, {Kansas}, {Kentucky}, and {Kansas, Kentucky}. These four sets are the elements of the power set of S. ∎

1.5. Remark. *Specifying the elements of a set.* In Example 1.4, we specified a set both by listing its elements explicitly and by describing it as a particular subset of a larger set (the set of all states). When we want to specify a set S consisting of the elements in a set A that satisfy a given condition, we write "$\{x \in A: \text{condition}\}$". We read this as "the set of x in A such that 'condition' holds for x". For example, the expression $S = \{x \in \mathbb{R}: ax^2 + bx + c = 0\}$ specifies S as the set of real solutions to the quadratic equation $ax^2 + bx + c = 0$. The first part of the descriptive notation for a set gives a name to the element being described and specifies a universe for it, the colon means "such that", and the last part states the required condition on the element. (We may omit the specification of the universe when the context makes it clear.) ∎

The next three definitions introduce notation and terminology for special sets that we will use throughout this book.

1.6. Definition. *Sets of integers.* We use $\mathbb{Z}$ to denote the set of integers, $\{\cdots, -2, -1, 0, 1, 2, \cdots\}$. When $a, b \in \mathbb{Z}$ with $a \le b$, we use $\{a, \ldots, b\}$ to denote $\{i \in \mathbb{Z}: a \le i \le b\}$. The set of *even numbers* is $\{2k: k \in \mathbb{Z}\}$. The set of *odd numbers* is $\{2k + 1: k \in \mathbb{Z}\}$. The *parity* of an integer states whether it is even or odd.

The set of natural numbers, $\mathbb{N}$, is $\{1, 2, 3, \cdots\}$. By convention, 0 is not a natural number. When $n \in \mathbb{N}$, we write $[n]$ for the set $\{1, \ldots, n\}$.

We have encountered the special set $[n]$ before; in the Penny Problem (Solution 1.2) we proved that the set of solutions is $\{[n]: n \in \mathbb{N}\}$.

1.7. Definition. *Cartesian product and k-tuples.* The *Cartesian product* of sets S and T, written $S \times T$, is the set $\{(x, y): x \in S, y \in T\}$. In words, $S \times T$ is the set of *ordered pairs* whose first entry is an element of S and whose second entry is an element of T.

A *list* with entries in A consists of elements of A (repetition allowed) in a specified order. A *k-tuple* is a list with k entries. We write A^k for the set of k-tuples with entries in A, so $A^k = \{(x_1, x_2, \ldots, x_k): x_i \in A\}$. We read "$x_i$" as "$x$ sub i".

The sets used to form a Cartesian product may have common elements. When $S = T = \mathbb{R}$, for example, $S \times T$ can be viewed as the set of all points in the plane, designated by horizontal and vertical coordinates, called the *Cartesian coordinates*. Ordered pairs are 2-tuples; hence $A^2 = A \times A$, and $\mathbb{R}^2$ denotes the set of points in the plane. The Cartesian product operation is named for René Descartes (1596-1650).

1.8. Definition. *Intervals.* When $a, b \in \mathbb{R}$ with $a \leq b$, the *closed interval* $[a, b]$ is the set $\{x \in \mathbb{R}: a \leq x \leq b\}$. The *open interval* (a, b) is the set $\{x \in \mathbb{R}: a < x < b\}$. Since we also use (a, b) to denote an ordered pair, we usually say "the interval (a, b)" to avoid ambiguity.

1.9. Remark. Since two sets are equal when they have the same elements, the statement "$A = B$" means precisely "$A \subseteq B$ and $B \subseteq A$". Solving a problem may involve letting S be the set of solutions and proving that S equals a set T described in some other way. This usually involves proving that $T \subseteq S$ and that $S \subseteq T$. In other words, every element of T is a solution and there are no other solutions. ■

1.10. Example. *Equality of sets.* We present four examples.

1) *The Penny Problem.* Let S be the set of solutions (lists unchanged by the operation), and let $T = \{[n]: n \in \mathbb{N}\}$. We proved that $S = T$ by proving both $T \subseteq S$ and $S \subseteq T$.

2) *The inequality $x^2 < x$.* Let $S = \{x \in \mathbb{R}: x^2 < x\}$, and let T be the interval $(0, 1)$. We claim that $S = T$. To prove this, we show that $(0, 1) \subseteq S$ and that $S \subseteq (0, 1)$. First suppose $x \in (0, 1)$. Since $x > 0$, we can multiply the known inequality $x < 1$ by x to obtain $x^2 < x$, so $x \in S$. Conversely, suppose $x \in S$. Since $x^2 < x$, we have $0 > x^2 - x = x(x - 1)$. This requires that x and $x - 1$ are nonzero and have opposite signs, which implies $x \in (0, 1)$.

3) *The inequality $xy \leq [(x + y)/2]^2$.* Let $T = \mathbb{R}^2$. Letting $S = \{(x, y) \in \mathbb{R}^2: xy \leq [(x + y)/2]^2\}$ yields $S \subseteq T$ by definition. We proved $T \subseteq S$ in Proposition 1.1 by starting with an inequality satisfied by all of $\mathbb{R}^2$ and operating on it in ways that did not change the set of solutions.

4) *The quadratic equation $ax^2 + bx + c = 0$.* Let S be the set of solutions to the equation, and let T be the set consisting of $(-b + \sqrt{b^2 - 4ac})/(2a)$ and $(-b - \sqrt{b^2 - 4ac})/(2a)$. When we proved that $S = T$, we could have explicitly proved both $T \subseteq S$ and $S \subseteq T$. Our reasoning was more efficient; we operated on the equation in ways that preserved the set of solutions. This produced a string of equalities of sets, starting with S and ending with T. ■

1.11. Definition. *Set operations.* Let A and B be sets. Their *union*, written $A \cup B$, consists of the elements of A together with the elements of B. Their *intersection*, written $A \cap B$, consists of the elements belonging to both A and B. Their *difference*, written $A - B$, consists of the elements of A that are not in B. Two sets are *disjoint* if their intersection is the empty set $\varnothing$. If a set A is contained in some universe U under discussion, then the *complement* A^c of A is the set of elements of U *not* in A.

1.12. Example. *Set operations.* When A and B are sets, we have $A \cap B = B \cap A$ and $A \cup B = B \cup A$, but generally $A - B \neq B - A$ (see Exercise 1.26). The sets $A - B$ and $B - A$ are disjoint, and their union is the difference between $A \cup B$ and $A \cap B$.

The set of even numbers and the set of odd numbers are disjoint, and their union is $\mathbb{Z}$. Within the universe of integers, the complement of the set of even numbers is the set of odd numbers. ∎

Pictures give life to mathematical concepts and illuminate the essential ideas behind them. We encourage the reader to draw pictures whenever they are helpful. The operations in Definition 1.11 become easy to grasp when we draw an appropriate picture. The diagrams describing set operations are named for John Venn (1834-1923), though he was not the first person to use them.

1.13. Remark. *Venn diagrams.* In a *Venn diagram*, an outer box represents the universe under consideration, and regions within the box correspond to sets. Non-overlapping regions correspond to disjoint sets. The four regions in the Venn diagram for two sets A and B represent $A \cap B$, $(A \cup B)^c$, $A - B$, and $B - A$. Since $A - B$ consists of the elements in A and not in B, we have $A - B = A \cap B^c$. Similarly, the diagram suggests that B^c is the union of $A - B$ and $(A \cup B)^c$, which are disjoint. Exercise 1.27 lists other elementary relationships. ∎

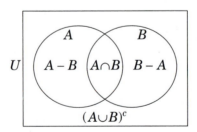

SUMMARY OF THE AXIOMS

We are ready to list our assumptions about the real numbers. We repeat that our purpose in listing them and their elementary arithmetic consequences is to state clearly the starting point for this book; these are the properties of numbers that we may use freely. Understanding what the real numbers are amounts to understanding the following definition, which will require some elaboration.

1.14. Definition. The set of real numbers, $\mathbb{R}$, with its structure of arithmetic and order, is the unique complete ordered field.

The definition of a complete ordered field appears in the next three Definitions. In Appendix A, we prove that there is a structure satisfying the definition of a complete ordered field and that the structure is unique. This justifies our definition of $\mathbb{R}$.

1.15. Definition. *Field Axioms.* A set S with operations $+$ and $\cdot$ and distinguished elements "0" and "1" is a *field* if the following properties hold for all $x, y, z \in S$.

A0: $x + y \in S$	M0: $x \cdot y \in S$	Closure
A1: $(x+y)+z = x+(y+z)$	M1: $(x \cdot y) \cdot z = x \cdot (y \cdot z)$	Associativity
A2: $x + y = y + x$	M2: $x \cdot y = y \cdot x$	Commutativity
A3: $x + 0 = x$	M3: $x \cdot 1 = x$	Identity
A4: given x, there is a y such that $x + y = 0$	M4: if $x \neq 0$, there is a y such that $x \cdot y = 1$	Inverses
	DL: $x \cdot (y + z) = x \cdot y + x \cdot z$	Distributive Law

The operations $+$ and $\cdot$ are called *addition* and *multiplication*. The elements 0 and 1 are the *additive identity element* and the *multiplicative identity element*.

The additive inverse of x is the *negative* of x, written as $-x$. The multiplicative inverse of x is the *reciprocal* of x, written as x^{-1}. The element 0 has no reciprocal. We often write xy for $x \cdot y$, and x^2 for $x \cdot x$. We may use parentheses to clarify the order of operations.

1.16. Definition. *Order Axioms.* An *ordered field* is a field F containing a positive set, where a *positive set* is a subset $P \subset F$ such that

P1: $x, y \in P$ implies $x + y \in P$ Closure under Addition
P2: $x, y \in P$ implies $xy \in P$ Closure under Multiplication
P3: $x \in F$ implies exactly one of
    $\{x = 0, x \in P, -x \in P\}$ Trichotomy

In an ordered field, we define $x < y$ to mean $y - x \in P$. The relations $\leq$, $<$, and $\geq$ have analogous definitions in terms of P.

From the closure of P under $+$ and $\cdot$, we conclude that $a < b$ and $c < d$ imply $a + c < b + d$, and that $0 < a < b$ and $0 < c < d$ imply $ac < bd$. An equivalent statement of the trichotomy property is that for all x, y, exactly one of the three statements $x < y$, $x = y$, and $x > y$ holds.

If S is a subset of a field F, then $\beta \in F$ is an *upper bound* for S if $x \leq \beta$ for all $x \in S$.

1.17. Definition. *Completeness Axiom.* An ordered field F is a *complete ordered field* if for every non-empty subset of F that has an upper bound there is a least upper bound in F.

Until Part IV, we do not need the completeness axiom for $\mathbb{R}$. The other axioms guarantee that arithmetic has its familiar properties. If a set with operations called addition and multiplication satisfies the field and order axioms, then it also has the properties 1.18-1.21 below. In Appendix A we build a structure and prove that it satisfies 1.15-1.17; all the properties of $\mathbb{R}$ follow from this. Note that $\mathbb{Q}$ also is an ordered field, and that the set $\mathbb{Z}$ of integers satisfies all the field and order axioms except the existence of multiplicative inverses.

Propositions 1.18-1.21 follow from the field and order axioms. We agree to assume them (except in Exercises 1.29-1.33). The axioms imply all properties of the real numbers. There are other equivalent formulations, especially for the order and completeness axioms, so the choice made for the axioms is a matter of taste.

1.18. Proposition. *Order properties.*

> O1: $x \leq x$ Reflexive Property
> O2: $x \leq y$ and $y \leq x$ imply $x = y$ Antisymmetric Property
> O3: $x \leq y$ and $y \leq z$ imply $x \leq z$ Transitive Property
> O4: at least one of $x \leq y$ and $y \leq x$ holds Total Ordering Property

1.19. Proposition. *Ordered field properties.*

> F1: $x \leq y$ implies $x + z \leq y + z$ Additive Order Law
> F2: $x \leq y$ and $0 \leq z$ imply $xz \leq yz$ Multiplicative Order Law

1.20. Proposition. *Elementary consequences of the field axioms.* In a field, the following properties hold for all elements x, y, z.

> a) $x + z = y + z$ implies $x = y$ e) $(-x)(-y) = xy$
> b) $x \cdot 0 = 0$ f) $xz = yz$ and $z \neq 0$ imply $x = y$
> c) $(-x)y = -(xy)$ g) $xy = 0$ implies $x = 0$ or $y = 0$
> d) $-x = (-1)x$

1.21. Proposition. *Elementary properties of an ordered field.* In an ordered field, the following properties hold for all elements x, y, z.

> a) $x \leq y$ implies $-y \leq -x$ e) $0 < 1$
> b) $x \leq y$ and $z \leq 0$ imply $yz \leq xz$ f) $0 < x$ implies $0 < x^{-1}$
> c) $0 \leq x$ and $0 \leq y$ imply $0 \leq xy$ g) $0 < x < y$ implies $0 < y^{-1} < x^{-1}$
> d) $0 \leq x^2$

Properties (a) and (b) of Proposition 1.21 tell us that multiplying an inequality by a negative number requires reversing the inequality.

EXERCISES

In this chapter, we discussed mathematical objects; in Chapter 2, we will discuss mathematical statements. As a warmup, these exercises begin with translations between mathematics and English involving only arithmetic.

Words like "determine", "show", "obtain", or "construct" include a request for proof; there are more than a dozen ways to say "prove".

1.1. (–) Suppose we have many tables and chairs. Let t be the number of tables, and let c be the number of chairs. Write down an inequality that means "We have at least four times as many chairs as tables."

1.2. (–) Consider the Celsius (C) and Fahrenheit (F) temperature scales.

C	0	5	10	15	20	25	30
F	32	41	50	59	68	77	86

Express the sentence "The temperature was 10° C and increased by 20° C" using the Fahrenheit scale.

1.3. (–) If a stock declines 20% in one year and rises 23% in the next, is there a net profit? What if it goes up 20% in the first year and down 18% in the next?

1.4. (–) On July 4, 1995, the *New York Times* reported that the nation's universities were awarding 25% more PhD degrees than the economy could absorb. The headline concluded that there was a 1 in 4 chance of underemployment. Here "underemployment" means having no job or having a job not requiring the PhD degree. What should the correct statement of the odds have been?

1.5. (–) A store offers a 15% promotional discount for its grand opening. A customer argues that the discount should be applied to the total after the 5% sales tax is added, expecting to save more money that way. The clerk believes that the law requires the discount to be applied first and then the tax computed on the resulting amount. Will it benefit the customer to apply the discount after the tax is added instead of before? Explain.

1.6. (–) A store offers an "installment plan" option, with no interest to be paid. There are 13 monthly payments, with the first being a "down-payment" that is half the size of the others, so payment is completed one year after purchase. If a customer buys a $1000 stereo, what are the payments under this plan?

1.7. We have two identical glasses. Glass 1 contains x ounces of wine; glass 2 contains x ounces of water ($x \geq 1$). We remove 1 ounce of wine from glass 1 and add it to glass 2. The wine and water in glass 2 mix uniformly. We now remove 1 ounce of liquid from glass 2 and add it to glass 1. Prove that the amount of water in glass 1 is now the same as the amount of wine in glass 2.

1.8. A digital 12-hour clock is broken in such a way that the reading for minutes and the reading for hours are always the same. Determine the minimum number of minutes between possible correct readings of the clock.

1.9. Three people check into a hotel and register for a room. The desk attendant charges them $30; each pays $10. The manager returns and says this was an overcharge, instructing the attendant to return $5. The attendant heads for the room with five $1 bills, but pockets $2 as a tip and returns $1 to each guest. Of the original $30 payment, each guest actually paid $9, and $2 went to the attendant. What happened to the "missing" dollar?

1.10. A census taker interviews a woman in a house. "Who lives here?" he asks. "My husband and I and my three daughters," she replies. "What are the ages of your daughters?" "The product of their ages is 36 and the sum of their ages is the house number." The census taker looks at the house number, thinks a moment, and says, "You haven't given me enough information to determine the ages." "Oh, you're right," she replies, "Let me also tell you that my eldest daughter is asleep upstairs." "Ah! Thank you very much!" What are the ages of the daughters? (Assume that the ages are positive integers; the problem requires a "reasonable" mathematical interpretation of its words.)

1.11. (+) Two mail carriers meet on their routes and have a conversation. A: "I know you have three sons. How old are they?" B: "If you take their ages, expressed in years, and multiply those numbers, the result will equal your age." A: "But that's not enough to tell me the answer!" B: "The sum of these three numbers equals the number of windows in that building." A: "Hmm...[pause]. But it's still not enough!" B: "My middle son is red-haired." A: "Ah, now it's clear!" How old are the sons? (Hint: the ambiguity at the earlier stages is needed to determine the solution for the full conversation.) (G.P. Klimov)

1.12. (!) What are the dimensions of a rectangular carpet with perimeter 48 feet and area 108 square feet? Given positive numbers p and a, under what conditions does there exist a rectangular carpet with perimeter p and area a?

1.13. (!) Use Proposition 1.1 to prove that the value of x maximizing $x(c - x)$ is $x = c/2$. For $a > 0$, use this to find the value of y maximizing $y(c - ay)$.

1.14. (!) Suppose that a and b are nonnegative real numbers with $a \le b$. Prove that $\sqrt{a} \le \sqrt{b}$. (Hint: first use Proposition 1.19(F2) (twice) to show that if $0 \le x \le y$, then $x^2 \le y^2$.)

1.15. (+) *Extensions of the AGM Inequality.*
    a) Prove that $4xyzw \le x^4 + y^4 + z^4 + w^4$. (Hint: start with Proposition 1.1 for $\{x, y\}$ and multiply it by the corresponding inequality for $\{z, w\}$.)
    b) Prove that $3abc \le a^3 + b^3 + c^3$. (Hint: in the inequaltiy of part (a), set w equal to the cube root of xyz.)

1.16. (!) Assuming only arithmetic (not the quadratic formula or calculus), prove that $\{x \in \mathbb{R}: x^2 - 2x - 3 < 0\} = \{x \in \mathbb{R}: -1 < x < 3\}$.

1.17. Let $S = \{(x, y) \in \mathbb{N}^2: (2 - x)(2 + y) > 2(y - x)\}$. Prove that $S = T$, where $T = \{(1, 1), (1, 2), (1, 3), (2, 1), (3, 1)\}$.

1.18. Let $S = [3] \times [3]$ (the Cartesian product of $\{1, 2, 3\}$ with itself). Let T be the set of ordered pairs $(x, y) \in \mathbb{Z} \times \mathbb{Z}$ such that $0 \le 3x + y - 4 \le 8$. Prove that $S \subseteq T$. Does equality hold?

1.19. Let $S = \{(x, y) \in \mathbb{R}^2 : (1 - x)(1 - y) \geq 1 - x - y\}$. Give a simple description of S involving the signs of x and y.

1.20. (!) Determine the set of ordered pairs (x, y) of nonzero real numbers such that $x/y + y/x \geq 2$. (Hint: the set is the union of two Cartesian products.)

1.21. Determine the set of solutions to the general quadratic inequality $ax^2 + bx + c \leq 0$, in terms of linear inequalities or intervals. (Use the quadratic formula; the complete solution involves many cases.)

1.22. (−) Suppose $a < b < c < d$ are real numbers. Express the set $[a, b] \cup [c, d]$ as the difference of two sets.

1.23. Let $S = \{x \in \mathbb{R} : x(x - 1)(x - 2)(x - 3) < 0\}$. Let T be the interval $(0, 1)$, and let U be the interval $(2, 3)$. Obtain a simple set equality relating S, T, U.

1.24. (!) Suppose n is a positive integer, and let $a_1, a_2, \ldots, a_n$ be real numbers such that $a_1 < a_2 < \cdots < a_n$. Let $(-\infty, a_1)$ denote the set $\{x \in \mathbb{R} : x < a\}$. Obtain a formula for the set $\{x \in \mathbb{R} : (x - a_1)(x - a_2) \cdots (x - a_n) < 0\}$ using the notation for intervals.

1.25. Suppose A and B are sets. Explain why the two sets $(A - B) \cup (B - A)$ and $(A \cup B) - (A \cap B)$ are always equal. Check this equality explicitly when A is the set of states in the United States whose names begin with a vowel and B is the set of states in the United States whose names have at most 6 letters.

1.26. (−) Given that A and B are sets, when is $A - B = B - A$ true?

1.27. (−) Suppose A, B, C are sets. Explain the relationships below using the definitions of set operations and containment and using Venn diagrams.

a) $A \subseteq A \cup B$, and $A \cap B \subseteq A$.   d) $A \subseteq B$ and $B \subseteq C$ imply $A \subseteq C$.
b) $A - B \subseteq A$.   e) $A \cap (B \cap C) = (A \cap B) \cap C$.
c) $A \cap B = B \cap A$, and $A \cup B = B \cup A$.   f) $A \cup (B \cup C) = (A \cup B) \cup C$.

1.28. Suppose F is a field consisting of two elements, 0 and 1, with $1 + 1 = 0$. Why is F not an ordered field?

1.29. Prove that every element in a field has exactly one additive inverse, and every nonzero element has exactly one multiplicative inverse.

1.30. Suppose x, y, z belong to a field. Use the field axioms to prove the properties below. Exercise 1.29 may be assumed, and each item may be used to prove items later in the list. (Comment: we assume that addition and multiplication are defined, so $x = y$ immediately implies $x + z = y + z$ and $xz = yz$.)

a) $x + z = y + z$ implies $x = y$ d) $-x = (-1)x$
b) $xz = yz$ and $z \neq 0$ imply $x = y$ e) $(-x)(-y) = xy$
c) $x \cdot 0 = 0$ f) $(-x)y = -(xy)$
     g) $xy = 0$ implies $x = 0$ or $y = 0$

1.31. Suppose x, y, z are elements of an ordered field. The order relation is defined by putting $x \leq y$ if $y - x$ is in the positive set or is 0. Using the order axioms and field properties, prove the following properties of the order relation.

a) $x \leq x$ Reflexive Property
b) $x \leq y$ and $y \leq x$ imply $x = y$ Antisymmetric Property
c) $x \leq y$ and $y \leq z$ imply $x \leq z$ Transitive Property
d) at least one of $x \leq y$ and $y \leq x$ holds   Total Ordering Property
e) $x \leq y$ implies $x + z \leq y + z$ Additive Order Law
f) $x \leq y$ and $0 \leq z$ imply $x \cdot z \leq y \cdot z$ Multiplicative Order Law

1.32. Suppose x, y, z are elements of an ordered field. Using the axioms and Exercises 1.29-1.31, prove the properties below. Each item listed may be used to prove later items.

a) $x \leq y$ implies $-y \leq -x$ e) $0 < 1$
b) $x \leq y$ and $z \leq 0$ imply $yz \leq xz$ f) $0 < x$ implies $0 < x^{-1}$
c) $0 \leq x$ and $0 \leq y$ imply $0 \leq xy$ g) $0 < x < y$ implies $0 < y^{-1} < x^{-1}$
d) $0 \leq x^2$

1.33. Prove that $(x + y)(z + w) = xz + xw + yz + yw$ when x, y, z, w are four elements of a field, naming the property used at each step of the proof.

1.34. (+) *Solution of the general cubic equation.* Consider the equation $ax^3 + bx^2 + cx + d = 0$ with $a \neq 0$ and $a, b, c, d \in \mathbb{R}$.

a) Determine appropriate constants s, t so that the change of variables $x = s(y + t)$ reduces solving this equation to solving the equation $y^3 + Ay + B = 0$, where A, B are constants.

b) Determine a constant r such that the change of variables $y = z + r/z$ in the equation for y reduces it to a quadratic equation in z^3.

c) Solve the resulting quadratic equation for z^3, and use the solution to solve the general cubic equation for x. (Comment: this method is tedious even for easy cubic equations, and it uses complex numbers even when all the roots are real. Nevertheless, it does produce the solutions. There is no formula for solving a general polynomial equation of degree 5 or higher.)

Chapter 2

Language and Proofs

Understanding mathematical reasoning requires familiarity with the logical meaning of words like "every", "some", "not", "and", "or", etc. We illustrate how these concepts arise by analyzing two mathematical problems whose solution indicates the importance of careful language. We therefore discuss the use of language in mathematics, considering word order, quantifiers, logical statements, and logical symbols. We close this chapter with examples of elementary techniques of proof.

TWO THEOREMS ABOUT EQUATIONS

We consider two problems about solving equations. We analyze the solution of pairs of linear equations, and we show that certain quadratic equations with integer coefficients have no solutions in rational numbers. Our discussion illustrates the use of language and the variety of techniques in proofs. In order to solve interesting problems here, we assume some easy intuitive facts that we will prove later.

2.1. Definition. A *linear equation* in two variables x and y is an equation $ax + by = r$, where the coefficients a, b and the constant r are real numbers. A *line* in $\mathbb{R}^2$ is the set of pairs (x, y) satisfying a linear equation whose coefficients a and b are not both 0.

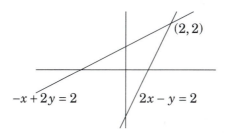

Geometric intuition suggests three possibilities when solving a pair of linear equations in two variables. If each equation describes a line, then the lines may intersect in one point, may be parallel, or may be the same line. These cases yield a unique solution, no solution, or infinitely many solutions, respectively. We can analyze this without relying on geometric intuition, because we have defined "line" using only properties of real numbers.

2.2. Theorem. Suppose $ax + by = r$ and $cx + dy = s$ are linear equations in two variables x and y. If $ad - bc \neq 0$, then there is a unique common solution. If $ad - bc = 0$, then there is no common solution or there are infinitely many, depending on the values of r and s.

Proof. If all four coefficients are zero, then there is no solution unless $r = s = 0$, in which case all pairs (x, y) are solutions. Otherwise, at least one coefficient is nonzero. By interchanging the equations and/or interchanging the roles of x and y, we may assume that $d \neq 0$. We can now solve the second equation for y, obtaining $y = (s - cx)/d$. By substituting this expression for y into the first equation and simplifying, we obtain $(a - \frac{bc}{d})x + \frac{bs}{d} = r$. Multiplying by d yields $(ad - bc)x + bs = rd$.

When $ad - bc \neq 0$, we may divide by $ad - bc$ to obtain $x = \frac{rd - bs}{ad - bc}$. Substituting this into the equation for y yields the unique solution

$$(x, y) = (\frac{rd - bs}{ad - bc}, \frac{as - rc}{ad - bc}).$$

When $ad - bc = 0$, the equation for x becomes $bs = rd$. If $bs \neq rd$, then there is no solution. If $bs = rd$, then for each x we obtain the solution $(x, y) = (x, (s - cx)/d)$; here there are infinitely many solutions. ∎

When $ad - bc \neq 0$, the equations define lines with one common point. When $ad - bc = 0$ and both equations describe lines, there may be no solution (parallel lines) or infinitely many solutions (the lines coincide). An equation does not describe a line if both its coefficients are 0; here there is no solution unless the equation is $0x + 0y = 0$, in which case the common solutions are the solutions to the other equation.

In the proof, avoiding division by 0 leads us to consider cases. No single solution formula holds for all pairs of linear equations; the form of the solution changes when $ad - bc = 0$, and even the statement of the solution requires careful attention to language.

Our next argument uses a fundamental method called "proof by contradiction"; we suppose that the desired conclusion is false and then derive a contradiction from this supposition. The method is particularly useful for proving statements of non-existence. By combining the method of proof by contradiction with several elementary observations, we prove next that quadratic equations with odd coefficients have no rational solutions. For this proof we assume the definition of a rational number and elementary properties of odd and even numbers.

2.3. Theorem. If a, b, c are odd integers, then $ax^2 + bx + c = 0$ has no solution in the set of rational numbers.

Proof. Suppose there *is* a rational solution x. We write this as p/q for integers p, q. We may assume that p/q is an expression for x "in lowest terms", meaning that p and q have no common integer factor larger than 1. From $ax^2 + bx + c$ we obtain $ap^2 + bpq + cq^2 = 0$ after multiplying both sides by q^2.

We obtain a contradiction by showing that $ap^2 + bpq + cq^2$ cannot equal 0. We do this by proving the stronger statement that it is odd. Because we expressed x as a rational number in lowest terms, p and q cannot both be even. If both are odd, then the three terms in the sum are all odd, since the product of odd numbers is odd. Since the sum of three odd numbers is odd, we have the desired contradiction in this case. If p is odd and q is even (or vice versa), then we have the sum of two even numbers and an odd number, which again is odd. In each case, the assumption of a solution leads to a contradiction. ∎

QUANTIFIERS AND LOGICAL STATEMENTS

Understanding a subject and writing clearly about it go together. We next discuss how well-chosen words and symbols combine to express mathematical ideas clearly. Our later development does not require the terminology of this chapter, but we hope that this discussion helps the reader become comfortable with reading and writing mathematics.

Using proof by contradiction requires understanding what it means for a statement to be false. Consider the sentence "Every classroom has a chair that is not broken". Without using negative words, can we write a sentence with the opposite meaning? This will be easy once we learn how logical operations are expressed in English.

2.4. Example. *Negation of simple sentences.* What is the negation of "All students are male"? Some would say, incorrectly, "All students are not male". The correct negation is "At least one student is not male". Similarly, it is not true that all integers are odd, but that does not mean that all integers are not odd. ∎

Common English usage tolerates ambiguities, because the listener can understand the intended meaning from context. In mathematics we must avoid such ambiguities.

2.5. Example. *Word order is crucial.* Consider the sentence "There is a real number y such that $x = y^3$ for every real number x". This seems to say that some number y is the cube root of all numbers, which is false. To say that every number has a cube root, we write "For every real number x, there is a real number y such that $x = y^3$".

In both English and mathematics, meaning depends on the order of words. Compare "Mary made Jane eat the food", "Mary, eat the food Jane made", "Eat, Mary; Jane made the food", and "Eat the food Mary Jane made". Compare also the two mathematical sentences "For every $\varepsilon > 0$ and $a \in \mathbb{R}$, there exists a $\delta > 0$ such that $|f(x) - f(a)| < \varepsilon$ whenever $|x - a| < \delta$" and "For every $\varepsilon > 0$, there exists a $\delta > 0$ such that $|f(x) - f(a)| < \varepsilon$ whenever $|x - a| < \delta$ and $a \in \mathbb{R}$" (Exercise 2.10).

The meaning of an English sentence may change depending on emphasis or context. What does "The bartender served two aces" mean? The answer depends on whether we are watching tennis or relaxing in a bar on an airbase (see Exercise 2.3b). Mathematics is broad enough to present similar difficulties; words such as "square" and "cycle" have several mathematical meanings. ∎

The fundamental issue in mathematics is whether mathematical statements are true or false. Before discussing proofs, we must agree on what assertions we accept as mathematical statements. We first require that such an assertion be a grammatically correct sentence, possibly involving words and mathematical symbols. Grammar eliminates both "food Mary Jane" and "1+ =".

Among the grammatically correct sentences, we accept as a mathematical statement any arithmetic assertion about specified sets and numbers where performing the computations determines whether the assertion is true or false. The sentences "$1 + 1 = 3$" and "$1 + 1 < 3$" are mathematical statements, even though the first is false. Similarly, "$(1 + 1)^{(4 \cdot 3)}$ is 96 more than 4000" is acceptable. We extend this computational criterion to more complicated operations and to objects defined using sets and numbers.

We also consider general assertions about many numbers or objects, such as "the square of each odd integer is one more than a multiple of 8". This statement is closely related to the sequence of statements "$1^2 = 1 + 0 \cdot 8$", "$3^2 = 1 + 1 \cdot 8$", "$5^2 = 1 + 3 \cdot 8$", $\cdots$. We can describe many related mathematical statements by introducing a *variable*. When the variable takes a specific value, the assertion becomes a mathematical statement. If $P(x)$ is a mathematical statement when x takes a specific value in a given set S, then we accept as mathematical statements the sentences below:

"For all x in S, the assertion $P(x)$ is true"
"There exists an x in S such that the assertion $P(x)$ is true"

2.6. Example. The sentence "$x^2 - 1 = 0$" by itself is not a mathematical statement, but it becomes one when we specify a value for x. The sentence "For all $x \in \{1, -1\}$, $x^2 - 1 = 0$" is a mathematical statement. This statement is true; there are two values of x to check, and each satisfies the conclusion. Also "For all $x \in \{1, 0\}$, $x^2 - 1 = 0$" is a mathematical statement; it is false. The sentence "There exists $x \in \{1, 0\}$ such that $x^2 - 1 = 0$" is a mathematical statement that is true. ∎

If it is not possible to assign "True" or "False" to an assertion, then it is not a mathematical statement. Consider the sentence "This statement is false;" call it P. If the words "this statement" in P refer to some other sentence Q, then P has a truth value. Without a context, one must assume that "this sentence" refers to P itself. In that case, P must be false if it is true, and true if it is false! Without a context, P makes no sense and hence has no truth value.

2.7. Remark. *Terminology and notation.* We use uppercase letters "$P, Q, R \cdots$" to denote (mathematical) statements. The truth or falsity of a statement is its *truth value*. Negating a statement reverses its truth value. We use ¬ to indicate negation, so "¬P" means "not P". If P is a false statement, then ¬P is a true statement.

In the statement "For all x in S, $P(x)$ is true", the variable x is *universally quantified*. In symbols, we write this as $(\forall x \in S)P(x)$ and say that $\forall$ is a *universal quantifier*. In the statement "There exists an x in S such that $P(x)$ is true", the variable x is *existentially quantified*. In symbols, we write this as $(\exists x \in S)P(x)$ and say that $\exists$ is an *existential quantifier*. The set of allowed values for a variable is its *universe*. ∎

Informally, we also refer to English words that express quantification as quantifiers. Typically, "every" and "for all" are universal quantifiers, while "some" and "there is" are existential quantifiers. Universal quantification can also be expressed by referring to an arbitrary element of the univers, as in "Let x be any integer," or "A student failing the exam will fail the course". Below we list common words used as quantifiers; the "helpers" don't always appear:

Universal (∀)	(helpers)	Existential (∃)	(helpers)
for all, for every	\<comma\>	for some	\<comma\>
if	then	there exists	such that
whenever		at least one	for which
every, any	satisfies	some	satisfies
a	must	has a	such that

2.8. Remark. *Order of quantifiers.* In conversation, a quantifier may appear after the expression it quantifies. "I drink whenever I eat" differs from "Whenever I eat, I drink" only in what is emphasized. Similarly, we easily understand the sentences "The AGM Inequality states that $(a + b)/2 \geq \sqrt{ab}$ for every pair a, b of positive real numbers" and "The value of $x^2 - 1$ is 0 for some x between 0 and 2". These quantifiers appear at the end for smoother reading. Error is unlikely when a sentence has one quantifier, but the order of quantification matters when a sentence has more than one.

We adopt a convention to avoid ambiguity. Consider the sentence "If n is even, then n is the sum of two odd numbers". Letting E and O denote the sets of even and odd integers, we write this as $(\forall n \in E)(\exists x, y \in O)P(n, x, y)$, where $P(n, x, y)$ is "$n = x + y$". In this format, the value chosen for a quantified variable remains fixed for later expressions but can be chosen in terms of variables quantified earlier. When we reach $(\exists x, y \in O)P(n, x, y)$ in the statement above, we treat "n" as a constant, already chosen. We use the same convention in mathematical English: place quantifications in proper order at the beginning of the sentence. The value of each variable is chosen independently of the subsequently quantified variables, viewing the previously quantified variables as constants. ∎

2.9. Example. *Interpretation of* $(\forall x)(\exists y)(\forall z)P(x, y, z)$. Suppose we play a game against an intelligent adversary. We want to obtain x, y, z such that $P(x, y, z)$ is true. The adversary wants to prevent this. We choose the existentially quantified variables; the adversary chooses the universally quantified ones. The adversary starts by picking x. After seeing x, we choose y. Having seen x and y, the adversary picks z. The adversary wins if this choice of z makes $P(x, y, z)$ false. We win (i.e., the sentence is true) if the adversary can't win. When we pick y, we do not know which z will be chosen; we must pick a y that works for all z. On the other hand, since x was chosen earlier, we have the freedom to pick a different y for each possible x. Universal quantification means that for each possible x, we can pick a winning y. ∎

2.10. Remark. *Negation of quantified statements.* After locating the quantifiers and variables, negating a statement involving quantifiers is easy. If it is false that $P(x)$ is true for every value of x, then there must be some value of x such that $P(x)$ is false, and vice versa. In notation, $\neg[(\forall x)P(x)]$ means the same as $(\exists x)(\neg P(x))$. Similarly, if it is false that there is some value of x for which $P(x)$ is true, then $P(x)$ is false for every value of x: $\neg[(\exists x)P(x)]$ means the same as $(\forall x)(\neg P(x))$. ∎

When using logical symbols, we may add matched parentheses or brackets to clarify grouping. We also simplified the expressions above by dropping the specification of the universe for the variables; we did not need to know the particular universes, only that they exist.

When negating quantified statements with specified universes, one must take care not to change the universe of potential values.

2.11. Example. *Negation involving universes.* The negation of "Every Good Boy Does Fine" (a mnemonic for reading music) is "some good boy does not do fine"; it says nothing about bad boys. The negation of "Every chair in this room is broken" is "Some chair in this room is not broken"; it says nothing about chairs outside this room.

Similarly, the negation of the statement $(\forall n \in \mathbb{N})(\exists x \in A)(nx < 1)$ is $(\exists n \in \mathbb{N})(\forall x \in A)(nx \geq 1)$. The negated sentence means that the set A has a positive lower bound that is the reciprocal of an integer. It does not mention values of n outside $\mathbb{N}$ or values of x outside A. ∎

2.12. Example. Let us rephrase "It is false that every classroom has a chair that is not broken". The quantifiers make it improper to cancel the "double negative"; the sentence "every classroom has a chair that is broken" has a different meaning. The original statement has a universal quantifier ("every") and an existential quantifier ("has a"). By successively rewording the negated quantifiers, we obtain first "There is a classroom that has no chair that is not broken" and then "There is a classroom in which every chair is broken". In symbols, the statements are $\neg[(\forall r \in R)(\exists c \in C(r))(\neg B(c))]$, then $(\exists r \in R)(\neg[(\exists c \in C(r))(\neg B(c))])$, and finally $(\exists r \in R)(\forall c \in C(r))B(c)$, all having the same meaning. ∎

COMPOUND STATEMENTS AND SETS

There are various ways to change or to combine logical statements to obtain new ones. For example, the negation of a logical statement is another logical statement. We also want to represent "and", "or", "if and only if", and "implies". For each choice of truth values for the component statements, the compound statement also has a truth value. We will also discuss the meaning of these combinations in terms of sets.

2.13. Definition. *Logical connectives.* In the following table, we define the operations named in the first column by the truth values specified in the last column.

Name	Notation	Meaning	When true
Negation	$\neg P$	"not P"	P false
Conjunction	$P \wedge Q$	"P and Q"	P, Q both true
Disjunction	$P \vee Q$	"P or Q"	at least one of P, Q true
Biconditional	$P \Leftrightarrow Q$	"P if & only if Q"	P, Q same truth value
Conditional	$P \Rightarrow Q$	"P implies Q"	unless P true & Q false

2.14. Remark. *Disjunctions.* The meaning of "or" in mathematics differs from its common usage in English. Answering "Yes" to "Are you going home or not?" causes annoyance despite being logically correct; in common English the word "or" means "one or the other but not both". In mathematics, this usage is called "exclusive or"; we reserve "or" for disjunction. Disjunction is mathematically more common than "exclusive or" because "and" and "or" act as quantifiers. A conjunction is true if *all* of its component statements are true; "and" is a universal quantifier. A disjunction is true if *at least one* of its component statements is true; "or" is an existential quantifier. ∎

In the conditional statement $P \Rightarrow Q$, we call P the *hypothesis* and Q the *conclusion*. The statement $Q \Rightarrow P$ is the *converse* of $P \Rightarrow Q$.

2.15. Example. *Conditionals.* The conditional is the only connective in Definition 2.13 for which the meaning of the compound statement changes when P and Q are interchanged. There is no general relationship between the truth values of $P \Rightarrow Q$ and $Q \Rightarrow P$. If P is the statement "2 is odd", and Q is the statement "4 is even", then $P \Rightarrow Q$ is true, but $Q \Rightarrow P$ is false. If R is the statement "4 is odd", then the statements $P \Rightarrow R$ and $R \Rightarrow P$ are both true. A conditional statement is false when and only when the hypothesis is true and the conclusion is false. It may be more helpful to read the conditional as "if-then" rather than "implies". Below we list a few of the ways to phrase $P \Rightarrow Q$ in English. ∎

> If P (is true), then Q (is true).
> Q is true whenever P is true.
> Q is true if P is true.
> P is true only if Q is true.
> P is a sufficient condition for Q.
> Q is a necessary condition for P.

Using connectives, we can build complicated logical expressions from elementary statements. We treat the elementary statements as variables from the universe {True, False}. For each choice of True/False for the variables, we can use the behavior of the connectives to compute the truth value of the full expression. The list of these outcomes is called a "truth table".

Two logical expressions X, Y are *logically equivalent* if they have the same truth value for each assignment of truth values to the variables representing statements. Equivalences allow us to rephrase a statement in a convenient way.

2.16. Remark. *Elementary logical equivalences.* We may substitute P for $\neg(\neg P)$ whenever we wish, and vice versa. Similarly, $P \vee Q$ is equivalent to $Q \vee P$, and $P \wedge Q$ is equivalent to $Q \wedge P$. Whenever P and Q are statements, we may substitute the expression in the right column below for the corresponding expression in the left column (or vice versa); they always have the same truth value.

1)	$\neg(P \wedge Q)$	$(\neg P) \vee (\neg Q)$
2)	$\neg(P \vee Q)$	$(\neg P) \wedge (\neg Q)$
3)	$\neg(P \Rightarrow Q)$	$P \wedge (\neg Q)$
4)	$P \Leftrightarrow Q$	$(P \Rightarrow Q) \wedge (Q \Rightarrow P)$
5)	$P \vee Q$	$(\neg P) \Rightarrow Q$
6)	$P \Rightarrow Q$	$(\neg Q) \Rightarrow (\neg P)$

We could verify these equivalences by manipulating symbols in truth tables, but we prefer to explain why the expressions in a row are equivalent by emphasizing the meaning of the English words used to state the connectives. When we view disjunction and conjunction as quantifiers over their component statements, for example, our understanding of negating quantified statements yields (1) and (2) above, which are called *de Morgan's laws* in honor of the logician Augustus de Morgan (1806-1871).

Equivalence (3) is the definition of the conditional. Both sides of (4) hold when P and Q have the same truth value, while both fail when P and Q have different truth values. The right side of (5) is false precisely when P fails and Q fails, which is when the left side is false. Each side of (6) fails precisely when P is true and Q is false. ∎

2.17. Remark. *Logical connectives and membership in sets.* Suppose $P(x)$ and $Q(x)$ are statements about a particular element x from a universe U. Suppose also that $A = \{x \in U: P(x) \text{ is true}\}$ and $B = \{x \in U: Q(x) \text{ is true}\}$. The statement $(\forall x \in U)(P(x) \Rightarrow Q(x))$ is equivalent to the statement $A \subseteq B$. When the universe for the variable x is understood, we may use the informal expression $P \Rightarrow Q$ for this (but note that $(\forall x \in U)(P(x) \Rightarrow Q(x))$ differs from $(\forall x \in U)P(x) \Rightarrow (\forall x \in U)Q(x)$ - Exercise 2.24). The converse $Q \Rightarrow P$ is equivalent to $B \subseteq A$. Hence the biconditional $P \Leftrightarrow Q$ is equivalent to $A = B$.

When we start with sets A, B and prove $A = B$ by proving that every element of A belongs to B and every element of B belongs to A, we are proving $P \Rightarrow Q$ and $Q \Rightarrow P$, where P is the statement of membership in A and Q is the statement of membership in B. Other logical connectives correspond to other set operations. ∎

$x \in A^c$	$\Leftrightarrow$	$\text{not } (x \in A)$	$\Leftrightarrow$	$\neg(x \in A)$
$x \in A \cup B$	$\Leftrightarrow$	$(x \in A \text{ or } x \in B)$	$\Leftrightarrow$	$(x \in A) \vee (x \in B)$
$x \in A \cap B$	$\Leftrightarrow$	$(x \in A \text{ and } x \in B)$	$\Leftrightarrow$	$(x \in A) \wedge (x \in B)$
$A \subseteq B$	$\Leftrightarrow$	$(\forall x \in A)(x \in B)$	$\Leftrightarrow$	$(x \in A) \Rightarrow (x \in B)$

We can translate any relationship between sets into a logical statement about membership, and vice versa. The two expressions tell the
same story in different languages, which must not be mixed. For example, $A \cap B$ is a set, not a statement; it has no truth value, and the notation "$(A \cap B)^c \Leftrightarrow A^c \cup B^c$" has no meaning.

2.18. Remark. *de Morgan's laws for sets.* In the language of sets, de
Morgan's laws become 1) $(A \cap B)^c = A^c \cup B^c$, and 2) $(A \cup B)^c = A^c \cap B^c$. We
verify (1) by translating both sides into a logical equivalence, leaving (2)
to Exercise 2.26. Given an element x, let P be the property $x \in A$, and
let Q be the property $x \in B$. Remarks 2.16 and 2.17 imply that

$$x \in (A \cap B)^c \Leftrightarrow \neg(P \wedge Q) \Leftrightarrow (\neg P) \vee (\neg Q) \Leftrightarrow (x \notin A) \vee (x \notin B)$$

Alternatively, a Venn diagram makes the identity apparent. ∎

ELEMENTARY PROOF TECHNIQUES

The business of mathematics is deriving consequences from
hypotheses; hence proofs in mathematics require verifying conditional
statements (implications). Biconditional statements and universally
quantified statements both lead to conditionals. Although sometimes
we can prove a biconditional by a chain of equivalences, as in Remark
2.18, we usually prove a biconditional by proving a conditional and its
converse, as suggested by Equivalence (4) in Remark 2.16. We also can
prove the truth of a universally quantified statement like $(\forall x \in A)Q(x)$
by proving the conditional statement "If $x \in A$, then $Q(x)$ is true".

2.19. Remark. *Elementary methods of proving* $P \Rightarrow Q$. The *direct
method* of proving that P implies Q is to assume that P is true and then
to apply mathematical reasoning to deduce that Q is true. When P is
"$x \in A$" and Q is "$Q(x)$", the direct method considers an *arbitrary* $x \in A$
and deduces that $Q(x)$ is true. This must not be confused with the
invalid "proof by example"; the proof must consider every member of A
as a possible instance of x.

Equivalence (6) in Remark 2.16 suggests another method. We call
$\neg Q \Rightarrow \neg P$ the *contrapositive* of $P \Rightarrow Q$. The equivalence between a conditional and its contrapositive allows us to prove $P \Rightarrow Q$ by proving
$\neg Q \Rightarrow \neg P$. This is the *contrapositive method.*

The meaning of the conditional (Equivalence (3) in Remark 2.16)
leads to a related method. We know that $P \Rightarrow Q$ is false if and only if P
is true and Q is false. Equivalences (3) and (1) together imply
$(P \Rightarrow Q) \Leftrightarrow \neg[P \wedge (\neg Q)]$. Hence we can prove $P \Rightarrow Q$ by proving that P
and $\neg Q$ cannot both be true. We do this by obtaining a contradiction
after assuming both P and $\neg Q$. This is the *method of contradiction* or

indirect proof. We summarize these methods below:

> **Direct proof**: Assume P, follow logical deductions, conclude Q.
> **Contrapositive**: Assume $\neg Q$, follow deductions, conclude $\neg P$.
> **Method of Contradiction**: Assume P and $\neg Q$, follow deductions, obtain a contradiction. ■

We begin with easy examples of the direct method, including statements used in proving Theorem 2.3.

2.20. Example. *If integers x and y are both odd, then $x + y$ is even.*
Suppose that x and y are odd. By the definition of "odd", there exist integers k, l such that $x = 2k + 1$ and $y = 2l + 1$. By the properties of addition and the distributive law, $x + y = 2k + 2l + 2 = 2(k + l + 1)$. This is twice an integer, so $x + y$ is even.

The converse of this conditional is false. When x, y are integers, it is possible that $x + y$ is even but x, y are not both odd. Compare this with the next example. ■

2.21. Example. *An integer is even if and only if it is the sum of two odd integers.* First we clarify what must be proved. Formally, the statement is $(\forall x \in \mathbb{Z})[(\exists k \in \mathbb{Z})(x = 2k) \Leftrightarrow (\exists y, z \in O)(x = y + z)]$, where O is the set of odd numbers. If $x = 2k$ is even, then $x = (2k - 1) + 1$, which expresses x as the sum of two odd integers. Conversely, suppose y and z are odd. By the definition of "odd", there exist integers k, l such that $y = 2k + 1$ and $z = 2l + 1$. Then $y + z = 2k + 1 + 2l + 1 = 2(k + l + 1)$, which is even. ■

2.22. Example. *If x and y are odd, then xy is odd.* If x and y are odd, then there are integers k, l such that $x = 2k + 1$ and $y = 2l + 1$. Now $xy = 4kl + 2k + 2l + 1 = 2(2kl + k + l) + 1$. Since this is one more than twice an integer, xy also is odd. ■

A special case of Example 2.22 is "x odd $\Rightarrow x^2$ odd". Here the conclusion is "There exists an integer m such that $x^2 = 2m + 1$". To prove an existential conclusion, we need only provide an *example*: in this case a value m (constructed in terms of x) such that the statement is true. For this reason, the direct method often succeeds when the conclusion of the desired implication is an existential statement.

2.23. Example. *An integer is even if and only if its square is even.* If n is even, then we can write $n = 2k$, where k is an integer. Now $n^2 = 4k^2 = 2(2k^2)$, proving that "n even" implies "n^2 even" by the direct method. For the converse, we want to prove "n^2 even implies n even", but this we have already done! Since integers are even or odd, the desired implication is the contrapositive of "n odd implies n^2 odd". ■

2.24. Remark. *Converse versus contrapositive.* Proving the biconditional statement $P \Leftrightarrow Q$ requires proving one statement from each column below. Each statement is the converse of the other statement in its row. Each statement is the contrapositive of the other statement in its column. Every conditional statement is equivalent to its contrapositive, so proving the two statements in the same column would be proving the same statement twice.

$$P \Rightarrow Q \qquad Q \Rightarrow P$$
$$\neg Q \Rightarrow \neg P \qquad \neg P \Rightarrow \neg Q$$

For example, consider "the product of two nonzero real numbers is positive if and only if they have the same sign". The axioms for real numbers allow us to prove that if x and y have the same sign, then xy is positive. We might then argue, "Now suppose that xy is negative. This implies that x and y have opposite signs." This accomplishes nothing; we have proved the contrapositive of the first conditional, not its converse. Instead, we must prove "If xy is positive, then x and y have the same sign" or "If x and y have opposite signs, then xy is negative".

We can interpret the first line of the display above as the direct method and the second line as the contrapositive method. To include the method of contradiction, we could add the line below. ∎

$$\neg(P \wedge \neg Q) \qquad \neg(Q \wedge \neg P)$$

The next example uses the contrapositive and illustrates that care must be taken to avoid unjustified assumptions.

2.25. Example. Consider the statement *"If $f(x) = mx + b$, then the values of f are different for distinct values of x."* We are being asked to prove that $x \neq y$ implies $f(x) \neq f(y)$. The direct method would consider $x < y$ and $x > y$ separately and obtain $f(x) < f(y)$ or $f(x) > f(y)$. This unsatisfying analysis by cases results from "not equals" being a messier condition than "equals". We can use the contrapositive to retain the language of equalities and reduce analysis by cases. If $f(x) = f(y)$, we obtain $mx + b = my + b$ and then $mx = my$. If $m \neq 0$, then we obtain $x = y$. If $m = 0$, we cannot divide by m, and in fact the statement is false. The difficulty is that m is a variable in the statement we want to prove, and we cannot determine its truth without quantifying m. The statement is true if and only if $m \neq 0$. ∎

The method of contradiction proves $P \Rightarrow Q$ by proving that P and $\neg Q$ cannot both hold, thereby proving that $P \Rightarrow Q$ cannot be false.

2.26. Example. *If a is less than every real number greater than b, then $a \leq b$.* The direct method goes nowhere, but when we say "suppose not", the light begins to dawn. The assumption $a > b$ (the negation of the conclusion $a \le b$) provides a way to construct a counterexample x to the hypothesis. Let $x = (a+b)/2$, the average of a and b. If $a > b$, then $x = (a+b)/2$ implies $a > x > b$. Now $x > b$ and a is not less than x. We have proved the contrapositive of the desired implication. ∎

In Example 2.26, the hypothesis P is a universally quantified statement about every real number x greater than b. The contrapositive is easy to prove because the negation of P is an existence statement; having assumed $a > b$, we need only construct a real number x for which P is false. Proving the negation of a universally quantified statement $(\forall x \in U)P(x)$ is the same as providing a *counterexample*, a value of x in U such that $P(x)$ is false.

2.27. Example. *Every list of numbers contains a number as large as its average.* Given numbers $y_1, \ldots, y_n$, let Y be their sum. The average z is defined to be Y/n. We can prove the claim directly or indirectly. For an indirect proof, we say "suppose the conclusion is false". This means $y_i < z$ for all y_i in the list. If we sum these inequalities, we obtain $Y < nz$, but this contradicts the definition of z, which yields $Y = nz$. Hence the assumption that each element was too small must be false.

For a direct proof, we construct the desired number. Let y^* be the largest number in the set. This is a likely candidate to be as large as the average; we prove that it works. Since $y_i \le y^*$ for all i, we compute $z = Y/n \le ny^*/n = y^*$. ∎

In Example 2.27, the "hypothesis" is that $\{y_i\}$ is a set of n numbers. We didn't derive its negation; we obtained a different contradiction. This is the method of contradiction. Like the contrapositive method, it begins by assuming $\neg Q$ when proving $P \Rightarrow Q$. We need not decide in advance whether to deduce $\neg P$ or to use both P and $\neg Q$ to obtain some other contradiction.

2.28. Example. *There is no largest real number.* If there is a largest real number z, then for all $x \in \mathbb{R}$, $z \ge x$. Since $z + 1$ is also a real number, this contradicts the inequality $z < z + 1$. ∎

The method of contradiction works well when the conclusion is a statement of non-existence or impossibility, because negating the conclusion provides an *example* to use, like p/q in the proof of Theorem 2.3 or z in Example 2.28. In one sense the method of contradiction ("indirect proof") has more power than the contrapositive, since we start with more information (P *and* $\neg Q$), but in another sense it is less satisfying, because we start with a situation that (we hope) cannot be true.

2.29. Remark. *The consequences of false statements.* Recall that a conditional statement is false only if the hypothesis is true and the conclusion is false. When there is no way for the hypothesis to be true, we say that the conditional follows *vacuously*. Similarly, every statement that is universally quantified over an empty set is true; when there are no dogs in the class, the statement "Every dog in the class has three heads" is true. In contrast, every statement that is existentially quantified over an empty set is false; when there are no dogs in the class, the statement "Some dog in the class has four legs" is false!

Returning to the conditional, we have argued that $P \Rightarrow Q$ is true whenever P is false. This explains why a proof containing a single error in reasoning cannot be considered "nearly correct"; we can derive any conclusion from a single false statement (see Exercise 2.19a). Bertrand Russell (1872-1970) once stated this in a public lecture and was challenged to start with the assumption that $1 = 2$ and prove that he was God. He replied, "Consider the set {Russell, God}. If $1 = 2$, then the two elements of the set are one element, and therefore Russell = God." ∎

We close by proving another biconditional. We state two identities involving three arbitrary sets; in the Venn diagrams, we place "•" in the regions corresponding to the sets discussed.

(1) 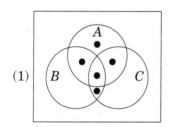   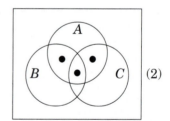 (2)

2.30. Proposition. Union and intersection distribute over each other. In particular, given arbitrary sets A, B, C, we have

1) $A \cup (B \cap C) = (A \cup B) \cap (A \cup C)$.
2) $A \cap (B \cup C) = (A \cap B) \cup (A \cap C)$.

Proof. We leave the proof of (2) to Exercise 2.27 and consider only (1). First suppose $x \in A \cup (B \cap C)$. If x is in A, then x is in both $A \cup B$ and $A \cup C$. If x is in both B and C, then x is in both $A \cup B$ and $A \cup C$. Hence $x \in A \cup (B \cap C)$ implies $x \in (A \cup B)$ and $x \in (A \cup C)$, which proves $A \cup (B \cap C) \subseteq (A \cup B) \cap (A \cup C)$.

To prove containment in the other direction, we suppose that $x \in (A \cup B) \cap (A \cup C)$. If x is in A, then x is in $A \cup (B \cap C)$. If x is not in A, then $(x \in (A \cup B)) \wedge (x \in (A \cup C))$ implies that x must be in both B and C. In either case, x is in A or in $B \cap C$. Hence $(A \cup B) \cap (A \cup C) \subseteq A \cup (B \cap C)$, and the two sets are equal. ∎

EXERCISES

2.1. (–) *The quadratic formula.* Suppose $a, b, c \in \mathbb{R}$ with $a \neq 0$. Assuming that $a^2 + bx + c$ can be factored as $a(x - r)(x - s)$ for real numbers r, s, setting x to be r or s shows that r and s are solutions of the quadratic equation $ax^2 + bx + c = 0$. Using the steps below, solve for r, s in terms of a, b, c to obtain another derivation of the quadratic formula.

a) By equating coefficients of corresponding powers of x, obtain the equations $r + s = -b/a$ and $rs = c/a$.

b) Prove that $(r - s)^2 = (b^2 - 4ac)/a^2$. (Hint: $(r - s)^2 = (r + s)^2 - 4rs$.)

c) From (a) and (b), obtain $r + s = -b/a$ and $r - s = \sqrt{b^2 - 4ac}/a$. Solve this system for r, s in terms of a, b, c.

d) Alternatively, we could use $r - s = -\sqrt{b^2 - 4ac}/a$ as the second equation. Show that this interchanges the roles of r and s and hence does not change the set of solutions.

2.2. (–) "In one year, my wife will be one-third as old as my house. In nine years, I will be half as old as my house. I am ten years older than my wife. How old are my house, my wife, and I?" Answer the question, describing clearly the mathematical equations that are needed.

2.3. (–) *Usage of language.*

a) The following sentence appeared on a restaurant menu: "Please note that every alternative may not be available at this time". Describe the unintended meaning. Rewrite the sentence to state the intended meaning clearly.

b) Give an example of an English sentence that has different meaning depending on inflection, pronunciation, or context.

2.4. (–) The negation of the statement "No slow learners attend this school" is:[†]

a) All slow learners attend this school.

b) All slow learners do not attend this school.

c) Some slow learners attend this school.

d) Some slow learners do not attend this school.

e) No slow learners attend this school.

2.5. Abraham Lincoln said, "You can fool all of the people some of the time, and you can fool some of the people all of the time, but you can't fool all of the people all of the time." Write this sentence in logical notation, negate the symbolic sentence, and state the negation in English.

2.6. In simpler language, describe the meaning of the following two statements and their negations. Which one implies the other, and why?

a) There is a number M such that, for every x in the set S, $|x| \leq M$.

b) For every x in the set S, there is a number M such that $|x| \leq M$.

2.7. Consider the sentence "For every integer $n > 0$ there is some real number $x > 0$ such that $x < 1/n$". Without using words of negation, write a complete sentence that means the same as "It is false that for every integer $n > 0$ there is some real number $x > 0$ such that $x < 1/n$". Which sentence is true?

[†]This appeared on the 1955 High School Mathematics Exam (C.T. Salkind, *Annual High School Mathematics Examinations 1950-1960*, Math. Assn. Amer. (1961), p37.)

2.8. Express each of the following statements as a conditional statement in "if-then" form or as a universally quantified statement. Also write the negation.

a) Every odd number is prime.

b) The sum of the angles of a triangle is 180 degrees.

c) Passing the test requires solving all the problems.

d) Being first in line guarantees getting a good seat.

e) Lockers must be turned in by the last day of class.

f) Haste makes waste.

g) I get mad whenever you do that.

h) I won't say that unless I mean it.

2.9. (!) Consider tokens that have some letter written on one side and some integer written on the other, in unknown combinations. The tokens are laid out, some with letter side up, some with number side up. Explain which tokens must be turned over to determine whether these statements are true:

a) Whenever the letter side is a vowel, the number side is odd.

b) The letter side is a vowel if and only if the number side is odd.

2.10. Which of the statements below implies the other?

a) For every $\varepsilon > 0$ and every real number a, there is a $\delta > 0$ such that $|f(x) - f(a)| < \varepsilon$ whenever $|x - a| < \delta$.

b) For every $\varepsilon > 0$, there is a $\delta > 0$ such that $|f(x) - f(a)| < \varepsilon$ whenever $|x - a| < \delta$ and a is a real number.

2.11. (!) Suppose I have a one-dollar bill, a ten-dollar bill, and a 100-dollar bill. Suppose I declare the following: "If you make a true statement, I will give you one of the bills. If you make a false statement, I will not give you anything." What should you say to maximize the amount I am forced to give you?

2.12. A fraternity has a rule for new members: each must always tell the truth or always lie. They know who does which. If I meet three of them on the street and they make the statements below, which ones (if any) should I believe?

A says: "All three of us are liars."

B says: "Exactly two of us are liars."

C says: "The other two are liars."

2.13. Three children are in line. From a collection of two red hats and three black hats, the teacher places a hat on each child's head. The third child sees the hats on the heads of the first two, the middle child sees the hat on the first child, and the first child sees no hats. The children, who reason carefully, are told to speak out as soon as they can determine the color of the hat they are wearing. After 30 seconds, the front child correctly names the color of her hat. Which color is it, and why?

2.14. (!) It is well known that $3^2 + 4^2 = 5^2$.

a) Is there any other triple of three consecutive natural numbers such that the square of one is the sum of the squares of the other two?

b) Is there any triple of three consecutive natural numbers such that the cube of one is the sum of the cubes of the other two?

2.15. Suppose x and y are integers. Determine whether each of the following statements is true.

    a) xy is odd if and only if x and y are odd.

    b) xy is even if and only if x and y are even.

2.16. (!) *Checkerboard problems.* (Hint: use the method of contradiction.)

    a) Suppose two opposite corner squares are deleted from an eight by eight checkerboard. Prove that the remaining squares cannot be covered exactly by dominoes (rectangles consisting of two adjacent squares).

    b) Suppose two squares from each of two opposite corners are deleted as indicated on the right below. Prove that the remaining squares cannot be covered exactly by copies of the "T-shape" and its rotations.

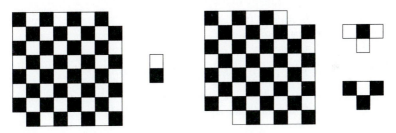

2.17. If a closet contains n different pairs of shoes, determine the minimum number of shoes that must be extracted to guarantee that at least one matching pair of shoes is obtained. What is the minimum number to guarantee that two matching pairs of shoes are obtained?

2.18. Using the equivalences discussed in Example 2.16, write a chain of symbolic equivalences to prove that $P \Leftrightarrow Q$ is logically equivalent to $Q \Leftrightarrow P$.

2.19. Suppose P and Q are statements. Prove that the following are true.

    a) $(Q \wedge \neg Q) \Rightarrow P$.   b) $P \wedge Q \Rightarrow P$.   c) $P \Rightarrow P \vee Q$.

2.20. Prove that the statements $P \Rightarrow Q$ and $Q \Rightarrow R$ imply $P \Rightarrow R$, and that the statements $P \Leftrightarrow Q$ and $Q \Leftrightarrow R$ imply $P \Leftrightarrow R$. (Comment: This is the justification for using a chain of equivalences to prove an equivalence.)

2.21. Prove that the logical expression S is equivalent to the logical expression $\neg S \Rightarrow (R \wedge \neg R)$, and explain the relationship between this equivalence and the method of proof by contradiction.

2.22. Consider the following sentence: "If a is a real number, then $ax = 0$ implies $x = 0$". Write this sentence symbolically using quantifiers, letting $P(a, x)$ be the assertion "$ax = 0$" and $Q(x)$ be the assertion "$x = 0$". Show that the implication is false, and find a small change in the quantifiers to make it true.

2.23. Let $P(x)$ be the assertion "x is odd", and let $Q(x)$ be the assertion "$x^2 - 1$ is divisible by 8". Determine whether the following statements are true:

    a) $(\forall x \in \mathbb{Z})(P(x) \Rightarrow Q(x))$.   b) $(\forall x \in \mathbb{Z})(Q(x) \Rightarrow P(x))$.

2.24. Let $P(x)$ be the assertion "x is odd", and let $Q(x)$ be the assertion "x is twice an integer". Determine whether the following statements are true:

a) $(\forall x \in \mathbb{Z})(P(x) \Rightarrow Q(x))$.   b) $(\forall x \in \mathbb{Z})(P(x)) \Rightarrow (\forall x \in \mathbb{Z})(Q(x))$.

2.25. Let $S = \{x \in \mathbb{R}: x^2 > x + 6\}$. Let $T = \{x \in \mathbb{R}: x > 3\}$. Determine whether the following statements are true, and interpret these results in words:

a) $T \subseteq S$.   b) $S \subseteq T$.

2.26. (–) Prove the second of de Morgan's laws: $(A \cup B)^c = A^c \cap B^c$.

2.27. Prove the second distributive law of sets:
$A \cap (B \cup C) = (A \cap B) \cup (A \cap C)$.

2.28. Prove the following identities about sets.

a) $A \cap [(A \cap B)^c] = A - B$
b) $A \cap [(A \cap B^c)^c] = A \cap B$
c) $(A \cup B) \cap A^c = B - A$.

2.29. If A, B, C are sets, prove that $A \cap (B - C) = (A \cap B) - (A \cap C)$.

2.30. (!) Suppose that A, B, C are sets. Prove that $(A \cup B) - C$ is a subset of $[A - (B \cup C)] \cup [B - (A \cap C)]$, but that equality need not hold.

2.31. Suppose statements P, Q, R depend on a variable x. Translate the statements below into the language of sets, and use this to explain why they are always true.

1) $(P \Leftrightarrow Q) \Leftrightarrow (\neg P \Leftrightarrow \neg Q)$.
2) $[(P \Rightarrow Q) \wedge (Q \Rightarrow R)] \Rightarrow (P \Rightarrow R)$. (Transitivity)
3) $[P \vee (Q \wedge R)] \Leftrightarrow [(P \vee Q) \wedge (P \vee R)]$. (Distributivity)
4) $[P \wedge (Q \vee R)] \Leftrightarrow [(P \wedge Q) \vee (P \wedge R)]$. (Distributivity)
5) $[(P \vee Q) \vee R] \Leftrightarrow [P \vee (Q \vee R)]$. (Associativity)
6) $[(P \wedge Q) \wedge R] \Leftrightarrow [P \wedge (Q \wedge R)]$. (Associativity)

2.32. (+) Suppose three circles in the plane intersect as shown below. Each region contains a token that is white on one side and black on the other. At each step, we can either (a) flip all four tokens inside one of the three circles, or (b) flip those tokens showing white inside one of the three circles to make all four tokens in that circle show black. From the starting configuration with all tokens black, can we reach the indicated configuration with all tokens black except for the token in the central region? (Hint: consider parity conditions.)

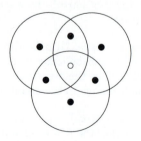

Chapter 3

Functions

We use functions to express relationships between sets. These sets may consist of numbers, geometric objects, or functions themselves. In this chapter, we present the definition of a function, exhibit examples used later in this book, and study general properties of functions.

A function may allow us to express one measurement in terms of another, thereby relating two scales.

3.1. Example. *Fahrenheit vs. Celsius.* Two common scales for measuring temperature are Fahrenheit and Celsius. The Celsius measurement can be converted into Fahrenheit temperature by multiplying by 9/5 and then adding 32. The Fahrenheit measurement can be converted into the corresponding Celsius temperature by subtracting 32 and then multiplying by 5/9. ■

The process of obtaining the Celsius value from the Fahrenheit value is reversible and hence provides a "one-to-one correspondence" between the values on the two scales. We explore this notion in our study of functions.

We can also determine the existence and uniqueness of solutions to equations using properties of functions. For what functions f can we always solve the equation $f(x) = b$? When is the solution unique?

These fundamental questions are generally difficult. In this chapter, we study them and answer them in some special cases. This will enable us to solve the following problem in Solution 3.37.

3.2. Problem. *A Special Polynomial.* Can $1 + x + x^2 + \cdots + x^{2k}$ ever equal 0? We will answer this by solving the more general problem of determining when $x^{2k} + x^{2k-1}y + \cdots + xy^{2k-1} + y^{2k}$ can be 0. ■

DEFINITIONS AND EXAMPLES

We view functions as procedures or "machines" that take inputs and return outputs.

3.3. Definition. A *function* f from a set A to a set B assigns to each $a \in A$ a single element $f(a)$ in B. We read the notation $f \colon A \to B$ as "f maps A into B" or "f mapping A to B". The set A is the *domain* of f; the set B is the *target*. We say that f is *defined on A*. A function is *real-valued* if its target is contained in $\mathbb{R}$.

If $a \in A$, then $f(a)$ is the *image* of a under f. The *image* under f of a set $C \subseteq A$, written $f(C)$, is $\{f(a) \colon a \in C\}$. The *image* of f is the image of its domain. Two functions f, g are *equal*, written $f = g$, if they have the same domain and target and satisfy $f(x) = g(x)$ for all x in their domain.

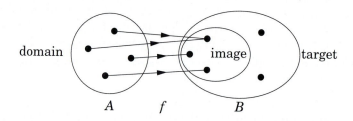

$$A \qquad f \qquad B$$

Comments on terminology. When f is a real-valued function, $f(x)$ is a number. Although it is sometimes convenient to use a formula to name a function, such as by writing "x^2" instead of "the function f defined by $f(x) = x^2$", it is more precise to distinguish between a function and its values.

Many authors define the "range" of a function to be the image of its domain. Others use "range" for the target set. To avoid the ambiguity, we do not use the word "range". The target of f contains the image of f, but the two sets may be different, so the concepts have different names. For example, when $f \colon \mathbb{R} \to \mathbb{R}$ is defined by $f(x) = x^2$, the target is $\mathbb{R}$, but the image is the set of nonnegative real numbers.

We often define a concept A by saying that A occurs "if" some property holds. For example, a number is even "if" it is twice an integer; in stating the definition we mean that the set of even integers is the set of doubled integers. Hence the "if" in a *definition* means "if and only if"; we say "if" because the concept or set being defined has no name until the definition is complete.

3.4. Remark. *The meaning of "well-defined".* The statement "f is well-defined" means that the rule given for computing f assigns to each element of the domain exactly one element. For example, there may be several ways to compute the image of an element via a "machine" f. In order that f be a function, the result must not depend on which computation is chosen. For example, consider the machine defined on $\mathbb{R}$ by $f(x) = x + 2$ when $x > 0$ and $f(x) = (x^2 - 4)/(x - 2)$ when $x < 2$. When $0 < x < 2$, this specifies two ways to compute $f(x)$. The results agree, so $f: \mathbb{R} \to \mathbb{R}$ is well-defined. See Exercise 3.3 for additional examples. ∎

Comment on notation. The concise notation $f: A \to B$ has become universal in mathematics. Notational conventions that have evolved over many years belong to the language of mathematics and facilitate rapid communication of ideas. The symbols $\mathbb{N}, \mathbb{Z}, \mathbb{Q}, \mathbb{R}$ are examples; no mathematician would say "Let $\mathbb{R}$ denote the set of integers." We use many such conventions in this book. For example, we often discuss elements and sets by using lower-case letters to denote the elements and upper-case letters to denote the sets. We use P, Q, R to denote mathematical statements because "P" suggests words like "proposition", "predicate", or "property". We usually use i, j, k, l, m, n to denote integers; the phrase "Let n be a real number between 0 and 1" sounds strange. This convention is so common that the scientific programming language FORTRAN "implicitly" defined as integers all variables having names beginning with these letters.

We next present definitions and notation for some useful functions.

3.5. Definition. Given $x \in \mathbb{R}$, the *absolute value* of x, written as $|x|$, is defined by

$$|x| = \begin{cases} x & \text{if } x \geq 0 \\ -x & \text{if } x < 0 \end{cases}$$

Given $x, y \in \mathbb{R}$, the *triangle inequality* for x, y is the inequality $|x + y| \leq |x| + |y|$.

The absolute value function maps $\mathbb{R}$ to $\mathbb{R}$; its image is the set of nonnegative real numbers. We think of $|x|$ as describing the distance from x to 0, and more generally $|x - y|$ describes the distance between x and y. The triangle inequality is a property of the absolute value function that will be crucial in Part IV; Exercise 3.4 requests a proof.

3.6. Definition. The *floor* of x, written $\lfloor x \rfloor$, is the largest integer less than or equal to x. The *ceiling* of x, written $\lceil x \rceil$, is the least integer greater than or equal to x.

The floor function and ceiling function map $\mathbb{R}$ to $\mathbb{Z}$; note that $\lfloor -x \rfloor = -\lceil x \rceil$ and $\lceil -x \rceil = -\lfloor x \rfloor$. We use the modern notation for floor and ceiling; in number theory, the old tradition was $[x]$ for $\lfloor x \rfloor$ and $\{x\}$ for

$\lceil x \rceil$. The new style became prevalent in computer science; when digital computers perform integer division by truncation, they use the floor function. We adopt this style because there are many other uses for [] and {}, while $\lfloor \, \rfloor$ and $\lceil \, \rceil$ are almost impossible to misinterpret.

3.7. Remark. *Notation for summation and product.* Often we wish to sum the values of a function f over a particular subset of the domain. The character used to denote summation is "Σ", the uppercase Greek "sigma". When a, b are integers with $a \le b$, the value of $\Sigma_{i=a}^{b} f(i)$ is the sum of the numbers $f(i)$ over the integers i satisfying $a \le i \le b$. Here i is called the *index of summation*, and the formula $f(i)$ is the *summand*. We also write $\Sigma_{x \in S} f(x)$ to indicate the more general situation of summing over the elements of a set S. When no limits of summation are specified, as in $\Sigma_j x_j$, we sum over all values of the index where the summand makes sense. When the summand has only one symbol that can vary, we may also omit the subscript on the summation symbol, as in Σx_i. These comments apply also to indexed products, where we use the uppercase Greek "pi", as in $\Pi_{i=1}^{n} i = 1 \times 2 \times \cdots \times n$. ∎

3.8. Example. *Three easy examples.*

$$\Sigma_{i=0}^{10} \lfloor i/2 \rfloor = 0 + 0 + 1 + 1 + 2 + 2 + 3 + 3 + 4 + 4 + 5 = 25.$$

$$\Pi_{j=3}^{6} \frac{j}{j+1} = \frac{3}{4} \cdot \frac{4}{5} \cdot \frac{5}{6} \cdot \frac{6}{7} = \frac{3}{7}.$$

$$\Sigma_{i=-n}^{n} i^3 = 0 \quad \text{for} \quad n \in \mathbb{N}. \quad ∎$$

3.9. Definition. A *sequence* is a function with domain $\mathbb{N}$. A function from $\mathbb{N}$ to S is a "sequence of elements of S". We describe sequences by listing the images in order, using the domain element as a subscript. We write $\langle x \rangle$ or $\{x_n\}$ for the sequence with values $x_1, x_2, x_3, \cdots$. We also consider functions with domain $\mathbb{N} \cup \{0\}$ to be sequences, corresponding to starting the indices at 0.

3.10. Definition. If B is a set where arithmetic is defined, and f, g are functions from A to B, then their *sum* $f + g$ is the function h defined by $h(x) = f(x) + g(x)$, and their *product* $f \cdot g$ is the function h defined by $h(x) = f(x) \cdot g(x)$. Pointwise combinations are defined similarly for other arithmetic operations.

We use the same notation for sequences; if $\langle a \rangle, \langle b \rangle$ are sequences, then $\langle a \rangle + \langle b \rangle$ is the sequence $\langle c \rangle$ such that $c_n = a_n + b_n$ for all $n \in \mathbb{N}$.

3.11. Definition. A *(real) polynomial* in x is a function $f \colon \mathbb{R} \to \mathbb{R}$ defined by $f(x) = \Sigma_{i=0}^{k} c_i x^i$, where k is some nonnegative integer and $c_0, \ldots, c_k$ are real numbers called the *coefficients*. The *degree* of f is the largest index d such that $c_d \neq 0$. The polynomial whose coefficients are all 0 has no degree.

We often think of a polynomial as its list of coefficients; we add polynomials f and g by adding corresponding coefficients (taking coefficients of powers above the degree to be zero). By the distributive law, the resulting list of coefficients defines the same polynomial as the pointwise sum $f + g$. We will prove in Chapter 4 that two polynomials are the same function only if they have the same coefficients.

Given an expression like $x^n - y^n$, we can treat x or y as a constant and obtain a polynomial in the other variable. We can also treat both x and y as variables and think of the expression as a polynomial in two variables. A *(real) polynomial* in the variables $x_1, \ldots, x_k$ is a finite sum of terms of the form $c\Pi_{i=1}^{k} x_i^{n_i}$, where each n_i is a nonnegative integer and $c \in \mathbb{R}$. Each such term is a *monomial*. The function f defined by

$$f(x, y, z) = x^2 + y^2 + z^2 + 2xy + 2xz + 2yz$$

is a polynomial in three variables.

GRAPHS OF FUNCTIONS

The "graph" of a function is a set that contains a complete description of the function and leads to a natural pictorial representation. The word "graph" stems from the Greek word for "picture". Mathematicians use "graph" in various ways, usually for an abstract structure that has a helpful pictorial representation.

3.12. Definition. The *graph* of a function $f \colon S \to T$ is the set of ordered pairs $\{(x, f(x)) \colon x \in S\}$.

3.13. Remark. *Pictures of graphs.* Suppose f is a real-valued function defined on a subset of $\mathbb{R}$. We draw two copies of $\mathbb{R}$ as horizontal and vertical axes, associating the horizontal axis with the domain and the vertical axis with the target. The graph of f is then a set of points in the plane. A set S of points in the plane is the graph of a function if and only if it contains at most one element (x, y) for each real number x; in other words, each vertical line intersects S at most once. ∎

We use "graphing a function f" or "drawing the graph of f" to mean plotting the elements of the graph in a picture of the plane that has enough detail to be meaningful. The advantages of graphing a function include: 1) it introduces geometric interpretations of properties we define for functions, 2) the visual presentation of the graph may suggest properties of the function, and 3) the visual presentation may *suggest* a method of proof.

3.14. Example. *Identity functions*. The *identity* function on a set A is the function $f: A \to A$ defined by $f(x) = x$ for all $x \in A$. Depending on whether f is the identity function on $\mathbb{N}$, $\mathbb{Z}$, or $\mathbb{R}$, the graph of f would be drawn as in one of the pictures below. ∎

3.15. Definition. A subset of $\mathbb{R}$ is *bounded* if it has both a lower bound and an upper bound. A real-valued function f is *bounded* if its image is a bounded subset of $\mathbb{R}$.

In terms of quantifiers, $f: A \to \mathbb{R}$ is bounded if there exist $\alpha, \beta \in \mathbb{R}$ such that $x \in A$ implies $\alpha \leq f(x) \leq \beta$. To show that f is bounded, it suffices to find a number $M \in \mathbb{R}$ such that $x \in A$ implies $|f(x)| \leq M$. The meaning of *unbounded* is $(\forall M \in \mathbb{R})(\exists x \in A)(|f(x)| > M)$.

3.16. Definition. A function $f: A \to \mathbb{R}$ with $A \subseteq \mathbb{R}$ is *increasing* if $x < x'$ implies $f(x) < f(x')$. A function $f: A \to \mathbb{R}$ with $A \subseteq \mathbb{R}$ is *nondecreasing* if $x < x'$ implies $f(x) \leq f(x')$. The definitions for *decreasing* and *nonincreasing* are obtained by changing $<$ to $>$ and $\leq$ to $\geq$. A function is *monotone* on A if it is nondecreasing on A or if it is nonincreasing on A.

The properties "increasing" and "nondecreasing" are sometimes called *strictly increasing* and *weakly increasing*, respectively. Similarly, a function is *strictly monotone* on A if it is increasing on A or if it is decreasing on A. We introduce the word "monotone" to avoid repetition; many results apply both to nondecreasing and to nonincreasing functions (see Proposition 3.34). A function that is increasing on one interval and decreasing on another is not monotone.

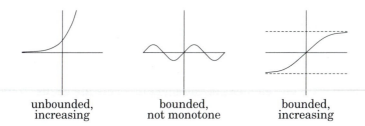

unbounded,   bounded,   bounded,
increasing   not monotone   increasing

3.17. Definition. A *fixed point* of a function $f: A \to A$ is an element $x \in A$ such that $f(x) = x$.

3.18. Remark. *Geometric interpretations.* Consider the graph of a function having domain and target in $\mathbb{R}$. The function is increasing if and only if for every horizontal line intersecting its graph, the graph is above that line to the right of the intersection and below it to the left. The function is bounded if and only if every point in the graph lies in the band between some pair of horizontal lines. The function has a fixed point if and only if the line $\{(x, x)\}$ through the origin intersects the graph. The first function graphed above has no fixed point; the other two have 0 as a fixed point. (They may have other fixed points, depending on the scales of the axes.) ■

BIJECTIONS

We return to the problem of solving equations. We want to know what conditions on a function $f: A \to B$ guarantee that for every $b \in B$, there exists exactly one solution to the equation $f(x) = b$.

3.19. Definition. A function $f: A \to B$ is a *bijection* if for every $b \in B$ there is exactly one $x \in A$ such that $f(x) = b$.

3.20. Example. *Linear equations in two variables.* Given constants $a, b, c, d \in \mathbb{R}$, let $f: \mathbb{R}^2 \to \mathbb{R}^2$ be defined by $f(x, y) = (ax + by, cx + dy)$. In Theorem 2.2, we proved that for each $(r, s) \in \mathbb{R}^2$, the pair of equations $ax + by = r$ and $cx + dy = s$ has a unique solution if and only if $ad - bc \neq 0$. In other words, we proved that the function f is a bijection if and only if $ad - bc \neq 0$. ■

3.21. Example. *Pairing up spouses.* Let M be the set of men at a party, let W be the set of women, and let $A = M \cup W$. If the attendees consist entirely of married couples, then we can define a function $f: M \to W$ by letting $f(x)$ be the spouse of x. For each woman b at this party, there is exactly one man x in M such that $f(x) = b$. Hence f is a bijection from M to W. We could also define $h: A \to A$ by letting $h(x)$ be the spouse of x; this is a bijection from A to itself. ∎

A bijection f establishes a *one-to-one correspondence* between two sets; it matches them up, with each element of the target being the image of exactly one element of the domain. This allows us to define a function from the target of f to its domain that reverses the effect of f.

3.22. Definition. If f is a bijection from A to B, then the *inverse* of f is the function $g: B \to A$ such that, for each $b \in B$, $g(b)$ is the unique element $x \in A$ such that $f(x) = b$. We write f^{-1} for the function g.

3.23. Remark. *If f is a bijection and g is the inverse of f, then g is also a bijection and f is the inverse of g.* This follows immediately from the interpretation of a bijection as a one-to-one correspondence. The pairing up of the sets can be viewed from either end; in one direction the map is f, and in the other it is g. Thus $(f^{-1})^{-1} = f$. ∎

3.24. Example. The conversion formula from Celsius temperature to Fahrenheit temperature is $f(x) = (9/5)x + 32$; this defines a bijection from $\mathbb{R}$ to $\mathbb{R}$. The bijection f has an inverse function g, defined by $g(y) = (y - 32)(5/9)$. We have $g(f(x)) = x$ for all x, but also $f(g(y)) = y$ for all y. When the bijection g is applied to true physical temperatures, the domain is $\{y \in \mathbb{R}: y \geq -273.15\}$.

When interpreting physical measurements, care is needed in converting from one scale to another. It is commonly believed that "normal body temperature" is 98.6 degrees Fahrenheit, which equals 37 degrees Celsius, exactly. Body temperature was originally studied under the Celsius scale. We presume that "37 degrees" is accurate to the nearest degree Celsius. It is inappropriate to state the normal Fahrenheit body temperature to the accuracy suggested by 98.6. ∎

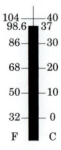

3.25. Definition. Suppose $f: A \to B$ and $y \in B$. The *inverse image* of y under f, written $I_f(y)$, is the set $\{x \in A: f(x) = y\}$.

Some authors write $f^{-1}(y)$ for $I_f(y)$; this can cause confusion. For example, if $f: \mathbb{R} \to \mathbb{R}$ is defined by $f(x) = x^2$, then $I_f(4) = \{2, -2\}$, but one sometimes sees $f^{-1}(4) = \pm 2$ even though f does not have an inverse.

The two conditions that a function must satisfy to be a bijection can be considered independently.

3.26. Definition. A function $f: A \to B$ is *injective* if for each $b \in B$, there is at most one $x \in A$ such that $f(x) = b$. A function $f: A \to B$ is *surjective* if for each $b \in B$, there is at least one $x \in A$ such that $f(x) = b$. Injective functions are *injections*, and surjective functions are *surjections*.

3.27. Remark. *Geometric interpretation of injection and surjection.* If $f: \mathbb{R} \to \mathbb{R}$, then f is injective if and only if every horizontal line intersects the graph of f at most once, and f is surjective if and only if every horizontal line intersects the graph of f at least once. ∎

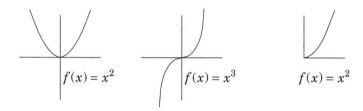

$f(x) = x^2$ $f(x) = x^3$ $f(x) = x^2$

3.28. Example. *Injections and surjections.* A function that is both injective and surjective is a bijection (Definition 3.21); we may say that it "is bijective". The function $f: \mathbb{R} \to \mathbb{R}$ defined by $f(x) = x^3$ is both injective and surjective and hence is bijective.

The function $f: \mathbb{R} \to \mathbb{R}$ defined by $f(x) = x^2$ is neither injective nor surjective. It maps -1 and 1 to the same value, so it is not injective. No negative number is in its image, so it is not surjective. A function from $\mathbb{R}$ to $\mathbb{R}$ cannot be surjective unless it is unbounded above and below.

When $P = \{x \in \mathbb{R} : x > 0\}$, the function $f: P \to P$ defined by $f(x) = x^2$ *is* a bijection, as is the function $f: A \to A$ defined by this rule when $A = \{x \in \mathbb{R} : 0 \le x \le 1\}$. ∎

3.29. Remark. *Schematic interpretation of injection and surjection.* To visualize a function $f: A \to B$ as a "mapping", we draw a region representing A and a region representing B, and from each $x \in A$ we draw an arrow to $f(x)$ in B. Each element of A is the tail of exactly one arrow; this follows from the definition of a function. The function is injective if each element of B is the head of at most one arrow, meaning that there is no "collapsing" of elements. The function is surjective if each element of B is the head of at least one arrow, meaning that no element of the target is "missed". The leftmost picture below illustrates a bijection; the function and its inverse are described by moving forward or backward along the arrows depicting the one-to-one correspondence. ■

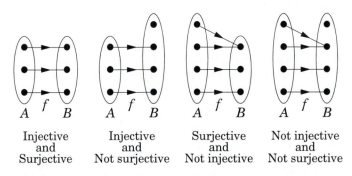

| Injective and Surjective | Injective and Not surjective | Surjective and Not injective | Not injective and Not surjective |

3.30. Remark. *Proofs for injective and surjective functions.* Although we can interpret injections and surjections schematically or geometrically, we return to the definitions to write proofs. Below we list symbolically the meanings of "injective", "surjective", and their negations.

A proof that $f: A \to B$ is injective shows that f does not map two elements of A to a single element of B. In other words, "every element of B is the image of at most one element under f" (**not** "every element maps to a unique element", which is the definition of *function*). We consider pairs $x, x' \in A$ and prove that $f(x) = f(x')$ implies $x = x'$ or that $x \neq x'$ implies $f(x) \neq f(x')$. To prove that f is not injective, it suffices to exhibit a pair $x, x' \in A$ with $x \neq x'$ and $f(x) = f(x')$.

When we prove that $f: A \to B$ is surjective, the definition again guides the proof: For all $y \in B$, we must prove that there exists $x \in A$ such that $f(x) = y$. Usually, we construct an example of such an x (in terms of y). To prove that f is not surjective, we provide a single example $y \in B$ that is not in the image of f. ■

Injective	Not Injective
$(\forall y \in B)[I_f(y)$ has at most one element]	$(\exists y \in B)[I_f(y)$ has at least two elements]
$(\forall x, x' \in A)[x \neq x' \Rightarrow f(x) \neq f(x')]$	$(\exists x, x' \in A)[x \neq x'$ and $f(x) = f(x')]$
$(\forall x, x' \in A)[f(x) = f(x') \Rightarrow x = x']$	

Surjective	Not Surjective
$(\forall y \in B)[I_f(y)$ is non-empty]	$(\exists y \in B)[I_f(y)$ is empty)
$(\forall y \in B)(\exists x \in A)[f(x) = y]$	$(\exists y \in B)(\forall x \in A)[f(x) \neq y]$

3.31. Example. When $A = \{x \in \mathbb{R}: 0 \leq x \leq 1\}$, the function $f: A \times A \to A$ defined by $f(x, y) = xy$ is surjective but not injective. It is surjective because each $z \in A$ is the image of the pair $(z, 1)$. It fails to be injective because $f(x, 0) = 0$ for all $x \in A$ (in other words, $I_f(0) = A$).

In Example 2.25, we discussed the equation $mx + b = mx' + b$, concluding that this requires $x = x'$ if $m \neq 0$. Hence the function $f: \mathbb{R} \to \mathbb{R}$ defined by $f(x) = mx + b$ is injective when $m \neq 0$. ∎

3.32. Definition. A *binary n-tuple* is an n-tuple whose entries are elements of $\{0, 1\}$. The *membership function* is the function defined on the power set of $[n]$ that assigns to the subset $S \subseteq [n]$ the binary n-tuple $(b_1, \ldots, b_n)$ such that $b_i = 1$ if $i \in S$ and $b_i = 0$ if $i \notin S$.

Given lights labeled $1, \ldots, n$, we can specify a subset of $[n]$ by turning the corresponding lights on. The membership function records whether each light is on or off. Below we list the resulting binary n-tuples when $n = 3$. We next prove that the membership function is a bijection to the set of binary n-tuples. This enables us to view subsets of $[n]$ as binary n-tuples and vice versa, translating statements about one context into statements about the other (see also Exercise 3.20).

lights on:	$\varnothing$	$\{1\}$	$\{2\}$	$\{3\}$	$\{1, 2\}$	$\{1, 3\}$	$\{2, 3\}$	$\{1, 2, 3\}$
image:	$(0,0,0)$	$(1,0,0)$	$(0,1,0)$	$(0,0,1)$	$(1,1,0)$	$(1,0,1)$	$(0,1,1)$	$(1,1,1)$

3.33. Proposition. The membership function is a bijection from the power set of $[n]$ to the set of binary n-tuples.

Proof. To prove that the membership function f is injective, suppose that S and T are two distinct subsets of $[n]$. Because they are distinct, some element $i \in [n]$ belongs to one of $\{S, T\}$ but not the other. Hence position i has a 1 in one of $\{f(S), f(T)\}$ and a 0 in the other. We have proved that $S \neq T$ implies $f(S) \neq f(T)$, so f is injective.

To prove that f is surjective, suppose b is a binary n-tuple; we construct a set S such that $f(S) = b$. For each $i \in [n]$, we put $i \in S$ if $b_i = 1$ and $i \notin S$ if $b_i = 0$. With this construction, the ith position in $f(S)$ is b_i, and indeed $f(S) = b$. ∎

Comment on terminology. Every function $f: A \to B$ maps A "into" B. Surjections are sometimes called "onto functions", because everything in B is "mapped onto"; we avoid this term to prevent confusion between "into" and "onto". Injections are sometimes called "one-to-one functions", because an injection establishes a one-to-one correspondence between its domain and its image. Since the image need not equal the target, we avoid this term to prevent confusion with the "one-to-one correspondence" between A and B when f is bijective.

The geometric interpretation of injection suggests that every increasing real-valued function is injective.

3.34. Proposition. Suppose $f: A \to B$ with $A, B \subseteq \mathbb{R}$. If f is a strictly monotone function, then f is injective.

Proof. Suppose f is increasing; the proof for decreasing functions is essentially the same. If $x, x' \in A$ and $x \neq x'$, then $x < x'$ or $x > x'$. In the first case, $f(x) < f(x')$; in the second case, $f(x) > f(x')$. Either implies $f(x) \neq f(x')$. Since x, x' were chosen arbitrarily from A, f satisfies the definition of an injection. ∎

The injectivity of increasing functions allows us to solve Problem 3.2. We need an elementary formula that we will apply again in Theorem 4.14 to study zeros of polynomials.

3.35. Lemma. If $x, y \in \mathbb{R}$ and $n \in \mathbb{N}$, then
$$x^n - y^n = (x - y)(x^{n-1} + x^{n-2}y + \cdots + xy^{n-2} + y^{n-1})$$

Proof. Using the distributive law (see Proposition 4.9), we multiply out the product on the right; below we write this with the terms using the factor of x on the first line and the terms using the factor of $-y$ on the second line. Canceling the like terms in the "columns" yields $x^n - y^n$, which completes the proof.

$$
\begin{array}{ccccccccc}
x^n & + & x^{n-1}y & + & \cdots & + & x^2 y^{n-2} & + & xy^{n-1} \\
    & - & x^{n-1}y & - & \cdots & - & x^2 y^{n-2} & - & xy^{n-1} & - & y^n
\end{array} \quad \blacksquare
$$

3.36. Remark. *Shifting the index.* We can also write the proof of Lemma 3.35 in summation notation. Substituting j for $i+1$ below is renaming the index of summation. We then rename j as i. Performing the two steps at once is called "shifting the index of summation".

$$(x - y)\left(\sum_{i=0}^{n-1} x^{n-1-i} y^i\right) = \sum_{i=0}^{n-1} (x^{n-i} y^i - x^{n-1-i} y^{i+1}) = \left(\sum_{i=0}^{n-1} x^{n-i} y^i\right) - \left(\sum_{i=0}^{n-1} x^{n-1-i} y^{i+1}\right)$$

$$= \left(\Sigma_{i=0}^{n-1} x^{n-i} y^i\right) - \left(\Sigma_{j=1}^{n} x^{n-j} y^j\right) = \left(\Sigma_{i=0}^{n-1} x^{n-i} y^i\right) - \left(\Sigma_{i=1}^{n} x^{n-i} y^i\right)$$

$$= x^n + \left(\Sigma_{i=1}^{n-1} x^{n-i} y^i\right) - \left(\Sigma_{i=1}^{n-1} x^{n-i} y^i\right) - y^n = x^n - y^n \quad \blacksquare$$

With Lemma 3.35, we can solve Problem 3.2. First consider $x^2 + xy + y^2 = 0$; we seek solutions other than $(0,0)$. When x and y have the same sign, the left side is positive and there is no solution. Hence we may assume that one of $\{x, y\}$ is positive and the other is negative. By adding xy to both sides, we obtain $x^2 + 2xy + y^2 = xy$ and then $(x + y)^2 = xy$. The right side is negative and the left side is not; again this is a contradiction. Hence there are no solutions other than $(0,0)$. Problem 3.2 generalizes this; we use the injectivity of the function that maps x to x^{2k+1}.

3.37. Solution. *A Special Polynomial.* Given a positive integer k, we prove that the pair $(x, y) = (0, 0)$ is the only solution of the equation

$$x^{2k} + x^{2k-1}y + \cdots + xy^{2k-1} + y^{2k} = 0. \qquad (*)$$

When we set $y = 1$, this tells us that $x^{2k} + \cdots + x + 1$ is never 0.

First we show that the function f defined by $f(x) = x^{2k+1}$ is increasing and therefore injective. Since the product of an odd number of negative factors is negative, we have $x^{2k+1} < 0 < y^{2k+1}$ if $x < 0 < y$. Hence it suffices to show that $x^{2k+1} < y^{2k+1}$ if $x < y$ and x, y have the same sign. Since $0 < x < y$ implies $x^2 < y^2$, the desired inequality is obtained as a product of $k + 1$ inequalities when x, y are positive. (Technically, this uses induction on k, a technique presented formally in Chapter 4.) A similar argument, using $(-1)^{2k+1} = -1$ and Proposition 1.21a, proves that $x^{2k+1} < y^{2k+1}$ when $x < y < 0$ (Exercise 3.6).

Suppose (x, y) is a solution to $(*)$, and multiply both sides by $(x - y)$. By Lemma 3.35, we obtain $x^{2k+1} - y^{2k+1} = 0$. Since exponentiation by an odd power is injective, this requires that $x = y$. When $x = y$, the polynomial becomes $\Sigma_{i=0}^{2k} x^{2k-i}y^i = (2k + 1)x^{2k}$. Since nonzero numbers have nonzero product, this requires $x = 0$. Therefore, the only solution to $(*)$ is $(0, 0)$. ∎

For readers familiar with calculus, we briefly pause to discuss these properties for some of the transcendental functions of calculus. The next two examples may be skipped without loss of continuity.

3.38. Example. The function defined by $f(x) = e^x$ maps $\mathbb{R}$ to $\mathbb{R}$. It is increasing and hence injective, but it is not surjective; its image is the set of positive real numbers. By restricting the target of e^x to the set of positive real numbers, we obtain a bijection. The natural logarithm function, its inverse, is a bijection from the set of positive real numbers to $\mathbb{R}$. The function defined by $f(x) = \sin x$ from $\mathbb{R}$ to the interval $[-1, 1]$ is surjective but not injective. By restricting the domain of $\sin x$ to $[-\pi/2, \pi/2]$, we obtain a bijection to the interval $[-1, 1]$. ∎

Suppose that $f : [a, b] \to [u, v]$ is a bijection. The picture below interprets this geometrically; as x varies from a to b, $y = f(x)$ varies from u to v. If x measures a physical quantity, then y measures it on another scale. The essence of f is to change variables from x to y.

3.39. Example. *Limits of integration under change of variables.* In calculus, we often compute the integral $\int_a^b f(x)dx$ by introducing a new variable via a transformation $y = g(x)$. The function g must be injective on the interval $[a, b]$, and therefore g is a bijection from $[a, b]$ to $g([a, b])$. As x varies from a to b, y varies from $g(a)$ to $g(b)$. We rewrite the integral in terms of y, using the bijection.

For example, suppose $f(x) = x^3 + 1$. This function is a bijection from the interval $[0, 2]$ to the interval $[1, 9]$. Letting $y = f(x)$ leads to $\int_0^2 (x^3 + 1)^5 3x^2 dx = \int_1^9 y^5 dy = (1/6)(9^6 - 1)$. Similarly, because $y = \sin x$ defines a bijection from $\{x \in \mathbb{R}: -\pi/2 \le x \le \pi/2\}$ to $\{y \in \mathbb{R}: -1 \le y \le 1\}$, we compute $\int_{-\pi/2}^{\pi/2} \sin x \cos x \, dx = \int_{-1}^{1} y \, dy = 0$. ∎

COMPOSITION AND FUNCTIONAL DIGRAPHS

Whenever we have a function whose target is contained in the domain of a second function, we can build a compound function by applying the two functions successively. This yields a function from the domain of the first function into the target of the second.

3.40. Definition. *Composition.* If $f: A \to B$ and $g: B \to C$, then the *composition* of g with f is a function h defined by $h(x) = g(f(x))$ for every $x \in A$. When h is the composition of g with f, we write $h = g \circ f$. The composition maps the domain of f into the target of g, as illustrated below.

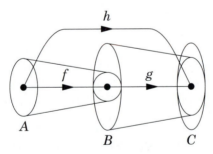

3.41. Example. Suppose $f: \mathbb{R} \to \mathbb{R}$ and $g: \mathbb{R} \to \mathbb{R}$ are defined by $f(x) = x - 2$ and $g(x) = x^2 + x$. Then $g \circ f$ is defined by

$$(g \circ f)(x) = g(f(x)) = (x - 2)^2 + (x - 2) = x^2 - 3x + 2.$$

On the other hand, $(f \circ g)(x) = f(g(x)) = x^2 + x - 2$. ∎

3.42. Example. If $f: A \to B$ is a bijection and $g = f^{-1}$, then $g \circ f$ is the identity function on A, and $f \circ g$ is the identity function on B. Exercise

3.23 considers whether f must be a bijection if one of $g \circ f$ or $f \circ g$ is an identity function. ∎

3.43. Proposition. The composition of two injections is an injection. The composition of two surjections is a surjection. The composition of two bijections is a bijection. Furthermore, if f, g are bijections (so $g \circ f$ is a bijection), then $(g \circ f)^{-1} = f^{-1} \circ g^{-1}$ (the *Inverse Composition Formula*).

Proof. (Exercise 3.24). ∎

Example 3.41 shows that the composition of functions from a set to itself is not generally commutative; $f \circ g$ need not equal $g \circ f$. On the other hand, composition of functions is always associative. We can form a composition $h \circ g \circ f$ by composing $h \circ g$ with f or by composing h with $g \circ f$. These two compositions always yield the same function, which justifies dropping the parentheses.

3.44. Proposition. Suppose $f: A \to B$, $g: B \to C$, and $h: C \to D$. Then
$$h \circ (g \circ f) = (h \circ g) \circ f \text{ (the associative property of composition).}$$

Proof. Let $\alpha = g \circ f$ and $\beta = h \circ g$. We show that $\alpha \circ h$ and $\beta \circ f$ are the same function. Consider their values at $x \in A$. Using only the definition of composition, we have $(h \circ \alpha)(x) = h(\alpha(x)) = h((g \circ f)(x)) = h(g(f(x))) = (h \circ g)(f(x)) = \beta(f(x)) = (\beta \circ f)(x)$. Since the functions $h \circ \alpha$ and $\beta \circ f$ have the same image at each element of the domain and have the same target, they must be the same function. ∎

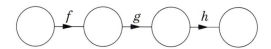

For a function with the same domain and target, we can study repeated composition of the function with itself. The *n*th *iterate* of $f: A \to A$ is the function f^n obtained by composing f with itself n times. In particular, $f^2 = f \circ f$, and we say that $f^1 = f$.

3.45. Example. *Rotation by 90 degrees.* Let $f: \mathbb{R}^2 \to \mathbb{R}^2$ be the function that rotates the plane by 90 degrees counterclockwise. The formula for f is $f(x, y) = (-y, x)$. Iterating f four times returns (x, y) to itself. ∎

$$(-y, x) \longleftarrow (x, y)$$
$$\uparrow \qquad\qquad \uparrow$$
$$(-x, -y) \longrightarrow (y, -x)$$

We next introduce a visual presentation that is useful for studying iteration of functions.

3.46. Definition. The *functional digraph* of a function $f: A \to A$ consists of a point for each element of A and, for each $x \in A$, an arrow from the point representing x to the point representing $f(x)$. These points are *vertices*. An arrow from a vertex to itself is a *loop*. A list of vertices $a_1, \ldots, a_k$ is a *cycle* of length k if there is an arrow from a_i to a_{i+1} for $1 \le i \le k-1$ and an arrow from a_k to a_1.

The functional digraph of f differs from the picture in Remark 3.29 because it uses only one copy of A. By the definition of a function, each vertex in a functional digraph is the tail of exactly one arrow. The function is injective if every vertex is the head of at most one arrow, and it is surjective if every vertex is the head of at least one arrow. A function from a set to itself has a fixed point if and only if its functional digraph has a loop. A loop is a cycle of length 1.

3.47. Example. *The functional digraph for the Penny Problem.* In the Penny Problem (Solution 1.2), we defined a function on the set of nondecreasing lists of positive integers. Since the function does not change the total number of pennies, we can study it on the set S_n of nondecreasing lists of positive integers summing to n. Below we illustrate the functional digraph that arises when $n = 5$; S_5 consists of 7 lists. As proved in Example 1.2, there is no fixed point among lists with sum 5, and thus the functional digraph has no loop. It does have a cycle of length 3. ∎

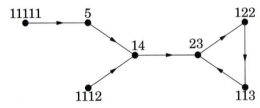

The functional digraph makes it easy to study what happens when we repeatedly compose a function with itself. In the example above we always wind up in the cycle of length 3.

3.48. Definition. A *permutation* of a set S is a bijection from S to itself. If $|S| = [n]$, then we can describe a permutation of S as an *arrangement* of S, listing the elements of S in some order. The *2-line notation* for a permutation f lists the elements of $[n]$ on the top line and the corresponding elements on the bottom line; the corresponding arrangement is the bottom line, in order. A *cycle* in a permutation is a set of elements that form a cycle in its functional digraph.

3.49. Example. *The functional digraph of a permutation.* Consider the arrangement $3\,6\,1\,4\,2\,5\,8\,9\,7$ of [9]; this is the permutation $f\colon [9] \to [9]$ with 2-line notation $\left(\begin{smallmatrix}1\,2\,3\,4\,5\,6\,7\,8\,9\\3\,6\,1\,4\,2\,5\,8\,9\,7\end{smallmatrix}\right)$. The visual presentation of this function using the model of Remark 3.29 appears on the left below. On the right we draw the functional digraph. In the functional digraph of a permutation, each vertex is the tail of one arrow and is the head of one arrow. Hence the digraph of a permutation always consists of disjoint "cycles". These are called the *cycles* of the permutation, and we can capture all the information about the function by using the *cycle description* $(13)(4)(265)(789)$ for the permutation. ∎

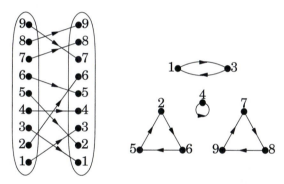

OPERATORS (optional)

We close this chapter by describing ways to obtain new functions from given functions. Such a procedure is a function whose domain and target are themselves sets of functions. To avoid confusion, we use the word "operator" to describe a function defined on a set of functions.

3.50. Example. The simplest example of an operator is the identity operator, mapping each function f to itself. Translation by a constant a is another example. Given $f\colon \mathbb{R} \to \mathbb{R}$, we let $T_a f$ be the function defined by $(T_a f)(x) = f(x + a)$. The "machine" T_a takes a function as its input and returns a function as its output. When $a = 0$, the translation operator is the identity operator. Similarly, the multiplication operator M_b takes f to the function $M_b f$ defined by $(M_b f)(x) = f(bx)$. When $b = 1$, the multiplication operator is the identity operator. Translation and multiplication have natural interpretations in terms of the graph of the function f (Exercise 3.32). ∎

3.51. Example. Let S be the set of polynomials in one variable. Given the polynomial f defined by $f(x) = \Sigma_{i=0}^{k} a_i x^i$, let Df denote the polynomial whose value at x is $\Sigma_{i=1}^{k} a_i i x^{i-1}$. The operator D is a function $D: S \to S$ (for readers familiar with calculus, D is the differentiation operator on these functions). The operator D is surjective; the polynomial with coefficients $\{a_k\}$ is the image of the polynomial in which the coefficient of x^0 is 0 and the coefficient of x^k is a_{k-1}/k for $k \geq 1$. The operator D is not injective; the polynomials f, g defined by $f(x) = x + 1$ and $g(x) = x + 2$ have the same image.

We define another operator $J: S \to S$. If $f(x) = \Sigma_{i=0}^{k} a_i x^i$, let Jf denote the polynomial whose value at x is $\Sigma_{i=0}^{k} a_i x^{i+1}/(i+1)$. If $Jf = Jg$, then term-by-term comparison of coefficients shows that $f = g$; hence J is injective. On the other hand, J is not surjective, because there is no polynomial f such that Jf is a nonzero polynomial of degree 0.

We can compose operators. We have $D(J(f)) = f$ for all $f \in S$, but $J(D(f))$ does not generally equal f. For example, if $f(x) = x^2 + 3$, then $J(D(f))$ is the function g defined by $g(x) = x^2$. ∎

EXERCISES

3.1. (–) At what temperatures do the following events occur?
   a) The Fahrenheit and Celsius values of the temperature are equal.
   b) The Fahrenheit value is the negative of the Celsius value.
   c) The Fahrenheit value is twice the Celsius value.

3.2. (!) Let $A = \{\text{January}, \text{February}, \dots, \text{December}\}$. Given $x \in A$, let $f(x)$ be the number of days in x. Does f define a function from A to $\mathbb{N}$?

3.3. (–) Determine whether the rules given below define functions from $\mathbb{R}$ to $\mathbb{R}$.
   a) $f(x) = |x - 1|$ if $x < 4$ and $f(x) = |x| - 1$ if $x > 2$.
   b) $f(x) = |x - 1|$ if $x < 2$ and $f(x) = |x| - 1$ if $x > -1$.
   c) $f(x) = ((x+3)^2 - 9)/x$ if $x \neq 0$ and $f(x) = 6$ if $x = 0$.
   d) $f(x) = ((x+3)^2 - 9)/x$ if $x > 0$ and $f(x) = x + 6$ if $x < 7$.
   e) $f(x) = \sqrt{x^2}$ if $x \in \mathbb{Z}$ and $f(x) = x$ if $x < 1$.

3.4. Prove the triangle inequality: $|x + y| \leq |x| + |y|$ for all $x, y \in \mathbb{R}$.

3.5. Suppose $x \in \mathbb{Z}$ and $k \in \mathbb{N}$. Prove that $\lceil x/k \rceil = \lfloor (x+k-1)/k \rfloor$ and $\lfloor x/k \rfloor = \lceil (x-k+1)/k \rceil$, where $\lceil y \rceil$ and $\lfloor y \rfloor$ denote the ceiling and floor of y.

3.6. Prove that $x^{2k+1} < y^{2k+1}$ if $x < y < 0$, where $k \in \mathbb{N}$.

3.7. Suppose $f: A \to B$, and suppose that C, D are subsets of A and X, Y are subsets of B. Determine which inclusion relationships must hold for the following pairs of sets.

   a) $f(C \cup D)$ and $f(C) \cup f(D)$.   c) $I_f(X \cup Y)$ and $I_f(X) \cup I_f(Y)$.
   b) $f(C \cap D)$ and $f(C) \cap f(D)$.   d) $I_f(X \cap Y)$ and $I_f(X) \cap I_f(Y)$.

3.8. (!) Suppose $f: \mathbb{R}^2 \to \mathbb{R}^2$ is defined by $f(x, y) = (ax - by, bx + ay)$, where a, b are fixed parameters with $a^2 + b^2 \neq 0$.
   a) Prove that f is a bijection.
   b) Find a formula for f^{-1}.
   c) Give a geometric interpretation of f for the case $a^2 + b^2 = 1$. (Describe the effect f has on geometric figures in the plane.)

3.9. (–) Let A be the set of days in the week. Given $x \in A$, let $f(x)$ be the number of letters in the English name of x. Does f define an injective function from A to $\mathbb{N}$?

3.10. Which of the following functions from $\mathbb{R}$ to $\mathbb{R}$ are injective, and which are surjective? For each that is not a bijection, find an interval $S \subseteq \mathbb{R}$ (containing more than a single point) such that the function is a bijection from S to S.
   a) $f(x) = x^3 - x + 1$.
   b) $f(x) = \cos(\pi x/2)$.

3.11. Suppose that f and g are surjective functions from $\mathbb{Z}$ to $\mathbb{Z}$, and suppose that $h = fg$ is the pointwise product of f and g (not the composition). Must h also be surjective? Give a proof or a counterexample.

3.12. (–) Suppose $f: \mathbb{R} \to \mathbb{R}$, and suppose that there are positive constants c, α such that, for all $x, y \in \mathbb{R}$, $|f(x) - f(y)| \geq c|x - y|^{\alpha}$. Prove that f is injective.

3.13. (–) Suppose $f: \mathbb{R} \to \mathbb{R}$ is a quadratic polynomial. Prove that f is not surjective. Give (with proof) an explicit example of a cubic polynomial $f: \mathbb{R} \to \mathbb{R}$ that is not injective.

3.14. (+) Determine which cubic polynomials are injections from $\mathbb{R}$ to $\mathbb{R}$. (Hint: this is easy if calculus is allowed. To do it without using calculus, first apply geometric arguments to reduce the problem to the case $x^3 + rx$. Comment: every cubic polynomial is a surjection from $\mathbb{R}$ to $\mathbb{R}$, but proving this requires the methods of Part IV.)

3.15. Consider three functions f, g, h mapping $\mathbb{R}$ to $\mathbb{R}$, defined by

$$f(x) = x/(1 + x^2), \quad g(x) = x^2/(1 + x^2), \quad h(x) = x^3/(1 + x^2).$$

   a) Determine which of these functions are injective.
   b) Prove that f and g are not surjective.
   c) Graph all three functions. (Comment: the graph of h should indicate that h is surjective, but proving this requires the methods of Part IV.)

3.16. Suppose $f: \mathbb{R} \to \mathbb{R}$.
   a) Prove that f can be expressed in a unique way as the sum of two functions g and h such that $g(-x) = g(x)$ for all $x \in \mathbb{R}$ and $h(-x) = -h(x)$ for all $x \in \mathbb{R}$. (Hint: find a formula for $g(x)$ and a formula for $h(x)$ in terms of the function f.)
   b) In the case that f is a polynomial, determine the functions g and h in terms of the coefficients of f.

3.17. (!) Suppose a, b, c, d are given real numbers, and $f: \mathbb{R}^2 \to \mathbb{R}^2$ is defined by $f(x, y) = (ax + by, cx + dy)$. Prove that f is injective if and only if f is surjective.

3.18. Suppose $f: A \to B$ is a bijection, where A and B are subsets of $\mathbb{R}$. Prove that if f is an increasing function, then f^{-1} is an increasing function.

3.19. Suppose $f: \mathbb{R} \to \mathbb{R}$. Let S be the set of functions g for which there exist positive constants $c, a \in \mathbb{R}$ such that $|g(x)| \le c|f(x)|$ for all $x > a$. State precisely what it means for a function g not to be in S. (Comment: the set S arises in computer science and in calculus to compare the "order of growth" of functions; the common notation for S is $O(f)$.)

3.20. (!) Let A be the set of binary n-tuples that have exactly k ones. Let B be the set of paths in the plane from the point $(0, 0)$ to the point $(k, n - k)$ consisting of steps of length 1 taken upward or to the right. Obtain a bijection from A to B. ("Obtain" means define a function on A and prove that the images belong to B and that the function is a bijection. As an example, when $(n, k) = (3, 1)$, the set A is $\{001, 010, 100\}$, and the set B consists of the three paths below.)

3.21. (–) Suppose F is an ordered field. Define f by $f(x) = -x$, and define g by $g(x) = x^{-1}$. Prove that f is a bijection from F to itself and that g is a bijection from $F - \{0\}$ to itself.

3.22. (–) Suppose $f(x) = x - 1$ and $g(x) = x^2 - 1$. Find formulas for $f \circ g$ and $g \circ f$.

3.23. (!) Suppose $f: A \to B$ and $g: B \to A$. Answer each question below by providing a proof or a counterexample.
    a) If $f(g(y)) = y$ for all $y \in B$, does it follow that f is a bijection?
    b) If $g(f(x)) = x$ for all $x \in A$, does it follow that $f(g(y)) = y$ for all $y \in B$?

3.24. (!) Prove the following statements about composition of functions.
    a) The composition of two injections is an injection
    b) The composition of two surjections is a surjection.
    c) The composition of two bijections is a bijection.
    d) If $f: A \to B$ and $g: B \to C$ are bijections, then $(g \circ f)^{-1} = f^{-1} \circ g^{-1}$.

3.25. Suppose $f: A \to A$. Prove that if $f \circ f$ is injective, then f is injective.

3.26. Suppose $f: A \to B$ and $g: B \to A$. Prove the following statements.
    a) If $f \circ g$ is the identity function on B, then f is surjective.
    b) If $g \circ f$ is the identity function on A, then f is injective.

3.27. (!) Suppose $f: A \to B$, $g: B \to C$, and $h = g \circ f$. For each statement below, give a proof or a counterexample.
    a) If h is injective, then f is injective.
    b) If h is injective, then g is injective.
    c) If h is surjective, then f is surjective.
    d) If h is surjective, then g is surjective.

3.28. Let S_n be the set of nondecreasing lists with sum n. Let $f: S_n \to S_n$ be the function defined on S_n by the operation in the Penny Problem (Solution 1.2).

a) Draw the functional digraph of the function of f when $n = 6$.

b) Determine all values of n such that f is injective. Determine all values of n such that f is surjective.

3.29. How many permutations of [4] have no fixed point? How many permutations of [5] have no fixed point?

3.30. (!) Suppose $a = a_1, \ldots, a_n$ is a permutation of $[n]$ (an arrangement of the elements of $[n]$ in a row). Let $f(a)$ be the number of pairs $i, j \in [n]$ such that $i < j$ and $a_i > a_j$. Let b be a permutation obtained from a by switching a pair of elements in the arrangement. Prove that $f(a)$ and $f(b)$ have opposite parity, meaning that one is odd and the other is even. For example, $f(4213) = 4$ and $f(1243) = 1$.

3.31. (+) Suppose that a, b are nonzero real numbers, and define $f: \mathbb{R} \to \mathbb{R}$ by $f(x) = 1/(ax + b)$ for $x \neq -b/a$ and $f(-b/a) = (-1/b) - (b/a)$. Determine the set of ordered pairs (a, b) such that the functional digraph of f has a 3-cycle. Solve the analogous problem when $f(x) = \frac{cx + d}{ax + b}$ and $ad \neq bc$.

3.32. Suppose that $f: \mathbb{R} \to \mathbb{R}$, and define the functions $T_a f$ and $M_b f$ by $(T_a f)(x) = f(x + a)$ and $(M_b f)(x) = f(bx)$. Determine how the graph of f should be modified to obtain the graphs of $T_a f$ and $M_b f$. For $M_b f$, consider the cases $b > 0$, $b = 0$, and $b < 0$.

Chapter 4

Induction

Many mathematical problems involve only integers; computers perform operations in terms of integer arithmetic. The natural numbers enable us to solve problems by working one step at a time. In this chapter, we develop the principle of mathematical induction, a fundamental technique of proof. We will use it to solve many problems, including the following three.

4.1. Problem. *The Checkerboard Problem.* Counting squares of sizes one-by-one through eight-by-eight, an ordinary eight-by-eight checkerboard has 204 squares. How can we obtain a formula for the number of squares of all sizes on an n-by-n checkerboard? ∎

4.2. Problem. *The Handshake Problem.* Consider a party consisting of n married couples. Suppose no person shakes hands with his or her spouse, and the $2n - 1$ people other than the host shake hands with different numbers of people. With how many people does the hostess shake hands? ∎

4.3. Problem. *Sums of Consecutive Integers.* Which natural numbers are sums of consecutive smaller natural numbers? For example, $30 = 9 + 10 + 11$ and $31 = 15 + 16$, but 32 has no such representation. ∎

THE PRINCIPLE OF INDUCTION

The set of *natural numbers*, written $\mathbb{N}$, is the set $\{1, 2, 3, \dots\}$. In listing the elements, we have agreed on an ordering for them. The first natural number is called 1. After each natural number k comes a specific next one, written $k + 1$ and called the *successor* of k. The successor function is a bijection from $\mathbb{N}$ to $\mathbb{N} - \{1\}$.

Under our definition, 0 is not a natural number. For some pur- poses it would be convenient to include 0; the same results can be proved, but some statements and proofs would need to be rephrased. Also there is no natural number called ∞. The symbol ∞ appears often in calculus and in Part IV of this book. Although the symbol is conve- nient, understanding its usage requires a careful study of limits.

Our understanding of the natural numbers includes one more property, called the Well-Ordering Property. If we assume the axioms of $\mathbb{R}$ (1.15-1.17), then we can obtain all properties of $\mathbb{N}$ as consequences. In Appendix A, however, we follow the historical and more appealing order of beginning with $\mathbb{N}$ and constructing $\mathbb{R}$ from it. Beginning with $\mathbb{N}$ requires assuming the Well-Ordering Property or some equivalent assertion as an axiom.

4.4. Axiom. (The Well-Ordering Property of the Natural Numbers)
Every non-empty subset of $\mathbb{N}$ has a least element (with respect to the usual ordering of $\mathbb{N}$).

We take the Well-Ordering Property as a fundamental Axiom, unassailable and unjustifiable, but intuitively believable. Neither the integers nor the nonnegative real numbers have this property under their "usual" orderings. A non-empty subset of $\mathbb{N}$ must have a least ele- ment, but it need not have a largest element.

The Well-Ordering Property implies the principle of induction, a technique for proving that all the statements in a sequence of mathema- tical statements are true. Conversely, the principle of induction implies the Well-Ordering Property (Exercise 4.40), so we could take either as the axiom. Theorem 4.5 is a formal statement of the principle of induc- tion whose proof follows easily from the Well-Ordering Property. Corol- lary 4.6 rephrases this in the way we generally use induction.

4.5. Theorem. (Principle of Mathematical Induction) Let T be a subset of $\mathbb{N}$. If the following two properties hold, then $T = \mathbb{N}$.
1) $1 \in T$.
2) If $k \in T$, then $k + 1 \in T$.

Proof. We use proof by contradiction. Consider the complementary set T' consisting of the natural numbers not in T. If $T \neq \mathbb{N}$, then T' is non- empty. By the Well-Ordering Property, T' contains a least element, say m. By property (1), $m \neq 1$, and hence $m - 1 \in \mathbb{N}$. Since m is the least ele- ment of T' and $m - 1$ is less than m, we have $m - 1 \in T$. Now property (2) implies $m \in T$, which contradicts $m \notin T$. ∎

4.6. Corollary. (Principle of Mathematical Induction) Suppose we have a sequence of mathematical statements $P(1), P(2), \cdots$ (one for each natural number). If the following two properties hold, then for every $n \in \mathbb{N}$, $P(n)$ is true.

1) $P(1)$ is true.
2) If $P(k)$ is true, then $P(k+1)$ is true.

Proof. Let $T = \{n \in \mathbb{N} : P(n) \text{ is true}\}$; we want to prove that $T = \mathbb{N}$. By Theorem 4.5, the hypotheses (1) and (2) imply that T is all of $\mathbb{N}$, and hence each $P(n)$ is true. ∎

When we give a proof by induction, proving property (1) of Corollary 4.6 is called the *basis step*, and proving property (2) is called the *induction step*. Our first example is a summation formula.

4.7. Proposition. For all $n \in \mathbb{N}$, $\sum_{i=1}^{n} i = n(n+1)/2$.

Proof. For each $n \in \mathbb{N}$, let $P(n)$ be the statement "$\sum_{i=1}^{n} i = n(n+1)/2$".
Basis Step: Since $1 = 1 \cdot 2/2$, the statement $P(1)$ is true.
Induction Step: From the hypothesis that $P(k)$ is true, we obtain

$$\sum_{i=1}^{k+1} i = \left(\sum_{i=1}^{k} i\right) + (k+1) = \frac{k(k+1)}{2} + (k+1)$$

$$= (k+1)\left(\frac{k}{2} + 1\right) = \frac{(k+1)(k+2)}{2},$$

and hence $P(k)$ implies $P(k+1)$.
By the principle of induction, the formula holds for every $n \in \mathbb{N}$. ∎

Induction works particularly well for Proposition 4.7 because the summation for $n = k+1$ is obtained from the summation for $n = k$ by adding one additional term. This approach fails for a summation like $\sum_{i=1}^{n} n^i$, where the terms in the summation for $n = k+1$ do not appear in the summation for $n = k$ (see Exercise 4.19). Thus we may need techniques other than induction to evaluate summations.

4.8. Remark. *Geometric argument for summation.* When still in grade school, Karl Friedrich Gauss (1777-1855) gave a direct proof that $\sum_{i=1}^{n} i = n(n+1)/2$. We list two copies of the sum, one above the other:

$$
\begin{array}{ccccccccc}
1 & + & 2 & + & 3 & + & \cdots & + & n \\
n & + & n-1 & + & n-2 & + & \cdots & + & 1
\end{array}
$$

For each i, the sum of the ith column is $i + (n + 1 - i) = n + 1$. There are n columns, so $2\sum_{i=1}^{n} i = n(n+1)$.

This argument has a "geometric" interpretation. Consider $n(n+1)$ points in a grid with n columns of size $n + 1$. Counted by columns, there are $n(n+1)$ points. We can also group the points into two disjoint subsets with columns of sizes 1 through n. In Chapter 5 we will discuss the "size" of sets and this technique of "counting two ways".

We can also express the argument using summations. In the computation below, we twice rename the index of summation on the second copy of the sum, replacing i by j and then j by $n + 1 - i$.

$$2\Sigma_{i=1}^{n} i = \Sigma_{i=1}^{n} i + \Sigma_{j=1}^{n} j = \Sigma_{i=1}^{n} i + \Sigma_{i=1}^{n}(n + 1 - i)$$
$$= \Sigma_{i=1}^{n}(n + 1) = n(n + 1) \quad \blacksquare$$

We accept that $\Sigma_{i=1}^{n}(n + 1) = n(n + 1)$; summing n terms, each equal to k, is the same as multiplying k by n. Appendix A explains how the definition of multiplication rests on induction. Thus the argument of Remark 4.8 indirectly uses induction.

We next prove two elementary facts about real numbers. We use induction to generalize statements about two objects to analogous statements about n objects. The proof is a useful model for proofs by induction. Before presenting it, we pause to list several reasons for proving seemingly obvious statements.

Some such statements have turned out to be false; others have been hard to prove. We can have no confidence in complicated results unless we can prove the elementary statements used in their proofs. Proving elementary statements also allows us to concentrate on the technique of proof without being distracted by complications. This book includes many such proofs to help the student master the techniques.

Understanding the proof of an obvious statement may enable us to prove less obvious statements by the same technique. This may also reveal the limitations of the argument and the difficulties that arise in extending the result to a more general context. Both "obvious" statements in the next proposition extend to infinite series; this requires the notion of limits (Chapter 14). In contrast, the "obvious" statement that the sum of n numbers is independent of the order of summation (Exercise 4.22) does not extend to infinite series! (See Exercise 14.36.)

4.9. Proposition. Suppose that $\langle a \rangle$ and $\langle b \rangle$ are sequences of real numbers and that $n \in \mathbb{N}$.
1) If $c \in \mathbb{R}$, then $\Sigma_{i=1}^{n} ca_i = c\Sigma_{i=1}^{n} a_i$.
2) If $a_i \leq b_i$ for all $i \in \mathbb{N}$, then $\Sigma_{i=1}^{n} a_i \leq \Sigma_{i=1}^{n} b_i$.

Proof. For each $n \in \mathbb{N}$, let $P(n)$ and $Q(n)$ denote these two conclusions, respectively. We prove each claim by induction.

1) The distributive law (Axiom 1.15DL) states that multiplication by a real number distributes over a sum of *two* real numbers: $x(y+z) = xy + xz$. We use induction to prove $P(n)$ for all $n \in \mathbb{N}$.

Basis Step: the statement $P(1)$ is "$ca_1 = ca_1$", which is true.

Induction Step: We use both the hypothesis that $P(k)$ is true and the distributive law to compute

$$\Sigma_{i=1}^{k+1} ca_i = (\Sigma_{i=1}^{k} ca_i) + ca_{k+1} = (c\Sigma_{i=1}^{k} a_i) + ca_{k+1}$$
$$= c[(\Sigma_{i=1}^{k} a_i) + a_{k+1}] = c\Sigma_{i=1}^{k+1} a_i.$$

This computation shows that $P(k)$ implies $P(k+1)$.

By the principle of induction, $P(n)$ is true for all n.

2) In our discussion of the order axioms (Definition 1.16), we defined $x < y$ to mean $y - x > 0$, and then we observed from the closure of the positive set that $a < b$ and $c < d$ implies $a + c < b + d$. This leads to a proof by induction that $Q(n)$ holds for all $n \in \mathbb{N}$.

Basis Step: the statement $Q(1)$ is "$a_1 \leq b_1$"; true by hypothesis.

Induction Step: If $Q(k)$ is true, then $\Sigma_{i=1}^{k} a_i \leq \Sigma_{i=1}^{k} b_i$, and we can use this and our ability to sum two inequalities to compute

$$\Sigma_{i=1}^{k+1} a_i = (\Sigma_{i=1}^{k} a_i) + a_{k+1} \leq (\Sigma_{i=1}^{k} b_i) + b_{k+1} = \Sigma_{i=1}^{k+1} b_i.$$

This computation shows that $Q(k)$ implies $Q(k+1)$.

By the principle of induction, $Q(n)$ is true for all n. ∎

This proof of the distributive law completes the formal proof of Lemma 3.35. Next we apply Lemma 3.35 to prove an important formula that can also be proved by induction (Exercise 4.17).

4.10. Corollary. (The Geometric Sum). If $q \in \mathbb{R}$, $q \neq 1$, and n is a non-negative integer, then

$$\Sigma_{i=0}^{n-1} q^i = \frac{q^n - 1}{q - 1}.$$

Proof. In the expression $(x - y)(\Sigma_{i=0}^{n-1} x^{n-1-i} y^i) = x^n - y^n$ of Lemma 3.35, set $x = q$ and $y = 1$ (and let $j = n - 1 - i$) to obtain $(q - 1)\Sigma_{j=0}^{n-1} q^i = q^n - 1$. Since $q \neq 1$, we can divide both sides by $q - 1$. ∎

4.11. Example. *A knockout tournament.* The NCAA basketball tournament starts with 64 teams. How many games are played to produce a winner? In the first round, there are 32 games. The 32 winners play 16 games in the second round. The subsequent rounds have 8, 4, 2, 1 games, respectively. By the geometric sum, the total number of games is $1 + 2 + 4 + 8 + 16 + 32 = \Sigma_{i=0}^{5} 2^i = 2^6 - 1 = 63$. We can also obtain this result by observing that every team other than the winner must lose exactly one game, and hence there must be exactly 63 games. The two sides of the equality give different methods for counting the games. ∎

When we express a claim to be proved by induction as a sequence of statements indexed by n, we say that the proof is "by induction on n" and call n the *induction parameter*. The induction step proves the conditional statement "$P(k)$ is true implies $P(k+1)$ is true", and the hypothesis of this conditional is called the *induction hypothesis*. Somewhere in the proof of the induction step, we say something like "by the induction hypothesis", which is the hypothesis that $P(k)$ is true. If we have not used the induction hypothesis, then we have not written a proof by induction.

Our next example of induction will be applied in Theorem 5.7. We begin to relax the formal template for proofs by induction to show the flexibility of the technique.

4.12. Proposition. If $n \in \mathbb{N}$ and $q \geq 2$, then $n < q^n$.

Proof. We use induction on n. For the basis step, we have $1 < q$ by the hypothesis on q, so the claim holds when $n = 1$. For the induction step, suppose that the claim holds when $n = k$, meaning that $k < q^k$. Using the induction hypothesis for the step of strict inequality, we compute

$$k+1 \leq k+k = 2k \leq qk < q \cdot q^k = q^{k+1}.$$

Hence the claim also holds when $n = k+1$, which completes the proof of the induction step. ∎

In proving the induction step, we do not assume the statement that $P(n)$ is true for all n; this is what we are trying to prove! We "assume" that the desired statement $P(n)$ is true for $n = k$ only to prove that $P(k+1)$ will follow if $P(k)$ is true. We prove for all $k \in \mathbb{N}$ that $P(k)$ and $\neg P(k+1)$ cannot both be true. Writing the proof in terms of an arbitrary natural number k encodes all the proofs at once, each proof enlarging the set of values of n for which $P(n)$ is known (starting from the value(s) proved in the basis step).

To visualize the process of induction, consider an infinite sequence of upright dominoes, one domino for each natural number. If any domino falls, it knocks over the next; this is the "induction step". The principle of induction says that if also the first domino falls (the "basis step"), then all the dominoes fall, because there is no first domino that doesn't fall. The proof that the first domino falls cannot be omitted.

4.13. Example. $n = n+1$ *(!?)*. The induction step $P(k) \Rightarrow P(k+1)$ is a conditional statement. A conditional statement fails only if the hypothesis is true and the conclusion is false. From the hypothesis that $k = k+1$, we can easily derive $k+1 = k+2$. This proof of the induction step is valid, but we have not proved that $n = n+1$ for all $n \in \mathbb{N}$, because the basis step is false: $1 \neq 2$. The basis step cannot be omitted. ∎

The next example illustrates a similar error.

4.14. Example. *All people have the same sex (!?).* We try to prove by induction on n that the people in every set of n people have the same sex. Certainly all the people in a set consisting of one person have the same sex, so the claim holds for $n = 1$. Now suppose the claim holds for $n = k$, and let $S = \{a_1, \ldots, a_{k+1}\}$ be a set of $k + 1$ people. If we delete a_1, then we have a set T of k people. If we delete a_2, then we have another set T' of k people. By the induction hypothesis, the people in T have the same sex, and the people in T' have the same sex. Now a_1 and a_2 have the same sex as a_{k+1}, so the people in S all have the same sex.

The problem is that the proof of the induction step is invalid when going from $n = 1$ to $n = 2$. In this case, the sets T and T' have no common element, so we cannot use the induction hypothesis to conclude that a_1 and a_2 have the same sex. ∎

We must always make sure that the *proof* of the induction step is valid for each value of the parameter where it is needed. The next example underscores this point and introduces the observation that we can start a proof by induction at any value. In particular, we can use induction to prove that $P(n)$ holds for $\{n \in \mathbb{N} : n \geq n_0\}$ by replacing $P(1)$ by $P(n_0)$ in the basis step. This is equivalent to treating $n - n_0$ as the induction parameter. Indeed, we may use any integer-valued function of the variables in our problem as an induction parameter.

4.15. Example. *Polynomial inequality.* In general, cubic polynomials "grow faster" than quadratics (see Chapter 14); let us try to prove that $n^3 + 20 > n^2 + 15n$ for all $n \in \mathbb{N}$. For the basis step, $21 > 16$, so the inequality holds when $n = 1$. For the induction step, suppose that it holds when $n = k$; we want $(k + 1)^3 + 20 > (k + 1)^2 + 15(k + 1)$. Using the induction hypothesis $k^3 + 20 > k^2 + 15k$, we compute

$$(k + 1)^3 + 20 = (k^3 + 3k^2 + 3k + 1) + 20 = (3k^2 + 3k + 1) + (k^3 + 20)$$

$$> (3k^2 + 3k + 1) + (k^2 + 15k) = (k^2 + 2k + 1) + (15k + 15) + 2k^2 + k - 15.$$

To prove that the final expression is at least $(k + 1)^2 + 15(k + 1)$, we need only verify that $k(2k + 1) \geq 15$.

But wait! The inequality $n^3 + 20 > n^2 + 15n$ fails when n is 2 or 3. The problem is that our argument for the induction step uses $k(2k + 1) \geq 15$, which is not valid when $k = 1$, so we cannot use it to go from $n = 1$ to $n = 2$.

All is not lost. When $n = 4$, we have $n^3 + 20 = 84 > 76 = n^2 + 15n$, so we can use $n = 4$ as a basis step. Now we only need the induction step to be valid when $k \geq 4$. In this range, we have $k(2k + 1) \geq 4 \cdot 9 > 15$, and we have proved the inequality for natural numbers at least 4. ∎

In the proof of the induction step in Example 4.15, it is tempting to start with the desired inequality $(k+1)^3 + 20 > (k+1)^2 + 15(k+1)$ and reduce it to a statement known to be true. A proof written in this manner must start with words such as "it suffices to show that" to make it clear that the desired statement is not being assumed, and words along the way must indicate that this statement is being obtained from the statement that follows it rather than the other way around. Direct computation usually presents the proof more efficiently, but the reduction method may aid in discovering it.

APPLICATIONS

Applications of induction occur throughout mathematics. We give several applications here, including the solutions of the Checkerboard Problem (Problem 4.1) and the Handshake Problem (Problem 4.2), a characterization of when polynomials are equal, and a geometric problem about cutting regions into smaller pieces.

To solve the Checkerboard Problem, we will need to sum the squares of the first n natural numbers. A consequence of this proposition is the perhaps surprising statement that $n(n+1)(2n+1)/6$ is an integer for each $n \in \mathbb{N}$.

4.16. Proposition. For all $n \in \mathbb{N}$, $\Sigma_{i=1}^{n} i^2 = n(n+1)(2n+1)/6$.

Proof. We use induction on n. Basis Step: when $n = 1$, the sum is 1, and the value of the formula is $1 \cdot 2 \cdot 3/6 = 1$, so the formula is correct. Induction step: we prove that the formula holds when $n = k+1$ if it holds when $n = k$. By the induction hypothesis, we have $\Sigma_{i=1}^{k} i^2 = k(k+1)(2k+1)/6$, and hence

$$\Sigma_{i=1}^{k+1} i^2 = \frac{k(k+1)(2k+1)}{6} + (k+1)^2 = (k+1)[\frac{2k^2+k}{6} + (k+1)]$$

$$= (k+1)(2k^2 + 7k + 6)/6 = (k+1)(k+2)(2k+3)/6.$$

The last expression is the desired formula when $n = k+1$, which completes the induction step. By the principle of induction, the formula holds for all n. ∎

4.17. Solution. *The Checkerboard Problem.* In the n-by-n checkerboard, there is one n-by-n square, and there are n^2 one-by-one squares. In general, there are $k \cdot k = k^2$ squares with sides of length $n - k + 1$, for $1 \le k \le n$. Hence the total number is $\Sigma_{k=1}^{n} k^2 = n(n+1)(2n+1)/6$, by Proposition 4.16. For $n = 8$, the value is $8 \cdot 9 \cdot 17/6 = 204$. ∎

Induction may provide an easy proof of a formula, but it does not help when we do not know the formula in advance. We may try to guess the formula from the first few values; if we guess right, then induction may provide a proof. In Chapter 9, we will develop summation methods that do not require knowing the answer in advance.

Up to now, we have given the induction parameter a new name when proving the induction step, emphasizing that the induction step uses the truth of one instance to prove the next and does not assume its truth for all instances. In the induction step the name of the induction parameter is a "dummy variable", with no meaning outside the induction step. Like an index of summation that has no meaning outside the sum, an induction parameter can be renamed.

When writing induction proofs, we may avoid giving this dummy variable a new name. Suppose we want to prove $P(n)$ for all $n \in \mathbb{N}$ by induction on n. Instead of "$P(k) \Rightarrow P(k+1)$ whenever $k \geq 1$", we may phrase the induction step as "$P(n-1) \Rightarrow P(n)$ whenever $n > 1$". Renaming the dummy parameter k as $n-1$ allows the conclusion of the induction step to look like the statement of the full theorem without changing the structure of a proof by induction.

Functions f and g with the same domain and target are equal when $f(x) = g(x)$ for all x. To justify the method of "equating corresponding coefficients" for equal polynomials, we need to know that polynomials are equal only if their corresponding coefficients are equal. The *zeros* of a function f are the solutions to the equation $f(x) = 0$. Recall that the zero polynomial does not have a degree.

4.18. Theorem. Every polynomial of degree d has at most d zeros.

Proof. Suppose f is a polynomial of degree d. By the definition of a polynomial, this means that $f(x) = \Sigma_{i=0}^{d} c_i x^i$, with d a nonnegative integer and $c_d \neq 0$. We prove by induction on d that f has at most d zeros. If $d = 0$ (basis step), then $f(x) = c_0 \neq 0$ for all x, and f has no zeros.

For the induction step, suppose $d \geq 1$, and suppose a is a zero of f. Since $f(x) = c_0 + \Sigma_{i=1}^{d} c_i x^i$ and $f(a) = 0$, we compute

$$f(x) = f(x) - f(a) = c_0 - c_0 + \Sigma_{i=1}^{d} c_i(x^i - a^i).$$

By Lemma 3.35, for $i \geq 1$ we have $x^i - a^i = (x-a)\Sigma_{j=1}^{i-1} x^{i-1-j} a^j$. This is a product of polynomials in x having degrees 1 and $i-1$. Hence for each x we can factor $(x-a)$ from each term in the sum $f(x) = \Sigma_{i=1}^{d} c_i(x^i - a^i)$ (this uses Proposition 4.9) to write $f(x) = (x-a)h(x)$, where h is a polynomial in x of degree $d-1$ (and hence not the zero polynomial). Since nonzero numbers have nonzero product, the only zeros of f are the zeros of $x-a$ and the zeros of h. Since $x - a = 0$ implies $x = a$, the first polynomial has exactly one zero. By the induction hypothesis, h has at most $d-1$ zeros. We conclude that f has at most d zeros. ∎

4.19. Corollary. Two real polynomials are equal if and only if their corresponding coefficients are equal.

Proof. Let f and g be the polynomials. If corresponding coefficients are equal, then $f(x) = g(x)$ for all $x \in \mathbb{R}$, because they have the same formula. Conversely, suppose the function values are always equal. Let h be the difference of f and g, so $h(x) = f(x) - g(x)$ for all $x \in \mathbb{R}$. Every real number is a zero of h, so Theorem 4.18 implies that h cannot be a polynomial of degree $d \geq 0$ and therefore must be the zero polynomial. Therefore, for each i the coefficients of x^i in f and in g are equal. ∎

Corollary 4.19 can be phrased in a slightly stronger form with the same proof: If polynomials f and g are equal at more than d places, where d is the maximum of the degrees of f and g, then they are the same polynomial. Contrast this with Exercise 7.28.

In the induction step of a proof by induction, we consider an arbitrary instance of the hypotheses for one value of the parameter, and we find an instance for a smaller value of the parameter in order to apply the induction hypothesis. In the proof of Theorem 4.18, we factored out $x - a$ to obtain a polynomial of smaller degree to which we could apply the induction hypothesis. Finding an appropriate smaller problem may take some effort. In the next example, the smaller instance emerges by stripping away pieces of the larger instance in an interesting way.

4.20. Solution. *The Handshake Problem.* Let a *handshake party* be a party with n married couples where no spouses shake hands with each other and the $2n - 1$ people other than the host shake hands with different numbers of people. We prove that in every handshake party, the hostess shakes hands with exactly $n - 1$ people.

If no one shakes with his or her spouse, then each person shakes with between 0 and $2n - 2$ people. Since the $2n - 1$ numbers are distinct, they must be 0 through $2n - 2$. The figure below illustrates for $n \in \{1, 2, 3\}$ the situation that is forced; each pair of encircled points indicates a married couple (host and hostess are rightmost), and two points are connected by a curve if and only if those two people shook hands.

  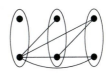

We prove the claim by induction on n. If $n = 1$, then the hostess shakes with 0 (which equals $n - 1$), because the host and hostess don't shake. Now suppose $n > 1$; the induction hypothesis states that the claim holds for a handshake party with $n - 1$ couples.

Let P_i denote the person (other than the host) who shakes with i

people. Since P_{2n-2} shakes with all but one person, P_0 must be the person omitted. Hence P_0 is the spouse of P_{2n-2}. Furthermore, this married couple is not the host and hostess, since $S = \{P_0, P_{2n-2}\}$ does not include the host. Everyone not in S shakes with exactly one person in S, namely P_{2n-2}. If we delete S to obtain a smaller party, then we have $n-1$ couples remaining (including the host and hostess), no person shakes with a spouse, and each person shakes with one fewer person than in the full party. Hence in the smaller party the people other than the host shake hands with different numbers of people.

By deleting the set S, we obtain a handshake party with $n-1$ couples (in the illustration, deleting the leftmost couple in the picture for $n = 3$ yields the picture for $n = 2$). We can thus apply the induction hypothesis to conclude that, outside of the couple S, the hostess shakes with $n-2$ people. Since she also shakes with P_{2n-2}, in the full party she shakes with $n-1$ people. ∎

In this proof, we cannot discard an arbitrary married couple to obtain the smaller configuration. We must find a couple S such that everyone outside S shakes with exactly one person in S. Only then will we be able to apply the induction hypothesis to the smaller party, because only then will we know that it satisfies the hypothesis about distinct numbers of handshakes. A similar problem arises if we start with a party of n couples in which the hostess shakes with $n-1$ people, and we add a couple S in which one person shakes with everyone else and the other person shakes with no one. This produces a party of $n+1$ couples satisfying the distinctness condition, in which the hostess shakes with n people, but it does not prove the induction step, because we have not shown that every party of $n+1$ couples satisfying the distinctness condition arises in this way. When we began with an arbitrary party of the larger size and proved that P_0 must be the spouse of the person shaking the most hands, we avoided this difficulty. An inductive argument that fails to consider all instances of the new (larger) size falls into what we call the *induction trap*.

Often one must explore a few small values to find a pattern. Eventually a uniform way of getting from one value of the parameter to the next emerges; describing and explaining that process for the general case becomes the proof of the induction step. Although they may aid in finding the proof, extra basis values should not appear in the final presentation of the proof.

Nevertheless, sometimes we must verify more than $P(1)$ in the basis step. This happens when the proof of the induction step needs the truth of the claim for more than just the preceding value. If we need both $P(n-1)$ and $P(n)$ to prove $P(n+1)$, then we must check $P(n)$ for the first two values of n in the basis step. Because the proof of the induction step uses two previous values, it does not apply for the second smallest value of n. This occurs in the next example and in many proofs using recurrence relations (Chapter 12).

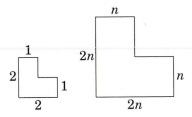

4.21. Solution. *The L-Tiling Problem.* A child has a large number of L-shaped tiles as illustrated on the left above. Is it possible to form the large similar area on the right with non-overlapping copies of this tile?

Let the large region be R_n, let the small region be L, and call a partition of a region into copies of L an L-*tiling*. When $n = 1$, R_n consists of exactly one L, so R_1 has an L-tiling. On the left below decompose R_2 explicitly (this suggests a simple inductive proof that R_n has an L-tiling whenever n is a power of 2 - Exercise 4.36).

Proving that every R_n has an L-tiling requires more effort. We can find a copy of R_{n-1} inside R_n by removing a strip of unit width along the topmost, leftmost, bottommost, and rightmost edges. This does not help; the usual induction hypothesis tells us that R_{n-1} can be tiled, but if $n \geq 3$ we cannot extend this to a tiling of R_n because the outer strip has no L-tiling.

To fix this flaw, we use an outer strip of width 2 and obtain an L-tiling of R_n from an L-tiling of R_{n-2}. Since R_1 has an L-tiling, this completes the proof for odd n, but to handle the even cases we also must consider R_2 in the basis. Below we explicitly tile R_2 and also illustrate that every 2 by $3k$ rectangle has an L-tiling.

For the induction step, consider R_n, where $n \geq 3$. The induction hypothesis provides an L-tiling of the inner region R_{n-2}, so to complete the proof it suffices to tile the outer strip. We use copies of R_2 in the corners and 2 by $3k$ rectangles along the sides; we already showed that these have L-tilings. We tile the outer band in one of three ways, depending on whether 3 divides n, $n - 1$, or $n - 2$. To prove that the decomposition works in each case, we need only verify that the length of the long side of each rectangle used is a multiple of 3. For clarity in the pictures, we list only these lengths; the short sides all equal 2. The three cases occur when $n \geq 3$, $n \geq 4$, and $n \geq 5$, respectively, so the lengths of the rectangles are all nonnegative multiples of 3. Verifying this completes the induction step. ∎

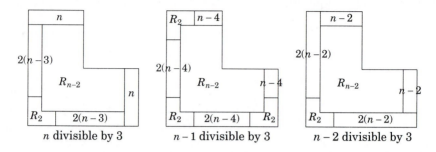

n divisible by 3   n − 1 divisible by 3   n − 2 divisible by 3

STRONG INDUCTION

A proof by induction asserts that there is no smallest natural number for which the claim fails. This observation leads to a variation called "strong induction", used when the proof of $P(k)$ in the induction step needs the hypothesis that $P(i)$ is true for every value of i before k. By assuming more in the induction hypothesis, we make the induction step a weaker implication. Because this weaker implication still suffices to complete the proof, we call the method "strong" induction.

4.22. Theorem. ("Strong" Principle of Mathematical Induction) Suppose we have a sequence of mathematical statements $P(1), P(2), \cdots$ (one for each natural number). If the following two properties hold, then for every $n \in \mathbb{N}$, $P(n)$ is true.
1) $P(1)$ is true.
2) If $P(i)$ is true for all $i < k$, then $P(k)$ is true.

Proof. As in Theorem 4.5, we use proof by contradiction. If $P(n)$ fails for some n, then the Well-Ordering Property guarantees that it fails for a smallest number, k. Statement 1 implies that k cannot be 1. Statement 2 implies that k cannot be larger than 1. ∎

Strong induction can be phrased as ordinary induction. Given a sequence $P(1), P(2), \cdots$ of mathematical statements, let $Q(n)$ be the statement that all of $\{P(1), \ldots, P(n)\}$ are true. Proving $Q(n)$ by ordinary induction on n is equivalent to proving $P(n)$ by strong induction on n. We can use strong induction to prove a statement for all nonnegative integers by changing statement (1) to "$P(0)$ is true".

Strong induction yields short proofs of many algebraic properties of natural numbers. As an example, we generalize our earlier discussion of parity. Since 1 equals 2^0, we consider 1 to be a power of 2.

4.23. Proposition. Every natural number n can be expressed as the product of an odd number and a power of 2. These numbers are uniquely determined.

Proof. We use strong induction. Basis step: the number 1 equals $1 \cdot 2^0$, and this representation is unique. Induction step: suppose $n > 1$. If n is odd, then $n = n \cdot 2^0$; the representation is unique because n does not have 2 as a factor. If n is even, then $n = 2m$ for some smaller natural number m. By the induction hypothesis, $m = s \cdot 2^t$ for some odd s and $t \geq 0$. Now $n = s \cdot 2^{t+1}$. Since m is even, every representation of m as an odd number times a power of two uses a positive power of 2. If m has two such representations, then $m/2$ also has two such representations, which contradicts the uniqueness part of the induction hypothesis; hence again the representation of m is unique. ∎

Using Proposition 4.23, we determine which natural numbers can be written as a sum of consecutive smaller natural numbers.

4.24. Example. *Sums of Consecutive Positive Integers.* For $r \geq 1$, we can write the odd number $n = 2r + 1$ as $r + (r + 1)$. When n is twice an odd number $2r + 1$, we can then write $n = (r - 1) + (r) + (r + 1) + (r + 2)$. This works whenever $r - 1 \geq 1$ and fails when $n = 2$. When $n = 6$, we have $r - 1 = 0$ and can drop this 0 to obtain $6 = 1 + 2 + 3$.

When $n = 4(2r + 1)$, we can write

$$n = (r - 3) + (r - 2) + (r - 1) + (r) + (r + 1) + (r + 2) + (r + 3) + (r + 4).$$

This works whenever $r - 3 \geq 1$. When $r - 3 = -1$, we can drop the first three terms $(-1) + (0) + (1)$ to write n as a sum of five consecutive integers. When $r - 3 = 0$, we drop the 0.

This suggests the general procedure used in the proof below. We illustrate it for 11, 22, 44, 88, which are powers of 2 times the odd number $11 = 2 \cdot 5 + 1$. For $2^t 11$, we use 2^t pairs of numbers summing to 11. Thus we write $11 = 5 + 6$, $22 = 4 + 5 + 6 + 7$, $44 = 2 + 3 + \cdots + 9$, and $88 = -2 + (-1) + 0 + 1 + 2 + \cdots + 13$. We cancel the first five terms in the last expression to obtain $88 = 3 + 4 + \cdots + 13$. ∎

4.25. Solution. *Sums of Consecutive Positive Integers.* We prove that a natural number n is a sum of consecutive smaller natural numbers if and only if n is not a power of 2.

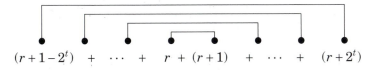

$$(r + 1 - 2^t) \quad + \quad \cdots \quad + \quad r \; + \; (r + 1) \quad + \quad \cdots \quad + \quad (r + 2^t)$$

If n is not a power of 2, then $n = 2^t(2r + 1)$ for some integers $t \geq 0$

and $r > 0$, by Proposition 4.23. Consider the 2^t numbers ending at r and the 2^t numbers starting at $r + 1$. Grouping these symmetrically around the middle yields 2^t pairs, each with sum $2r + 1$, so the overall sum is n. When $r + 1 - 2^t \le 0$, the numbers are not all positive. The two numbers in the middle are positive, so the number of positive terms exceeds the number of nonpositive terms by at least 2. In this case, the numbers $r + 1 - 2^t$ through $-(r + 1 - 2^t)$ have sum 0, and we delete them to express n as the sum of (at least two) consecutive natural numbers.

To prove the converse, suppose that n is the sum of p consecutive natural numbers starting with m. We will write n as an integer times an odd number larger than 1. When $p = 2$, n is odd. Otherwise, we use Proposition 4.7 to evaluate $\Sigma_{i=0}^{p-1} i$ and obtain

$$n \;=\; \Sigma_{i=0}^{p-1}(m+i) \;=\; mp + \Sigma_{i=0}^{p-1} i \;=\; mp + \frac{p(p-1)}{2} \;=\; \frac{p(2m+p-1)}{2}$$

Whether p is odd or even, exactly one of $\{p, 2m + p - 1\}$ is even and both exceed 2. Hence $n = \dfrac{p(2m+p-1)}{2}$ expresses n as the product of two integers, at least one of which is odd and larger than 1. We conclude that n is not a power of 2. ∎

EXERCISES

Words like "determine", "obtain", "construct", or "show" request proof.

4.1. (–) Find and prove a formula for $3 + 7 + 11 + \cdots + (4n - 1)$.

4.2. (–) Find and prove a formula for $1 + 5 + 9 + \cdots + (4n + 1)$.

4.3. (–) Find and prove a formula for $-1 + 2 - 3 + 4 - \cdots - (2n - 1) + 2n$.

4.4. (–) Find and prove a formula for $1 - 3 + 5 - 7 + \cdots + (4n - 3) - (4n - 1)$.

4.5. For $n \in \mathbb{N}$, prove that $\Sigma_{i=1}^{n}(-1)^i i^2 = (-1)^n \frac{n(n+1)}{2}$.

4.6. For $n \in \mathbb{N}$, prove that $\Sigma_{i=1}^{n} i^3 = (\frac{n(n+1)}{2})^2$.

4.7. For $n \in \mathbb{N}$, prove that $\Sigma_{i=1}^{n} i(i + 1) = \frac{n(n+1)(n+2)}{3}$.

4.8. (–) Suppose that $n \in \mathbb{N}$ and that $x_1, \ldots, x_{2n+1}$ are odd integers. Prove that $\Sigma_{i=1}^{2n+1} x_i$ is odd and that $\Pi_{i=1}^{2n+1} x_i$ is odd.

4.9. For $n \in \mathbb{N}$, prove that $\Sigma_{i=1}^{n} \frac{1}{(3i-2)(3i+1)} = \frac{n}{3n+1}$.

4.10. For $n \in \mathbb{N}$, find and prove a formula for $\Sigma_{i=1}^{n} \frac{1}{i(i+1)}$.

4.11. For $n \in \mathbb{N}$, find and prove a formula for $\Sigma_{i=1}^{n}(2i - 1)$.

4.12. For $n \in \mathbb{N}$, prove that $\Sigma_{i=1}^{n}(2i - 1)^2 = \frac{n(2n-1)(2n+1)}{3}$.

4.13. For $n \in \mathbb{N}$ and $n \geq 2$, find and prove a formula for $\Pi_{i=2}^{n}(1 - \frac{1}{i^2})$.

4.14. For $n \in \mathbb{N}$ and $n \geq 2$, find and prove a formula for $\Pi_{i=2}^{n}(1 - \frac{(-1)^i}{i})$.

4.15. Obtain a simple formula for the number of closed intervals with integer endpoints contained in the interval $[0, n]$ (including one-point intervals).

4.16. Consider a set of 20 boxes, each containing 20 balls. Suppose every ball weighs one pound, except that the balls in one box are all one ounce too heavy or all one ounce too light. A precise scale is available that can weigh to the nearest ounce (not a balance scale). By selecting some balls to place on the scale, explain how to determine in one weighing which is the defective box and whether its balls are too heavy or too light.

4.17. Suppose $q \neq 1$. Use induction on n to prove that $\Sigma_{i=0}^{n-1} q^i = (q^n - 1)/(q - 1)$.

4.18. Obtain a polynomial f such that $\Sigma_{i=2}^{n} x^k = f(x)/(x - 1)$.

4.19. For $n \in \mathbb{N}$, obtain a formula for $\Sigma_{i=1}^{n} n^i$. (Hint: do not use induction.)

4.20. (!) Starting with 0, two players alternately add 1, 2, or 3 to a single running total. The player who first brings the total to at least 1000 wins. Prove that the second player has a strategy to win against any strategy for the first player. (Hint: use induction to prove a more general statement.)

4.21. (!) Suppose that $x_1, \ldots, x_n$ are n numbers in the interval $[0, 1]$. Prove that $\Pi_{i=1}^{n}(1 - x_i) \geq 1 - \Sigma_{i=1}^{n} x_i$.

4.22. (!) Addition is defined as a function from $\mathbb{R} \times \mathbb{R}$ to $\mathbb{R}$; it sums only pairs of numbers. Use induction on n to prove that the sum of n numbers is independent of the order in which the numbers are added into the total. This justifies the use of summation notation for a sum of n numbers.

4.23. Given $f: A \to A$, the nth *iterate* of f, for $n \in \mathbb{N}$, is the function f^n defined by $f^1 = f$ and $f^{k+1} = f \circ f^k$ for $k \geq 1$. Suppose n, k are natural numbers with $k < n$. Prove that $f^n = f^k \circ f^{n-k}$. (Comment: Chapter 3 contains a more informal version of the definition of iteration.)

4.24. Suppose $f: \mathbb{R} \to \mathbb{R}$ is defined by $f(x) = a(x + b) - b$. Obtain an explicit formula for the nth iterate f^n (defined in Exercise 4.23). Express f and f^n in the language of Exercise 3.32.

4.25. Suppose $f: A \to B$ is a bijection and $g: B \to B$. Let h be the composition $h = f^{-1} \circ g \circ f$, so $h: A \to A$. Derive a formula in terms of f and g for the nth iterate of h. Compare this with Exercise 4.24.

4.26. (!) Suppose $f: \mathbb{R} \to \mathbb{R}$ satisfies $f(xy) = xf(y) + yf(x)$ for all $x, y \in \mathbb{R}$. Prove that $f(1) = 0$ and that $f(u^n) = nu^{n-1}f(u)$ for all $n \in \mathbb{N}$ and $u \in \mathbb{R}$.

4.27. (!) Determine the set of natural numbers that can be expressed as the sum of some nonnegative number of 3's and some nonnegative number of 10's.

4.28. (−) Let $f(n) = n^2 - 8n + 18$. For which $n \in \mathbb{N}$ is $f(n) > f(n - 1)$ true?

4.29. Prove that $5^n + 5 < 5^{n+1}$ for all $n \in \mathbb{N}$.

4.30. (!) Determine the set of positive real numbers x such that the inequality $x^n + x < x^{n+1}$ holds for all $n \in \mathbb{N}$.

4.31. For each of the following inequalities, determine the set of natural numbers n for which it holds.
    a) $3^n \geq 2^{n+1}$.
    b) $2^n \geq (n+1)^2$.
    c) $3^n > n^4$.
    d) $n^3 + (n+1)^3 > (n+2)^3$.

4.32. Construct a cubic polynomial such that the set of natural numbers where its value is at least 3 is $\{1\} \cup \{n \in \mathbb{N}: n \geq 5\}$.

4.33. The following statement is not valid: "Suppose x is a particular natural number. If T is a set of natural numbers such that 1) $x \in T$ and 2) $y \in T$ implies $y + 1 \in T$, then $T = \{y \in \mathbb{N}: y \geq x\}$." Explain why the statement is not valid, and change one symbol to correct it.

4.34. Suppose a is a nonzero real number. Find the flaw in the following "proof" that $a^n = 1$ for every nonnegative integer n. "Basis step: $a^0 = 1$. Induction step: $a^{n+1} = a^n \cdot a^n / a^{n-1} = 1 \cdot 1/1 = 1$."

4.35. *Partial fraction expansion.* Obtain real numbers A, B, r, s such that
$$\frac{1}{x^2 + x - 6} = \frac{A}{x-r} + \frac{B}{x-s}.$$

4.36. (!) *L-tilings.* Prove that R has an L-tiling in the following situations.
    a) R is a 2^k by 2^k chessboard with one corner square removed.
    b) R is a 2^k by 2^k chessboard with *any* single square removed.

4.37. (!) *The Coin-Removal Problem.* Suppose n coins are arranged in a row. Coins may be heads-up (H) or tails-up (T). We are allowed the following operation: remove an H and flip the coin or coins in the immediate neighboring position(s), if these are still present. For which arrangements of heads and tails can we remove all the coins? (When we remove a coin between two others, the others do NOT become neighbors.) For example, $THHHT$ succeeds, via $THHHT$, $H.THT, ..THT, ..H.H, ...H, ...$, but $THTHT$ fails.

4.38. (!) *The December 31 Game.* Two players alternately name dates. On each move, a player can increase the month or the day of the month but not both. The starting position is January 1, and the player who names December 31 wins. According to the rules, the first player can start by naming some day in January after the first or the first of some month after January. For example, (Jan. 10, Mar. 10, Mar. 15, Apr. 15, Apr. 25, Nov. 25, Nov. 30, Dec. 30, Dec. 31) is an instance of the game won by the first player. Derive a strategy that the first player can use to guarantee winning. (Hint: use strong induction to prove a more general statement.)

4.39. Let T be a subset of $\mathbb{N}$. The "method of infinite descent" says that if $\mathbb{N} - T$ contains no least element, then $T = \mathbb{N}$. Explain the relationship between the method of infinite descent and the principle of induction.

4.40. Derive the Well-Ordering Property for natural numbers from the principle of induction.

4.41. (!) Suppose that $f(x)$ is a polynomial of degree n and that the values $f(0), f(1), \ldots, f(n)$ are known. Describe a procedure for determining f, and state what is needed to justify that it works.

4.42. (!) Suppose f is defined by $f(x) = \Sigma_{i=0}^{n} c_i x^i$ and has zeros $\alpha_1, \ldots, \alpha_n$ such that $\alpha_i \neq 0$ for all i. Derive a formula for $\Sigma_{i=1}^{n}(1/\alpha_i)$ in terms of $c_0, \ldots, c_n$.

4.43. (!) In the village of Perfect Reasoning, each employer has an apprentice. At least one apprentice is a thief. To remedy this without embarrassment, the mayor proclaims the following true statements: "At least one apprentice in this town is a thief. Every thief is known a thief to everyone except his or her employer, and all employers reason perfectly. If n days from now you have concluded that your apprentice is a thief, you will come to the village square at noon that day to denounce your apprentice." The villagers gather at noon every day thereafter. If in fact $k \geq 1$ of the apprentices are thieves, when will they be denounced, and how do their employers reason? (Hint: study small values of k, and use induction to prove the pattern for all k.)

PART II

PROPERTIES OF NUMBERS

Chapter 5

Counting and Cardinality

In this chapter we discuss how to represent natural numbers and how to use them to solve counting problems. In both tasks, induction plays a fundamental role. We also use the notion of bijection to define precisely the "size" of a set. We introduce the binomial coefficients as the sizes of particular finite sets. Natural numbers, the notion of size, and binomial coefficients will be used to solve the following problems of existence and enumeration.

5.1. Problem. *The Weights Problem.* A balance scale has a left pan and a right pan; we can place objects in each pan and determine whether the total weight is the same on each side. Suppose 5 objects of known integer weight can be selected. How can we choose the weights to guarantee being able to check all integer weights from 1 through 121? Is it possible to choose 5 values to check more weights? Given a natural number $n \in [121]$, how should we place the known weights to check whether an unknown weight equals n? ∎

5.2. Problem. *Comparison of Poker Hands.* A poker hand consists of five cards from an ordinary deck of cards. Why is "three-of-a-kind" a higher-ranked poker hand than "two-pair", and why is a "flush" ranked higher than a "straight"? ∎

5.3. Problem. *Nonnegative Integer Solutions.* Suppose each resident of New York City has 100 coins in a jar. The coins come in five types (pennies, nickels, dimes, quarters, half-dollars). We consider two jars of coins to be "equivalent" if they have the same number of coins of each type. Is it possible that no two people have equivalent jars of coins? ∎

REPRESENTATION OF NATURAL NUMBERS

The most naive way to represent the number "one hundred" is by a collection of one hundred dots; it is hard even to count them. We can arrange the dots in a ten-by-ten square, but no geometric method gives convenient representations of all natural numbers. Representing large numbers is particularly difficult.

Roman numerals permit a reasonably concise description of large natural numbers, but they make arithmetic computations difficult. In Roman numerals, the numbers 1, 5, 10, 50, 100, 500, 1000, are represented by the symbols I, V, X, L, C, D, M. Other numbers are formed by strings of these symbols using complicated rules involving addition and subtraction of adjacent symbols. For example, 2 is written as II, 44 as XLIV, 88 as LXXXVIII, and 90 as XC; addition of 2 and multiplication by 2 are awkward operations!

The familiar decimal representation of integers is conducive to arithmetic computations, and fairly large numbers can be represented concisely. The decimal (base 10) representation of a natural number is a string of symbols from $\{0, 1, \ldots, 9\}$, encoding the expression of the number as a sum of multiples of powers of 10. Chemists, physicists, and astronomers often need very large numbers and express them in "scientific notation", a variant of decimal representation where only the significant digits and the order of magnitude are recorded (6.02×10^{23} is scientific notation for 602,000,000,000,000,000,000,000). Computer scientists use binary (base 2), octal (base 8), and hexadecimal (base 16) representations, where the string representing the number encodes its expression as a sum of multiples of powers of the base.

The appropriate method of representation depends on the problem being solved. In base q, there are q elementary symbols, representing the numbers 0 through $q - 1$. Computers use binary digits ("bits") because there are two alternatives for a switch: "on" or "off". In solving the Weights Problem (Problem 5.1), we will apply base 3 representation because there are three alternatives for how to use each weight.

5.4. Definition. Let q be a natural number greater than 1. A *q-ary expansion* of a natural number n is a list $a_m, \ldots, a_0$ such that each a_i is an integer in the set $\{0, \ldots, q - 1\}$ and such that $n = \Sigma_{i=0}^{m} a_i q^i$. The number q is the *base* of this representation. We use (q) as a subscript to indicate the base of a representation when q is not equal to 10. Representations in base 2, 3, or 10 are *binary*, *ternary*, or *decimal* representations, respectively.

In Theorem 5.7, we will prove that every natural number has a q-ary expansion and that it is unique except for leading zeros. In other words, there is only one q-ary expansion in which the first coefficient a_m

is nonzero. This allows us to replace "*a* q-ary expansion" by "*the* q-ary expansion".

5.5. Example. The ternary representations for the first ten natural numbers in order are 1, 2, 10, 11, 12, 20, 21, 22, 100, 101 (here we have suppressed the subscript notation for the base). The corresponding representations in base 4 are 1, 2, 3, 10, 11, 12, 13, 20, 21, 22. ∎

5.6. Example. Base 10 is the familiar base for representing numbers. For the natural number $354 = 3 \cdot 10^2 + 5 \cdot 10^1 + 4 \cdot 10^0$, the elements of the base 10 representation are $a_2 = 3$, $a_1 = 5$, $a_0 = 4$. We can also write $354 = 2 \cdot 5^3 + 4 \cdot 5^2 + 0 \cdot 5^1 + 4 \cdot 5^0$, expressed concisely as $2404_{(5)}$.

In order to write a number n in base q, we first determine the largest nonzero index m by finding the largest power of q that is at most n. The coefficient a_m is the largest multiple of q^m that can be subtracted from n without making it negative. We then repeat the procedure with what remains, which is smaller than q^m. For example, 5^4 is larger than 354, but 5^3 is not, so the 5-ary expansion of 354 starts with a_3. Since $5^3 = 125$ can be subtracted twice from 354, the expansion begins with $a_3 = 2$, followed by the representation of 104. By this procedure, we obtain $354_{(10)} = 2404_{(5)}$. In other bases, we have $354_{(10)} = 11202_{(4)} = 111010_{(3)} = 101100010_{(2)}$. ∎

The procedure described in Example 5.6 for finding a q-ary expansion works in general to prove that every natural number has a q-ary expansion. We use the geometric sum (Corollary 4.10) to prove that the expansion is unique.

5.7. Theorem. When q is a natural number greater than 1, every natural number has a unique q-ary expansion with no leading zeros.

Proof. First we prove by strong induction on n that n has a q-ary expansion $a_m, \ldots, a_0$ whenever $n < q^{m+1}$. We include the values of n with $1 \le n < q$ in the basis step, where we have the expansion $a_0 = n$. For the induction step, suppose that $n \ge q$ and that all natural numbers less than n have q-ary expansions. We will find a number m such that $q^m \le n < q^{m+1}$ and apply the induction hypothesis to a number n' less than q^m. From the inequality $q^n > n$ (Proposition 4.12), we conclude that there are powers of q larger than n. By the Well-Ordering Property, there is a least power of q larger than n; call this q^{m+1}, so $q^m \le n < q^{m+1}$. Among the multiples of q^m that are greater than n, let $(k+1)q^m$ be the smallest, so $kq^m \le n < (k+1)q^m$. Let $a_m = k$. Let $n' = n - kq^m$, so $0 \le n' < q^m \le n$. If $n' = 0$, then let $a_{m-1} = \cdots = a_0 = 0$. If $n' > 0$, then by the induction hypothesis the number n' has a q-ary expansion $a_{m-1}, \ldots, a_0$. In either case, combining $a_{m-1}, \ldots, a_0$ with

$a_m = k$ yields a q-ary expansion of n, because

$$\Sigma_{i=0}^{m} a_i q^i = a_m q^m + \Sigma_{i=0}^{m-1} a_i q^i = k q^m + n' = n.$$

To prove that there is only one q-ary expansion of n with nonzero first element, we prove the contrapositive: different q-ary expansions yield different natural numbers. Given a q-ary expansion $a = a_m, \ldots, a_0$, let $f(a) = \Sigma_{i=0}^{m} a_i q^i$, so that a is a q-ary expansion of $f(a)$. Suppose a and b are different q-ary expansions; we may add leading 0's to the lists to make them the same length without changing the value of f. Let k be the largest index where a, b differ; we may assume $b_k < a_k$. The contributions from indices exceeding k are the same, so we compare $a = \Sigma_{i=0}^{k} a_i q^i$ and $b = \Sigma_{i=0}^{k} b_i q^i$. Ignoring the contributions from powers lower than q^k yields $a \geq a_k q^k$. We obtain an upper bound on b by replacing the coefficients of powers lower than q^k by their maximum possible value, $q - 1$. Using this and then the geometric sum yields

$$b \leq b_k q^k + \Sigma_{i=0}^{k-1}(q-1)q^i \leq b_k q^k + (q^k - 1) < (b_k + 1)q^k \leq a_k q^k \leq a.$$

From $b < a$ we obtain $f(b) < f(a)$, and distinct q-ary expansions yield distinct natural numbers. ∎

The proof of uniqueness in Theorem 5.7 generalizes the familiar statement that m exceeds n if, in the highest-order position where their decimal representations differ, the digit for m is larger than the digit for n. Ternary representations lead to a solution of Problem 5.1.

5.8. Example. In order to understand the Weights Problem, we begin by considering the analogous question with fewer weights. In general, let k be the number of weights allowed. When $k = 2$, we do best by choosing $\{1, 3\}$; we can then test 2 by putting 1 and 3 on opposite sides and test 4 by putting them on the same side, along with testing 1 or 3 by using that weight alone.

Exploring a mathematical problem may involve both experimentation and insightful thinking. Here we may experiment to find that choosing $\{1, 3, 9\}$ for three known weights allows us to balance all unknown weights up to 13, and no other choice goes farther. This pattern suggests that choosing powers of 3 as the unknown weights may be a good idea. Insightful thinking may suggest this choice directly, since for each weight we have the three options of "left pan", "right pan", and "omit". Using powers of 3 allows us to exploit these three options fully.

Suppose $k = 5$. Using $\{1, 3, 9, 27, 81\}$, we can balance an unknown object A of weight 49 (for example) by using $\{9, 27\}$ and A on one side and using $\{1, 3, 81\}$ on the other side. The largest weight that can be balanced is $1 + 3 + 9 + 27 + 81 = 121$. We will prove that we can check all weights up to 121 and that this choice of five weights is best possible. ∎

5.9. Solution. *The Weights Problem.* We prove that the set $S_k = \{1, 3, \ldots, 3^{k-1}\}$ of known weights permits the checking of all integer weights from 1 through $(3^k - 1)/2$ on a balance scale, and that no other choice of k known weights permits more values to be checked.

Let $f(k) = (3^k - 1)/2$. We first prove that S_k permits us to balance A_n of integer weight n for $1 \le n \le f(k)$. When we place weights on the scale, placing them on the side opposite A_n contributes to balancing the weight n, while placing them on the same side as A_n provides a negative contribution. Therefore, it suffices to express n as $\Sigma_{i=0}^{k-1} b_i 3^i$, where each $b_i \in \{-1, 0, 1\}$, because then interpreting the values $-1, 0, 1$ for b_i to mean "same side as A_n", "off the balance", and "side opposite to A_n" yields an explicit configuration of the weights that balances A_n.

We find the desired numbers $\{b_i\}$ using the ternary expansion of the number $n' = n + f(k)$. The equation $n = \Sigma_{i=0}^{k-1} b_i 3^i$ holds if and only if the equation $n' = \Sigma_{i=0}^{k-1}(b_i + 1)3^i$ holds, because the geometric sum yields $(3^k - 1)/2 = \Sigma_{i=0}^{k-1} 3^i$. Since $n \le f(k)$, we have $n' \le 2f(k) = 3^k - 1$. Theorem 5.7 now guarantees a (unique) expression of n as $n = \Sigma_{i=0}^{k-1} a_i 3^i$ with each $a_i \in \{0, 1, 2\}$. Setting $b_i = a_i - 1$ yields the desired weighing of n.

We also must prove that no other set of weights can balance more values. We count the possible configurations: each weight can be placed on the left, on the right, or omitted, generating 3^k possible configurations. The configuration that omits all weights balances no nonzero weight. Of the remaining $3^k - 1$ configurations, each balances the same weight as the configuration obtained by switching the left pan and right pan. Hence at most $(3^k - 1)/2$ distinct values can be weighed. ∎

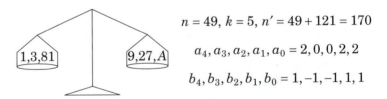

$n = 49,\ k = 5,\ n' = 49 + 121 = 170$

$a_4, a_3, a_2, a_1, a_0 = 2, 0, 0, 2, 2$

$b_4, b_3, b_2, b_1, b_0 = 1, -1, -1, 1, 1$

CARDINALITY

Enumerative problems generally involve counting the elements in a set. The precise meaning of this involves bijections. The definition agrees with intuition, and we have been using it implicitly. We have the notation $[k] = \{1, 2, \ldots, k\}$ for $k \in \mathbb{N}$; we also define $[0] = \varnothing$.

5.10. Definition. A set A is *finite* if there is a bijection from A to $[k]$ for some $k \in \mathbb{N} \cup \{0\}$. A set is *infinite* if there is no such bijection.

We state next a basic fact about the natural numbers, upon which counting is based. In a sense explained in Appendix A, its role is similar to that of the arithmetic properties of the natural numbers, which we have agreed to assume. Therefore, although the statement is easily proved by induction, we leave the proof to Appendix A.

5.11. Proposition. If there is a bijection $f\colon [m] \to [n]$, then $m = n$.

5.12. Corollary. If A is finite, then there is exactly one n for which there is a bijection from A to $[n]$.

Proof. By the definition of finiteness, such a number exists. Suppose bijections $g\colon A \to [m]$ and $h\colon A \to [n]$ exist. Because the composition of two bijections is a bijection (Proposition 3.43), the function $f = h \circ g^{-1}$ is a bijection from $[m]$ to $[n]$. By Proposition 5.11, $m = n$. ∎

5.13. Definition. The *size* of a finite set A, written $|A|$, is the unique value n such that there is a bijection from A to $[n]$. A set of size n is an *n-element set* or *n-set*.

"Size" is a function whose domain is the set of all finite sets and whose target is the set of nonnegative integers. Corollary 5.12 states that this function is well-defined in the sense of Remark 3.4. When we write $A = \{a_1, \ldots, a_n\}$ with the a_i's distinct, we are specifying a bijection from $[n]$ to A and stating that the size of A is n.

The notation for size is the same as the notation for absolute value; size measures "discrete distance" to A from the empty set, and absolute value measures distance to a number from 0. Since size applies only to sets and absolute value applies only to numbers, the context indicates whether size or absolute value is being used.

The definition of size using bijections enables us to justify many intuitive statements.

5.14. Corollary. If A and B are disjoint finite sets, then
$$|A \cup B| = |A| + |B|.$$

Proof. Suppose that $m = |A|$ and $n = |B|$. Given bijections $f\colon A \to [m]$ and $g\colon B \to [n]$, we define $h\colon A \cup B \to [m+n]$ by $h(x) = f(x)$ for $x \in A$ and $h(x) = g(x) + m$ for $x \in B$. Upon checking that h is a bijection (Exercise 5.8), the conclusion follows. ∎

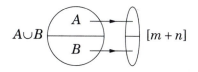

Deleting an element of an infinite set leaves another infinite set, but deleting an element of a non-empty finite set decreases its size by one. This observation enables us to prove statements about finite sets by induction on size. The next proof also fits the model of generalizing a fact about two objects to the corresponding fact about n objects (illustrated in Proposition 4.9).

5.15. Corollary. Every non-empty finite set of real numbers has both a maximum element and a minimum element.

Proof. We use induction on the size of the set. If $|A| = 1$, then the only element of A is both its maximum and its minimum. If $|A| > 1$, choose $x \in A$. The induction hypothesis yields a maximum M and minimum L for $A - \{x\}$. Compare x with these to find the maximum and minimum elements of A. ∎

We do not extend $|\ |$ to infinite sets. Nevertheless, we can use bijections to compare infinite sets.

5.16. Definition. An infinite set A is *countably infinite* (or *countable*) if there is a bijection from A to $\mathbb{N}$; otherwise an infinite set is *uncountably infinite* (or *uncountable*). Sets A and B *have the same cardinality* if there is a bijection from A to B.

Some authors define countable differently, by including finite sets among the countable sets. We adopt the more common convention that a countable set has the same cardinality as $\mathbb{N}$ and hence is infinite.

5.17. Proposition. The sets $\mathbb{N}$ and $\mathbb{Z}$ have the same cardinality.

Proof. To define a bijection $f: \mathbb{Z} \to \mathbb{N}$, we "spread out" the elements of $\mathbb{N}$ to provide room for the nonpositive integers. We define $f(x) = 2x$ if $x > 0$ and $f(x) = -2x + 1$ if $x \le 0$. This maps the positive integers to distinct even natural numbers and the nonpositive integers to distinct odd natural numbers, so f is injective. To prove that f is surjective, we observe that if n is even, then n is the image of the integer $x = n/2$, and if n is odd, then n is the image of the integer $x = (n-1)/(-2)$. ∎

The set $\mathbb{N}$ is properly contained in the set $\mathbb{Z}$, but they have the same cardinality. The set of rational numbers also has the same cardinality as $\mathbb{N}$ (Exercise 8.8), but $\mathbb{R}$ does not (it is uncountably infinite - Theorem 13.16). Since deleting an element reduces the size of a finite set, there is no bijection from a finite set to a proper subset of itself. Consequently, if there is a bijection from A to a proper subset of A, then A must be an infinite set.

Writing $A = \{a_1, a_2, \cdots\}$ with the a_i's distinct specifies a bijection from $\mathbb{N}$ to A. Proving that a set A is countable is equivalent to obtaining a sequence that contains each element of A exactly once. In the proof of Proposition 5.17, we generated such a sequence for $A = \mathbb{Z}$; this is the sequence of integers defined by the successive values of f^{-1}: 0, 1, -1, 2, -2, 3, -3, $\cdots$. Arguing that this sequence names every integer exactly once is another way to prove Proposition 5.17.

5.18. Theorem. The sets $\mathbb{N} \times \mathbb{N}$ and $\mathbb{N}$ have the same cardinality ($\mathbb{N} \times \mathbb{N}$ is countable).

Proof. View the ordered pairs $\{(i, j): i, j \in \mathbb{N}\}$ as points in the plane with positive integer coordinates. We list the ordered pairs in sequence by listing each successive diagonal in order. The pairs appear in order of increasing $i + j$, and the pairs with a fixed value of $i + j$ appear in increasing order by j. ∎

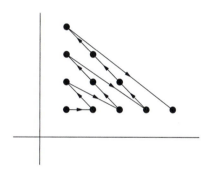

5.19. Example. *Another bijection from* $\mathbb{N} \times \mathbb{N}$ *to* $\mathbb{N}$. Define $f: \mathbb{N} \times \mathbb{N} \to \mathbb{N}$ by $f(m, n) = 2^{m-1}(2n - 1)$. By Proposition 4.23, every natural number is an odd number times a power of 2, so f is surjective. By the uniqueness result in the same proposition, f is also injective. ∎

Consider sets A and B and functions $f: A \to B$ and $g: B \to A$. If A and B are finite and f and g are injections, then f and g must also be bijections (Exercise 5.11). When A and B are not finite, the conclusion that f and g are bijections need not hold. For example, let $A = (0, 1)$ and $B = [0, 1]$, and define $f: A \to B$ and $g: B \to A$ by $f(x) = x$ and $g(x) = (3 - 2x)/4$. Then f and g are both injections, but neither is a bijection. Nevertheless, the existence of injections f and g always implies that A and B have the same cardinality.

5.20. Theorem. (Schroeder-Bernstein Theorem). If $f: A \to B$ and $g: B \to A$ are injections, then there exists a bijection $h: A \to B$, and hence A and B have the same cardinality.

Proof. We view A and B as disjoint sets, making two copies of common elements. For each element z of $A \cup B$, we define the *successor* of z to be $f(z)$ if $z \in A$, and $g(z)$ if $z \in B$. The *descendants* of z are the elements that can be reached by repeating the successor operation. We say that z is a *predecessor* of w if w is the successor of z. Because f and g are injective, every element of $A \cup B$ has at most one predecessor. The *ancestors* of z are the elements that can be reached by repeating the predecessor operation.

The *family* of z consists of z together with all its ancestors and descendants. We use the structure of families to define a one-to-one correspondence between A and B. The successor operation defines a function f' on $A \cup B$; below we show several possibilities for families using the functional digraph of f'.

First suppose that z is a descendant of z. Because every element has at most one predecessor, the family of z is finite (and forms a cycle in the functional digraph of f'). Also, the family alternates between A and B and thus has even size. For every $x \in A$ in such a family, we pair x with $f(x)$; because the family has even size, this is a one-to-one correspondence between its elements in A and its elements in B.

Otherwise, the family of z is infinite. Now the set of ancestors of z may be finite or infinite; if it is finite, then it has an *origin* that has no predecessor. If the family of z has an origin in B, then for every $x \in A$ in this family we pair x with its predecessor $g^{-1}(x)$; because the origin is in B this element exists. Otherwise, we pair x with its successor $f(x)$.

Because every element has at most one predecessor, the pairing we have defined is a one-to-one correspondence between the elements of A and the elements of B within each family. Since the families are pairwise disjoint, it is also a one-to-one correspondence between A and B. In more technical language, we have defined the function $h: A \to B$ by $h(x) = g^{-1}(x)$ when the family of x has an origin in B, and $h(x) = f(x)$ otherwise. The function h is the desired bijection. The statement about cardinality simply repeats Definition 5.16 of cardinality. ∎

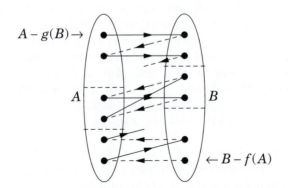

$A - g(B) \to$

A

B

$\leftarrow B - f(A)$

BINOMIAL COEFFICIENTS AND COUNTING

We next consider methods for counting arrangements of distinct objects and counting selections from finite sets. These problems introduce the factorial function and the binomial coefficients, quantities that appear throughout mathematics. We will use arrangements and selections as models when solving diverse problems.

Often we solve elementary counting problems by breaking them into subproblems. There are several ways to combine subproblems; the simplest leads to the *rule of sum*.

5.21. Definition. A *partition* of a set A is a collection of pairwise disjoint subsets of A whose union is A. The **rule of sum** states that if A is finite and $B_1, \ldots, B_m$ is a partition of A, then $|A| = \Sigma_{i=1}^{m} |B_i|$.

The rule of sum follows immediately from Corollary 5.14 by induction on m (Exercise 5.9).

The *rule of product* is a bit more subtle. Often we can describe a set by building its elements in stages. Although the choices available for the ith step may depend on the previous choices, the rule of product applies when the *number* of choices available for the ith step does not.

5.22. Example. A music practice room is available for only one hour during each weekday. In how many ways can three students sign up to use the room during the week? The first student picks one of the five days. The choices for the second student depend on the choice the first student makes, but in each case the number of choices remaining is four. Similarly, the third student always has three choices. There are therefore sixty possibilities. ■

5.23. Definition. A set T has a *k-step construction* if its elements can be formed using steps $S_1, \ldots, S_k$ such that the number of ways to perform step S_i does not depend on how steps $S_1, \ldots, S_{i-1}$ are performed. The **rule of product** states that if T has a k-step construction, then $|T|$ is the product of the number of ways to perform each step.

The rule of product follows from the rule of sum by induction on k (Exercise 5.17). Its most elementary interpretation is the observation that the size of the Cartesian product of finite sets A and B is $|A| \cdot |B|$. Our first application of the rule of product is to count the ways to arrange a set S of size n in a line. In Definition 3.48, we observed that these linear arrangements can be viewed as the bijections from $[n]$ to S or as permutations of the set S. We write $n!$, read as "n *factorial*", to mean $\Pi_{i=1}^{n} i = n(n-1) \cdots 1$.

5.24. Theorem. An n-element set has $n!$ permutations (arrangements).

Proof. We count the bijections from $[n]$ to S. We construct all possible bijections by choosing images one at a time. There are n ways to choose the image of 1. For each way we do this, there are $n-1$ ways to choose the image of 2. In general, after we have chosen the first i images, there are $n-i$ ways to choose the next image, no matter how we made the first i choices. By the rule of product, this yields the formula $\prod_{i=0}^{n-1}(n-i) = n!$ for the number of arrangements. ∎

By convention, we define $0! = 1$. In light of Theorem 5.24, we interpret this to mean that there is exactly one way to define a bijection from $\varnothing$ to $\varnothing$. This exemplifies a more general convention; the value of an empty sum is the additive identity, and the value of an empty product is the multiplicative identity. For example, we set $x^0 = 1$.

If k is smaller than $n = |B|$, then there are no bijections from $[k]$ to B, but we can count injections from $[k]$ to B. The injections are lists of k distinct elements of B in order. Again there are n ways to fill the first position, then $n-1$ ways to fill the next no matter how that was done, etc. This time we have only k factors, so $n(n-1)\cdots(n-k+1)$ is the answer. We can write this as $n!/(n-k)!$.

The injections are arrangements of k elements from B. We also consider selections of k elements from B, in which the order of the selected elements is unimportant.

5.25. Definition. A *selection* of k elements from $[n]$ is a k-element subset of $[n]$. The number of such selections is written $\binom{n}{k}$ and pronounced "n choose k".

If k is less than 0 or greater than n, then $\binom{n}{k} = 0$, because then there are no selections of k elements from $[n]$. When $0 \le k \le n$, the number has a simple formula.

5.26. Theorem. If n, k are integers with $0 \le k \le n$, then $\binom{n}{k} = \dfrac{n!}{k!(n-k)!}$.

Proof. We relate the problem of counting selections to the problem of counting arrangements using k of the elements. Instead of writing the elements one by one into the k positions, we can first choose the elements to be used and then permute them into the k positions. This counts the set of arrangements in two different ways, but it is the same set of arrangements, so we conclude that $n(n-1)\cdots(n-k+1) = \binom{n}{k}k!$. Dividing by $k!$ completes the proof. ∎

5.27. Solution. *Comparison of Poker Hands.* An ordinary deck has 52 distinct cards, so there are $\binom{52}{5} = 2{,}598{,}960$ distinct subsets with five cards. The cards come in 13 ranks of four cards each. They are also grouped into four "suits", with one card of each rank in each suit.

"Three-of-a-kind" means three cards of the same rank and one in each of two other ranks. "Two-pair" means two cards each in two ranks and the fifth card in some third rank. Three-of-a-kind can occur in $\binom{13}{1} \cdot \binom{4}{3} \binom{12}{2} \binom{4}{1} \binom{4}{1} = 54{,}912$ ways, since we pick the special rank, pick three cards from that rank, pick two other ranks, and pick one card from each of those. The rule of product applies, since the number of ways to make each choice does not depend on how the earlier choices were made. Two-pair can occur in $\binom{13}{2} \cdot \binom{4}{2} \cdot \binom{4}{2} \cdot \binom{44}{1} = 123{,}552$ ways; we pick the ranks for the pairs, pick the cards from those ranks, and pick the final card from the remaining ranks. The computation shows that three-of-a-kind is less than half as likely and hence is ranked higher.

A "flush" consists of five cards in one suit; a "straight" consists of one card each in five consecutive ranks, except that the "ace" can be considered either the lowest or the highest rank. A flush occurs in $4 \cdot \binom{13}{5} = 5148$ ways. A straight can begin at one of 10 possible ranks; thus it occurs in $10 \cdot 4^5 = 10240$ ways. The flush is rarer. ∎

The numbers $\binom{n}{k}$ are called the *binomial coefficients* due to their role in raising a sum to a power.

5.28. Theorem. (Binomial Theorem). $(x + y)^n = (x + y)(x + y) \cdots (x + y)$
$= \Sigma_{k=0}^{n} \binom{n}{k} x^k y^{n-k}$.

Proof. The proof interprets the process of multiplying out the factors. To form a term in the product, we must choose x or y from each factor; some factors contribute x, some y. The number of factors contributing x to the term is some integer k between 0 and n, and then the remaining $n - k$ factors contribute y. The number of terms of the form $x^k y^{n-k}$ is the number of ways to choose k of the factors to contribute x. Summing over k accounts for all the terms. ∎

5.29. Lemma. $\binom{n}{k} = \binom{n}{n-k}$.

Proof. Let S be the collection of k-element subsets of $[n]$. Counting them according to the elements selected yields $\binom{n}{k}$, by definition. Because the elements selected are determined by the elements omitted, we can also count the selections by counting the ways to omit $n - k$ elements, which equals $\binom{n}{n-k}$. ∎

We could also observe that the left side counts the k-element subsets of $[n]$, the right side counts the $n - k$-element subsets, and the operation of "complementation" establishes a bijection between these two collections. We could also observe that the formula for $\binom{n}{k}$ has the same value when k is replaced by $n - k$, so the bijection has an algebraic interpretation.

We include counting arguments as another weapon in our arsenal of proof techniques. A proof that interprets a formula as the size of a finite set is a *combinatorial proof*. The technique of *counting two ways* allows us to establish equality between two formulas by proving that both count the same set. We have already used this idea in Remark 4.8, Example 4.11, Solution 4.25, and Corollary 5.14. Counting two ways is closely related to proving equality of size by establishing a bijection; Lemma 5.29 can be proved in either way. Counting arguments may provide more information and deeper understanding than manipulation of formulas, but discovering them may require some cleverness.

5.30. Example. *Alternative models.* Several models are equivalent to the selection model; any of them can be used when making combinatorial arguments about binomial coefficients. In Proposition 3.33, we constructed a bijection (the "membership function") between the set of subsets of $[n]$ and the binary n-tuples; this takes subsets of size k into binary n-tuples with k ones. Whenever we discuss k-element subsets of $[n]$, we could alternatively discuss binary n-tuples with k 1's. ∎

We view n-tuples as arrangements with repetitions allowed. What happens when we consider selections with repetitions allowed? The next theorem permits us to solve Problem 5.3.

5.31. Theorem. With repetition allowed, there are $\binom{n+k-1}{k-1}$ ways to select n elements from k types of elements. This also equals the number of nonnegative integer solutions to $x_1 + \cdots + x_k = n$.

Proof. Each selection is determined by how many objects we select of each type. Since in this problem the order of selections does not matter, we can select the elements of type 1 first, then type 2, etc., keeping track of how many we selected of each type. Letting x_i be the number of objects of type i, the selections correspond to nonnegative integer solutions to $x_1 + \cdots + x_k = n$.

We can also model the selections as arrangements of n dots and $k - 1$ "slashes". If our selection has x_1 items of type 1, we can represent this by recording x_1 dots and marking the end with a "slash" before continuing to the next type. Doing this for each type of object forms an arrangement of dots and slashes. Below we illustrate this for $x_1 = 5$ from the first type, $x_2 = 2$ from the second, $x_3 = 0$ from the third, and $x_4 = 3$ from the last. Since the total number of objects selected is n, we have n dots and $k - 1$ slashes.

$$\bullet \ \bullet \ \bullet \ \bullet \ \bullet \ / \ \bullet \ \bullet \ / / \ \bullet \ \bullet \ \bullet$$

Conversely, given an arrangement of n dots and $k-1$ slashes, we can invert the process to obtain the number of each type chosen in the corresponding selection; the number of dots in the ith group is the number of objects of type i. This establishes a bijection between selections with repetition and arrangements of n dots and $k-1$ slashes. These arrangements are determined by choosing the locations for the slashes in a list of length $n+k-1$, so there are $\binom{n+k-1}{k-1}$ of them. ∎

This formula can also be written as $\binom{n+k-1}{n}$, so care must be taken to distinguish between the number of types and the number of elements being selected. It may be safer to remember the proof than to remember the formula, especially since the names "n" and "k" may be changed or exchanged in applications of the model.

5.32. Solution. *Non-negative Integer Solutions.* New York City has about 7 million residents. We are told that each resident has 100 coins in a jar, and that there are five types of coins. Two jars are "equivalent" if they have the same number of pennies, same number of nickels, and similarly for dimes, quarters, and half-dollars. When x_i denotes the number of coins of type i, the number of pairwise inequivalent jars of coins is the number of solutions to $x_1 + x_2 + x_3 + x_4 + x_5 = 100$ using nonnegative integers. By Theorem 5.31, this equals $\binom{104}{4} = 4{,}598{,}126$. Hence two people must have equivalent jars. ∎

For ease in stating the next corollary, we define a *monic monomial* with *degree d* in the m variables $x_1, \ldots, x_m$ to be a monomial of the form $\prod_{i=1}^{m} x_i^{d_i}$ such that $\Sigma d_i = d$.

5.33. Corollary. The number of monic monomials with degree d in the variables $x_1, \ldots, x_m$ is $\binom{d+m-1}{m-1}$.

Proof. The monomials correspond to the nonnegative integer solutions of $\Sigma_{i=1}^{m} d_i = d$. ∎

5.34. Example. *Monomials in a multinomial expansion.* Every monomial in the expansion of $(w + x + y + z)^3$ has degree 3. Ignoring the coefficients, we list the monomials below. By Corollary 5.33, there are $\binom{3+4-1}{4-1} = 20$ such monomials. In Chapter 9 we compute the formula for the coefficients, which are called the "multinomial coefficients". ∎

w^3	w^2x	w^2y	w^2z	wxy
x^3	x^2w	x^2y	x^2z	wxz
y^3	y^2w	y^2x	y^2z	wyz
z^3	z^2w	z^2x	z^2y	xyz

EXERCISES

In these problems, n denotes a positive integer. The phrase "count the" means "determine the number of" and requires a proof of the answer.

5.1. (–) Suppose that $(120102)_3$ and $(110222)_3$ are 3-ary expansions of two natural numbers. Compute the 3-ary expansion of their sum, using 3-ary arithmetic. Check the answer by converting each summand to base 10, adding, and converting back to base 3.

5.2. Suppose that $a_m, \ldots, a_0$ is the q-ary expansion of n. Describe a procedure for finding the q-ary expansion of $n + 1$, and prove that it works.

5.3. (!) Consider a balance scale plus k objects of known weights $1, 3, \ldots, 3^{k-1}$ (the first k powers of 3). Prove by induction on k that every unknown weight in the set $\{1, \ldots, (3^k - 1)/2\}$ can be balanced.

5.4. Consider a balance scale and positive integer weights $w_1 \le \cdots \le w_k$. Let $S_0 = 0$, and let $S_j = \Sigma_{i=1}^j w_i$ for $1 \le j \le k$. Prove that it is possible to balance every integer weight from 1 to S_k if and only if $w_j \le 2S_{j-1} + 1$ for $1 \le j \le k$.

5.5. (–) Many games involve rolling two dice, each having six sides numbered 1 through 6. Explain why x and $14 - x$ are equally likely to be the sum of the numbers facing up on the two dice.

5.6. The following problem appeared on a statewide exam for grade 10 in California. "A game involves two cubes with sides numbered 1 through 6. After throwing the two cubes, the smaller number is subtracted from the larger number to find the difference. If a player throws the cubes many times, what difference will probably occur most often? Provide a diagram and written explanation that you could use to explain to a friend."

5.7. (–) For which finite sets A does there exist a bijection from A to A that is different from the identity function on A?

5.8. (–) Prove that the function h in the proof of Corollary 5.14 is a bijection.

5.9. (–) Use Corollary 5.14 to prove that the size of the union of k pairwise disjoint finite sets is the sum of their sizes.

5.10. Suppose $f: A \to A$ and A is finite. Prove that f is injective if and only if f is surjective. Prove that this equivalence does not hold if A is infinite.

5.11. (!) Suppose A and B are finite, and $f: A \to B$.
  a) If f is injective, what is implied about the sizes of A and B?
  b) If f is surjective, what is implied about the sizes of A and B?
  c) Prove that if A and B are finite and $f: A \to B$ and $g: B \to A$ are injections, then $|A| = |B|$ and f and g are bijections.

5.12. Prove that the sets consisting of the natural numbers, the even natural numbers, and the odd natural numbers all have the same cardinality (they are countable).

5.13. The proof of countability of $\mathbb{N} \times \mathbb{N}$ in Theorem 5.18 specifies a sequence containing every ordered pair (i, j). Determine the index of the ordered pair (i, j) in this sequence, as a function of i and j. (This defines the bijection $f: \mathbb{N} \times \mathbb{N} \to \mathbb{N}$ explicitly.)

5.14. (!) Suppose $\{A_i: i \in \mathbb{N}\}$ is a sequence of sets, each of which is countable. Prove that the union $\cup_{i=1}^{\infty} A_i$ is a countable set.

5.15. (!) Suppose that $A = (0, 1)$ and $B = \{y \in \mathbb{R}: 0 \le y < 1\}$. Define $f: A \to B$ and $g: B \to A$ by $f(x) = x$ and $g(y) = (y+1)/2$. Obtain an explicit formula for the function h constructed in the proof of the Schroeder-Bernstein Theorem.

5.16. Construct an explicit bijection from the open interval $(0, 1)$ to the closed interval $[0, 1]$.

5.17. (!) Use the rule of sum to prove the rule of product. (Hint: use induction on the number k of steps used to form elements of the set T being counted, after expressing the elements of T as k-tuples.)

5.18. (–) Count the bijections from A to B, given that $|A| = |B| = n$.

5.19. Suppose that n, m, k are positive integers satisfying $n! + m! = k!$. Prove that $n = m = 1$ and $k = 2$.

5.20. Determine (with proof) the positive integers n such that $n! > 2^n$.

5.21. For $n \in \mathbb{N}$, find and prove a formula for $\Sigma_{k=1}^{n} k \cdot k!$.

5.22. (!) Prove that $(n^5 - 5n^3 + 4n)/120$ is an integer for all $n \in \mathbb{N}$.

5.23. Count the rectangles of all sizes formed using segments in a grid with m horizontal lines and n vertical lines. In the picture below, $m = 4$ and $n = 5$.

5.24. Let P be an n-sided polygon in the plane such that every segment joining pairs of vertices of P lies inside P; such segments are "diagonals" of P. Count the pairs of diagonals of P that cross.

5.25. In terms of binomial coefficients, count the (5-card) poker hands that have
    a) One pair (two cards of equal rank and no others of equal rank).
    b) Full house (two cards of equal rank and three cards of another rank).
    c) Straight flush (five cards in sequence from the same suit).

5.26. Suppose that x is an element of a set A of size $2n$. Count the n-element subsets of A that contain x and the n-element subsets of A that omit x. Use this to prove that $\binom{2n}{n} = 2\binom{2n-1}{n-1}$.

5.27. (!) Count the ways to group $2n$ distinct people into pairs. (The answer is 1 when $n = 1$ and is 3 when $n = 2$.)

5.28. Give a combinatorial proof that $\sum_{k=1}^{n} 2^{k-1} = 2^n - 1$.

5.29. Give a combinatorial proof that $\sum_{i=1}^{n-1} i = \binom{n}{2}$.

5.30. (+) Give a combinatorial proof that $\sum_{i=1}^{n} i(n-i) = \binom{n}{3}$.

5.31. (!) Suppose that $f_m \colon \mathbb{N} \to \mathbb{N}$ is defined by $f_m(n) = \sum_{k=0}^{m} \binom{n}{k}$. Prove that $f_m(n) = 2^n$ when $n \le m$. Find an n such that $f_m(n) \ne 2^n$. (Hint: count subsets.)

5.32. (!) Let A be the set of subsets of $[n]$ that have even size, and let B be the set of subsets of $[n]$ that have odd size.

a) Establish a bijection from A to B, thereby proving combinatorially $|A| = |B|$. (Such a bijection is suggested below for $n = 3$.)

b) Provide another proof of this equality by using the binomial theorem to prove that $|A| - |B| = 0$. Does this remain true if $n = 0$?

A	$\varnothing$	$\{2,3\}$	$\{1,3\}$	$\{1,2\}$
B	$\{3\}$	$\{2\}$	$\{1\}$	$\{1,2,3\}$

5.33. (+) Let A_n be the set of permutations of $[n]$. Let B_n be the set of n-tuples $(b_1, \ldots, b_n)$ such that $1 \le b_i \le i$ for each $i \in [n]$. Construct a bijection from A_n to B_n. (Hint: use induction on n, employing a bijection from A_{n-1} to B_{n-1} to construct a bijection from A_n to B_n. Below we illustrate this process for $n = 3$.)

A_3	321	231	213	312	132	123
B_3	111	112	113	121	122	123

5.34. A *partition of n* (where n is a positive integer) is a nonincreasing list of positive integers that sum to n (compare with the Penny Problem - Solution 1.2 and Example 3.47). For example, the partitions of 4 are 4, 31, 22, 211, 1111. The elements of the list are the "parts" of the partition.

a) List the partitions of 6.

b) Prove that the number of partitions of n with k parts equals the number of partitions of n with largest part k. (Hint: view the parts as piles of pennies or columns of dots.)

5.35. (+) By establishing a bijection, prove that the number of partitions of n into distinct parts equals the number of partitions of n into odd parts. For example, the partitions of 4 into distinct parts are 4 and 31, and the partitions of 4 into odd parts are 31 and 1111.

Chapter 6

Divisibility

We next study divisibility. We know already that each natural number has exactly one expression as an odd number times a power of two, but the odd number may have further factors.

6.1. Definition. If $a, b \in \mathbb{Z}$ with $a \neq 0$, and there is an integer m such that $b = ma$, then we say "a *divides* b", "b is *divisible by* a", "a is a *divisor* of b", and "a is a *factor* of b". We write "a divides b" as $a|b$. A natural number other than 1 is *prime* if its only positive factors are itself and 1; otherwise it is *composite*.

By convention, 1 is not prime; the first few primes are 2, 3, 5, 7, 11. We will prove that every natural number has a unique factorization into primes. We will also use divisibility and prime numbers to find the integer solutions to linear equations and to solve the following problems.

6.2. Problem. How can we find the greatest common divisor of two large numbers without factoring them? ∎

6.3. Problem. *The Dart Board Problem.* Suppose a dart board has regions associated with two values a and b, where a and b are natural numbers having no common divisor other than 1. What is the largest integer that cannot be achieved by summing the values of thrown darts? We must obtain an integer k such that $ma + nb = k$ has no solutions using nonnegative integers m, n, but $ma + nb = j$ does have such a solution whenever j is an integer larger than k. ∎

THE EUCLIDEAN ALGORITHM

In order to determine whether b divides a, we can start with a and subtract b repeatedly until what remains is less than b. Suppose this happens after k subtractions and what remains is r. Then $a = kb + r$, and b divides a if and only if $r = 0$. This process of obtaining k and r from the pair (a, b) is called the **division algorithm** (an *algorithm* is a procedure for performing a computation or construction). If we require $0 \le r \le b - 1$, then k and r are uniquely determined (Exercise 6.7). We say that *a has remainder r* upon division by b.

6.4. Definition. Given natural numbers a, b, the *greatest common divisor* $\gcd(a, b)$ is the largest natural number that divides both a and b. If $\gcd(a, b) = 1$, then a and b are *relatively prime* (they have no common prime divisor). By convention, $\gcd(n, 0) = n$.

We adopt the convention $\gcd(n, 0) = n$ because every natural number divides 0. There is a simple and efficient algorithm to compute $\gcd(a, b)$, thereby solving Problem 6.2. Exercise 12.31 considers the efficiency of the algorithm.

6.5. Definition. The input to the **Euclidean algorithm** is a pair (a, b) of integers with $a \ge b \ge 0$ and $a \ne 0$. A *step* of the algorithm takes a pair (a, b) and replaces it with the pair (b, c), where c is the remainder of a upon division by b (computed by the division algorithm). The Euclidean algorithm repeats this step until it produces a pair $(d, 0)$; the output is then the nonzero number d.

6.6. Example. *Application of the Euclidean Algorithm.* When we start with the numbers 68 and 40, the successive pairs are $(68, 40)$, $(40, 28)$, $(28, 12)$, $(12, 4)$, $(4, 0)$, and the output is 4. We say that the Euclidean algorithm *takes four steps* when applied to the pair $(68, 40)$. Note that $4 = \gcd(68, 40)$ and that $4 = 3 \cdot 68 - 5 \cdot 40$; this illustrates general properties of the output proved in the next theorem. ∎

6.7. Theorem. When the Euclidean algorithm is applied to the input (a, b), the output is $\gcd(a, b)$. Furthermore, $\gcd(a, b)$ can be expressed as $ma + nb$ for some $m, n \in \mathbb{Z}$.

Proof. The proof is by strong induction on b, the smaller entry of the input pair. For the basis step, we have $b = 0$. In this case, the output is a. This equals $\gcd(a, 0)$, and $a = 1 \cdot a + 0 \cdot 0$ expresses the greatest common divisor as an integer combination of a and b.

For the induction step, we have $a \geq b \geq 1$, and we assume that the Euclidean algorithm computes the gcd whenever the smaller input is less than b. The result of the first step is a pair (b, c) with $b > c \geq 0$, satisfying $a = kb + c$ for some $k \in \mathbb{N}$. The output of the algorithm when the input is (a, b) is the same as the output d when the input is (b, c). By the induction hypothesis, $d = \gcd(b, c)$. The induction hypothesis also supplies $m', n' \in \mathbb{N}$ such that $d = m'b + n'c$. It suffices to show that $\gcd(b, c) = \gcd(a, b)$ and to rewrite $m'b + n'c$ in terms of $\{a, b\}$. The latter is easy; since $d = m'b + n'c = m'b + n'(a - kb) = n'a + (m' - n'k)b$, we can set $m = n'$ and $n = (m' - n'k)$.

To prove that $\gcd(a, b) = \gcd(b, c)$, it suffices to show that a number is a common divisor of a and b if and only if it is a common divisor of b and c. This follows from the distributive law; if a number divides two of the three terms in the expression $a = kb + c$, then it must also divide the third. Hence the common divisors of $\{a, b\}$ are the same as the common divisors of $\{b, c\}$. ∎

The main part of the proof is showing that the gcd is unchanged by each step of the Euclidean algorithm. The process repeats until it produces a pair $(d, 0)$, for which d is obviously the gcd.

Since the division algorithm is performed by repeated subtraction, we could also find the greatest common divisor by repeatedly subtracting the smaller number from the larger and discarding the larger. Each such step replaces a pair (a, b) with $(a - b, b)$ or $(b, a - b)$. Using the division algorithm combines several subtractions and produces the greatest common divisor in larger steps. Exercise 6.8 considers a similar algorithm.

6.8. Example. *Finding the integer combination.* The inductive proof of the Euclidean algorithm provides a way to find integers m, n such that $ma + nb = \gcd(a, b)$. For example, consider the pair $(30, 13)$. We replace 30 by 4 using $30 - 2 \cdot 13 = 4$, then replace 13 by 1 using $13 - 3 \cdot 4 = 1$, and $\gcd(30, 13) = 1$. As we proceed, each new number is generated as an integer combination of the preceding ones, so each number can be expressed as an integer combination of the original pair, including the final number, which is the gcd. In this example, we have $1 = 13 - 3 \cdot 4 = 13 - 3(30 - 2 \cdot 13) = -3 \cdot 30 + 7 \cdot 13$. ∎

6.9. Corollary. Two integers a and b are relatively prime if and only if there exist integers m, n such that $1 = ma + nb$.

Proof. "Relatively prime" means $\gcd(a, b) = 1$, and by Theorem 6.7 the desired m, n exist. If $1 = ma + nb$, then a, b can have no common factor other than 1, because by the distributive law it would also divide 1. ∎

What numbers other than $\gcd(a, b)$ are integer combinations of a and b? When $d = ma + nb$, we can obtain kd by using km and kn as coefficients. The multiples of $\gcd(a, b)$ are the *only* numbers we can obtain. Let $d\mathbb{Z} = \{md\colon m \in \mathbb{Z}\}$ denote the set of integer multiples of d.

6.10. Theorem. If $S(a, b) = \{ma + nb\colon m, n \in \mathbb{Z}\}$ is the set of integer combinations of a and b, then $S(a, b) = d\mathbb{Z}$, where $d = \gcd(a, b)$.

Proof. We first prove that $d\mathbb{Z} \subseteq S(a, b)$. Because $d = \gcd(a, b)$, Theorem 6.7 guarantees integers m', n' such that $d = m'a + n'b$. If $k \in d\mathbb{Z}$, then k is expressible as rd, and the distributive and associative properties yield $rd = r(m'a + n'b) = (rm')a + (r'n)b$. Since $m = rm'$ and $n = rn'$, we conclude that $k \in S(a, b)$.

Conversely, we prove that $S(a, b) \subseteq d\mathbb{Z}$. Because $d = \gcd(a, b)$, there are integers s, t such that $a = sd$ and $b = td$. If $k \in S(a, b)$, then k is expressible as $ma + nb$, and we have $k = m(sd) + n(td) = (ms + nt)d$. Hence $k \subseteq d\mathbb{Z}$. ∎

We can use Theorem 6.10 to make statements about equations in integers. An equation where we seek integer solutions is called a *diophantine equation*, in honor of Diophantus (3rd century AD).

6.11. Example. *Impossibility of solutions.* The equation $6x + 15y = 79$ has no solution in integers. Such a solution would express 79 as an integer combination of 6 and 15. By Theorem 6.10, all such numbers are multiples of $\gcd(6, 15) = 3$, but 79 is not a multiple of 3. ∎

After finding one solution to the diophantine equation $ax + by = c$, we can easily find all solutions. We illustrate this by an example.

6.12. Example. *Description of all solutions.* What are the integer solutions of $6x + 15y = 99$? Let S denote this set. Since 99 is a multiple of $3 = \gcd(6, 15)$, Theorem 6.7 guarantees a solution. To find solutions, we first divide the equation by this gcd to obtain the *reduced* equation $2x + 5y = 33$; doing so does not change the set of solutions. Setting $x = -2$ and $y = 1$ produces 1 as an integer combination of the coefficients 2 and 5: $2(-2) + 5(1) = 1$. Multiplying this by 33 produces the solution $(x, y) = 33 \cdot (-2, 1) = (-66, 33)$ in S. (Had we not seen a solution to $2x + 5y = 1$, the Euclidean algorithm would have generated one.)

We can generate other solutions by increasing x and decreasing y, or vice versa. The amount by which we increase $2x$ must equal the amount by which we decrease $5y$. Hence it must be a multiple of both 2 and 5. Since 2 and 5 are relatively prime, we find all other solutions by changing x by a multiple of 5 while we change y in the other direction by that multiple of 2. Thus $S = \{(-66 + 5k, 33 - 2k)\colon k \in \mathbb{Z}\}$. ∎

PRIME FACTORIZATION

The Euclidean Algorithm implies further properties of divisibility.

6.13. Proposition. If a and b are relatively prime and a divides qb, then a divides q.

Proof. Since a, b are relatively prime, Corollary 6.9 allows us to write $1 = ma + nb$ for some integers m, n. Now $q = maq + nbq$. Since a divides each term on the right, a must also divide q. ∎

6.14. Corollary. If a prime p divides a product of k integers, then p divides at least one of the factors.

Proof. By induction on k; the statement is trivial when $k = 1$. For $k \geq 2$, let $n = \Pi_{i=1}^{k} b_i$, with $n' = \Pi_{i=1}^{k-1} b_i$. Then p divides $n = n'b_k$. If p divides b_k, the claim holds. Otherwise, $\gcd(p, b_k) = 1$, since p has no other divisors. Since p and b_k are relatively prime, Proposition 6.13 implies that p divides n'. By applying the induction hypothesis to n', we conclude that p divides one of $\{b_1, \ldots, b_{k-1}\}$. ∎

A *prime factorization* of n expresses n as a product of primes; the number of times each prime occurs in the product is its *multiplicity*. A prime that does not divide n has multiplicity 0 in every factorization. The next theorem guarantees both the existence of a prime factorization and its uniqueness (except for possibly reordering the factors).

6.15. Theorem. (Fundamental Theorem of Arithmetic). Every positive integer n has a prime factorization. Furthermore, the multiplicity of each prime is the same in every prime factorization of n.

Proof. We use strong induction on n. For $n = 1$, there are no prime factors. By convention, the product of the integers in an empty set is the multiplicative identity 1, so the basis step holds. For the induction step, suppose $n > 1$. Let S be the set of integers larger than 1 that divide n; this is non-empty, since $n \in S$. By the Well-Ordering Property, S has a smallest element p. Furthermore, p is prime, else p has a smaller prime factor that also divides n.

By Corollary 6.14, p appears in every list of primes (repetition allowed) whose product is n. Therefore, every prime factorization of n consists of p and a prime factorization of n/p. By the induction hypothesis, n/p has a unique prime factorization. Hence there is exactly one prime factorization of n, obtained by adding one to the multiplicity of p in the unique prime factorization of n/p. ∎

6.16. Corollary. If a and b are relatively prime, and a, b both divide n, then $ab|n$.

Proof. Exercise 6.10. ∎

THE DART BOARD PROBLEM

The Fundamental Theorem of Arithmetic has diverse applications to number theory, geometry, and cryptography; we do not consider these here. We can, however, apply Proposition 6.13 to solve the Dart Board Problem. The result is also known as Sylvester's Theorem, for James Joseph Sylvester (1814-1897).

Suppose that a and b are relatively prime positive integers and that k is a positive integer. First we present a geometric argument that suggests that $k = ma + nb$ must have a nonnegative integer solution (m, n) when k is large. Since a, b are relatively prime, the equation $k = ma + nb$ has integer solutions. We move from one to the next by adding b to m and subtracting a from n. Viewed as points in the plane, the solution pairs (m, n) lie along a line. There is a nonnegative integer solution for k if and only if the line for k contains an integer point in the first quadrant, which by definition is the set of points with both coordinates nonnegative.

The lines resulting from various choices of k are parallel. Below we sketch solution lines for $(a, b) = (3, 5)$ and $k = 1, 2, 4, 7$. These k are the positive integers not achievable as nonnegative integer combinations of a and b. The heavy dots indicate the integer points closest to the first quadrant on these lines. As k increases, the line cuts through more of the first quadrant; because the integer points have the same spacing on every line, making k large guarantees a solution. In terms of a and b, we can compute how large k must be made to guarantee the existence of a nonnegative integer solution.

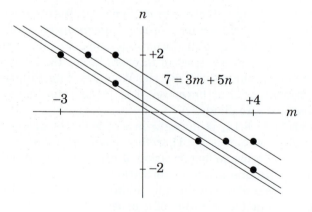

6.17. Solution. *The Dart Board Problem.* For relatively prime positive integers a and b, we prove that the largest integer that is not a nonnegative integer combination of a and b is $ab - a - b$. Define k to be *achievable* if $k = ma + nb$ has a nonnegative integer solution. We prove that $ab - a - b$ is not achievable and that every number larger than $ab - a - b$ is achievable.

First we prove that $ab - a - b$ is not achievable, by the method of contradiction. If $ab - a - b$ is achievable, then $ab - a - b = ma + nb$ for some nonnegative integers m, n. Thus $ab = (m + 1)a + (n + 1)b$. Since a and b are relatively prime, this implies that a divides $(n + 1)$ and that b divides $(m + 1)$. Since $m, n \geq 0$, this in turn implies that $n + 1 \geq a$ and $m + 1 \geq b$. These inequalities yield the contradiction

$$ab = (m + 1)a + (n + 1)b \geq 2ab.$$

Next, we prove "$k > ab - a - b \Rightarrow k$ is achievable" by proving the contrapositive: "k not achievable $\Rightarrow k \leq ab - a - b$". Suppose k is not achievable. Because $\gcd(a, b) = 1$, we can find integers r, s such that $1 = ra + sb$. Multiplying the equation by k yields $k = (kr)a + (ks)b$; this is an integer solution to $k = ma + nb$, but one coefficient is negative. Adding b to m and subtracting a from n produces another integer solution to the equation $k = ma + nb$. Since k is not achievable, there are no integer solutions in the first quadrant. Hence there are consecutive integer solutions with (m', n') in the second quadrant and $(m' + b, n' - a)$ in the fourth quadrant. Since these are integer solutions in these quadrants, they must satisfy $m' \leq -1$ and $n' - a \leq -1$. Now we can compute

$$k = m'a + n'b \leq (-1)a + (a - 1)b = ab - a - b. \quad \blacksquare$$

The Dart Board Problem also has an inductive solution. Exercise 4.26 asked for the set of achievable numbers when $a = 3$ and $b = 10$ (using only induction). Checking successive numbers reveals that 17 is not achievable but that 18,19,20 are. Once we have a consecutive achievable numbers, all larger numbers are achievable, since each larger number differs from one of these by a positive multiple of a.

This suggests proving inductively that numbers larger than $ab - a - b$ are achievable. We need only prove that the first a of these numbers are achievable. This illustrates that a conditional and its contrapositive may both have interesting (and quite different) direct proofs. The proof uses the following theorem, which we prove in a general form for application in the next chapter.

6.18. Theorem. If a, b are relatively prime and x is an integer, then x, $x + b$, $x + 2b, \ldots, x + (a - 1)b$ have different remainders when divided by a.

Proof. Suppose that $x + ib$ and $x + jb$ have the same remainder, which means that $x + ib = ka + r$ and $x + jb = la + r$ for some integers k, l, r with $0 \le r \le a - 1$. Subtracting the equations yields $(i - j)b = (k - l)a$. Since a divides $(k - l)a$, a must also divide $(i - j)b$. Since a and b are relatively prime, we conclude that a must divide $(i - j)$. Since i and j are nonnegative integers less than a, this implies that $i = j$. ∎

6.19. Solution. *The Dart Board Problem, alternative proof.* Assume that $a < b$, by symmetry. We proved that $ab - a - b$ is not achievable. We prove by induction that every larger integer is achievable. For the basis step, we will prove that the first a integers after $ab - a - b$. Given a consecutive achievable integers, the next a integers are also achievable, by increasing m (the coefficient of a) by one, and this proves the induction step. Let $S = \{z, \ldots, ab - b\}$ be the set consisting of the first a integers after $ab - a - b$.

  To prove the basis step, we first observe that all numbers in the set $T = \{0, b, 2b, \ldots, (a - 1)b\}$ are achievable, by using $m = 0$ and the first a nonnegative values of n. By setting $x = 0$ in Theorem 6.18, we know that the numbers in T have different remainders when divided by a. That also holds for S, since the numbers in S are consecutive. Each remainder appears exactly once in each set.

  Because $(a - 2)b = ab - b - b < ab - a - b$, all numbers in T other than $(a - 1)b$ are smaller than $ab - a - b$. By definition, all numbers in S are bigger than $ab - a - b$. We achieve each $y \in S$ by adding a nonnegative multiple of a to the number $x \in T$ that has the same remainder as y. Since they have the same remainder, they differ by a multiple of a. Also, the numbers in T are all smaller than the numbers in S except in the case $x = ab - b = (a - 1)b = y$, which is already achieved. ∎

$$
\begin{array}{ccccc}
 & & & z \quad y \quad ab-b & \leftarrow S \\
\hline
0 & b & 2b = x & (a-2)b \qquad (a-1)b & \leftarrow T
\end{array}
$$

EXERCISES

  American coins have values 1,5,10,25, and 50 cents, called pennies, nickels, dimes, quarters, and half-dollars, respectively.

6.1. (−) A person has the same (nonzero) number of each type of American coin. The total amount she has is a whole number of dollars. Determine the smallest such nonzero amount. Answer the same question assuming she has no pennies. Answer the same question assuming she has no pennies and no nickels.

6.2. (−) Suppose a parking meter contains the same number of dimes and quarters, in total a nonzero whole number of dollars. Determine the minimum number of coins.

6.3. (−) Suppose a parking meter can hold k quarters, $2k$ nickels, and $4k$ dimes. Find all k such that the amount of money when then meter is full is a whole number of dollars.

6.4. (−) Suppose a parking meter accepts only dimes and quarters and has twice as many dimes as quarters. If the total amount of money is a nonzero whole number of dollars, what is the smallest possible number of quarters?

6.5. (−) For each pair below, use the Euclidean algorithm to compute the greatest common divisor, and express the greatest common divisor as an integer combination of the two numbers.

         a) 464 and 637.   b) 610 and 987.

6.6. Suppose the function $f: \mathbb{N} \times \mathbb{N} \to \mathbb{N}$ is defined by $f(x, y) = 3^{x-1}(3y - 1)$. Is f surjective? Give a proof or a disproof.

6.7. (−) Suppose $b, a \in \mathbb{N}$. Prove that there is exactly one pair k, r of nonnegative integers such that $0 \le r \le a - 1$ and $b = ka + r$.

6.8. Suppose a and b are positive integers that are not both even. Prove that the following algorithm computes $\gcd(a, b)$. 1) If one of $\{a, b\}$ is even, divide it by 2 and iterate with the new pair. 2) If $\{a, b\}$ are both odd, replace the larger number by their difference. 3) When one number becomes 0, the nonzero number remaining is the gcd of the original pair. (Comment: this algorithm runs faster than the Euclidean algorithm.)

6.9. Theorem 5.7 develops one procedure for computing the q-ary expansion of a natural number n. Prove that the following inductive procedure also works, and use it to compute the expansion of 729 in base 5.

    1) If $1 \le n \le q - 1$, then the q-ary expansion of n is $a_0 = n$.
    2) If $n \ge q$, then let $n = kq + r$, where r is an integer in $\{0, \ldots, q - 1\}$, and let $b_m, \ldots, b_0$ be the q-ary expansion of k. The q-ary expansion of n is $a_{m+1}, \ldots, a_0$, where $a_0 = r$ and $a_i = b_{i-1}$ for $i > 0$.

6.10. Suppose that $\gcd(a, b) = 1$ and that $a|n$ and $b|n$. Prove that $ab|n$.

6.11. What is the smallest number of American coins (values may repeat) that enable a shopkeeper to make change for each amount from 1 cent through 99 cents? Is there a unique way of doing this with the minimum number of coins? Suppose coins can be made in any desired denomination; what then is the minimum number of coins needed (not the minimum number of values)?

6.12. If p is an integer greater than 3, prove that not all the numbers $p, p + 2, p + 4$ can be prime.

6.13. Prove that 3 divides $4^n - 1$, for every positive integer n. Prove that 6 divides $n^3 + 5n$, for every positive integer n.

6.14. Suppose $\langle a \rangle$ is a sequence such that $a_1 = 1$, $a_2 = 1$, and $a_{n+1} = a_n + 2a_{n-1}$ for $n \ge 2$. Prove that a_n is divisible by 3 if and only if n is divisible by 3.

6.15. If $n \in \mathbb{N}$, prove that $(n - 1)^3 + n^3 + (n + 1)^3$ is divisible by 9.

6.16. (!) Prove that $(2n)!/(2^n n!)$ is an odd number.

6.17. (!) Prove using contradiction that the set of prime numbers is not finite.

6.18. (!) Suppose n is a positive integer. Construct a set of n consecutive positive integers that are not prime. (Hint: determine a positive integer x such that x is divisible by 2, $x + 1$ is divisible by 3, $x + 2$ is divisible by 4, etc.)

6.19. (!) Suppose a, b, c are integers such that $a^2 + b^2 = c^2$.
  a) Is it always true that at least one of $\{a, b\}$ is even?
  b) If c is divisible by 3, prove that a and b are both divisible by 3.

6.20. Use the identity $\Sigma_{i=1}^{n}(2i - 1) = n^2$ to prove that there are infinitely many solutions to $a^2 + b^2 = c^2$ such that $c = b + 1$.

6.21. Suppose that abc is a 3-digit natural number (in base 10). Prove that the 6-digit number $abcabc$ has at least three distinct prime factors.

6.22. (+) *The Coconuts Problem.* Five suspicious sailors spend the day gathering a pile of coconuts. Exhausted, they postpone dividing the pile until the next morning. Suspicious, each decides to take his share during the night. The first sailor divides the pile into five equal portions plus one extra coconut, which he gives to a monkey. He takes one pile and leaves the rest in a single pile. The second sailor later does the same; again the monkey receives one leftover coconut. The third, fourth and fifth sailors also do this; each time a remainder of one goes to the monkey. In the morning they split the remaining coconuts into 5 equal piles, and each sailor gets his "share". (Each sailor knows some were taken, but none complains, since each is guilty!) What is the smallest possible number of coconuts in the original pile? (This problem appeared in the *Saturday Evening Post* on Oct. 9, 1926.)

6.23. (+) *The Postage Stamp Problem* (special case). Suppose the Post Office decides to issue stamps of two different values. Postage is one cent per ounce, and each envelope has space for s stamps. Being able to correctly post a one-ounce envelope requires that one of the stamp values be 1. The problem is to choose the other value m to maximize the integer n such that all weights in $[n]$ can be correctly posted.
  a) Prove that m should be at most $s + 1$.
  b) Prove that for each m satisfying $2 \le m \le s + 1$, the smallest integer weight that cannot be correctly posted has remainder $m - 2$ when divided by m.
  c) Use part (b) to prove that the best choice of m is $\lceil s/2 \rceil + 1$. (Comment: the more general problem in which d different values are allowed is unsolved.)

6.24. (+) Consider cards labeled $1, \ldots, 2n$. The cards are shuffled and dealt to two players A and B so that each gets n of the cards. Let x be the sum of the labels on the cards that have been played; initially, $x = 0$. Starting with A, play alternates between the two players. At each play, a player adds one of his or her remaining cards to x. The first player who makes x divisible by $2n + 1$ wins. Prove that for every deal, player B has a strategy to win. (Hint: prove that B can always make it impossible for A to win on the next move.)

6.25. (!) Suppose that x, y, k are nonnegative integers and that k is not a power of 2. Prove that $x^k + y^k$ is not prime. Conclude that if $2^n + 1$ is prime and n is not a power of 2, then n is prime.

6.26. Prove that if $2^n - 1$ is prime, then n is prime. (Hint: prove the contrapositive; if n is not prime, then $2^n - 1$ is not prime. Primes of the form $2^n - 1$ are called *Mersenne primes*; 35 such primes are known.)

6.27. (!) A natural number is *perfect* if it is the sum of the smaller natural numbers that divide it; 6,28 are the first two perfect numbers. Prove that if $2^n - 1$ is prime, then $2^{n-1}(2^n - 1)$ is perfect. (Hint: list the divisors and add them up. Euclid conjectured that all perfect numbers have this form, but no proof is known, although it is known these are the only even perfect numbers.)

6.28. *Pólya's proof for infinitude of primes.* Let $a_n = 2^{2^n} + 1$. Prove by induction that a_n divides $a_m - 2$ if $n < m$. Conclude that a_n and a_m have no common factors if $n \neq m$. Use this to prove there are infinitely many primes. (This method also proves that there are at least $\log_2 \log_2 N$ primes less than N.)

6.29. The royal treasury has 500 7-ounce weights, 500 13-ounce weights, and a balance scale. An envoy arrives with a bar of gold, claiming it weighs 500 ounces. Can the treasury determine whether the envoy is lying? If so, how? What if the weights are six-ounce and nine-ounce weights?

6.30. The *least common multiple (lcm)* of natural numbers a and b is the least natural number divisible by both. Prove that $\text{lcm}(a, b) \cdot \gcd(a, b) = a \cdot b$.

6.31. (–) For each diophantine equation below, find all solutions, if any exist.
a) $17x + 13y = 200$. b) $21x + 15y = 93$. c) $60x + 42y = 104$. d) $588x + 231y = 63$.

6.32. (–) Find all integer solutions to $70x + 28y = 518$. Determine how many solutions have both variables positive.

6.33. Find all integer solutions to $\frac{1}{60} = \frac{x}{5} + \frac{y}{12}$.

6.34. Suppose $a, b, c \in \mathbb{Z}$ and $d = \gcd(a, b)$. Suppose also that d divides c. Prove that the set of integer solutions to $ax + by = c$ is non-empty. Express the set of solutions in terms of one given solution and the parameters a, b, d.

6.35. Derive a formula for the exponent of the prime p in the prime factorization of the product $n!$. The formula should be a sum with finitely many terms.

6.36. Suppose p is a prime number.
a) Prove that p divides $\binom{p}{k}$ if $1 \leq k \leq p - 1$.
b) Prove that $n^p - n$ is divisible by p for every $n \in \mathbb{N}$. (Hint: use the binomial theorem and part (a) in a proof by induction.)

6.37. (+) *Ideals.* A set $S \subseteq \mathbb{Z}$ is an *ideal* in $\mathbb{Z}$ if S is non-empty and satisfies 1) If $a, b \in S$, then $a + b \in S$, and 2) If $a \in S$ and $n \in \mathbb{Z}$, then $na \in S$. Prove that every ideal in $\mathbb{Z}$ is $d\mathbb{Z}$ (the set of multiples of d) for some d. (Comment: an ideal consisting of the multiples of a single element is a *principal ideal*. This exercise strengthens Theorem 6.10 by showing that every ideal in $\mathbb{Z}$ is a principal ideal.)

6.38. *A "reciprocal" dart board problem.*

    a) Do there exist natural numbers m, n such that $7/17 = 1/m + 1/n$?

    b) (+) Suppose p is prime. For which $k \in \mathbb{N}$ do there exist $m, n \in \mathbb{N}$ such that $k/p = 1/m + 1/n$?

6.39. Suppose a jar contains some number of pennies, some number of nickels, and some number of dimes. Suppose the total value of the coins in cents is s, and the total number of coins is t. Determine the smallest s that permits more than one solution for some fixed t. Determine the largest s for which there is no solution using 10 coins.

6.40. Suppose there are two jars of jelly beans in a bear's cage, one with x beans and the other with y. Each jar has a lever. If a jar has at least two beans, then pressing its lever will give one bean from the jar to the bear and move one bean from it to the other jar; otherwise pressing the lever has no effect. Obtain necessary and sufficient conditions on the pair x, y so that the bear can successfully eat all the beans except one.

6.41. (+) Let S be a set of three positive integers. If r, s are members of the current set, with $r \leq s$, then these numbers can be replaced by $2r$ and $s - r$. Prove that every set S of three positive integers can be transformed by a sequence of these operations into a set that contains 0. (Hint: If x is the smallest number in S and y is the next smallest, prove that y can be expressed as $y = (2^n + a)x + b$, where $a < 2^n$ and $b < x$. Use this expression to prove the claim by strong induction on the minimum value in S.)

Chapter 7

Modular Arithmetic

In Chapter 6 we studied divisibility; now we study the remainders left by division. Parity describes the remainder when the divisor is 2; the odd integers are those having remainder 1. Considerations of parity are fundamental to atomic physics and computer science as well as to mathematics. We generalize parity by considering divisors other than 2; this leads to many applications and to another notion of arithmetic.

7.1. Problem. *Chinese Remainder Problem.* A general in ancient China wanted to count his troops. Suppose that when his soldiers were split into 3 equal groups there was one soldier left over, when split into 5 equal groups there were two left over, and when split into 7 equal groups there were four left over. What is the minimum number of soldiers that makes this possible?

More generally, suppose n_1, n_2, n_3 are three natural numbers that are pairwise relatively prime, and a_1, a_2, a_3 are integers with $0 \le a_i < n_i$ for each i. Describe all the integers that, for each i, have remainder a_i when divided by n_i. ∎

7.2. Problem. *The Newspaper Problem.* A math professor cashes a check for x dollars and y cents, but the teller inadvertently pays y dollars and x cents instead. After the professor buys a newspaper for 50 cents, the remaining money is twice the original value of the check. What was the value of the check? How does the solution change if the cost of the newspaper changes? ∎

7.3. Problem. *Primality Testing.* Is is possible to prove that a number is not prime without knowing any of its factors? ∎

EQUIVALENCE RELATIONS

The parity classes "even" and "odd" obey an arithmetic structure. Products of odd numbers are odd, the sum of two odds is even, etc. Indeed, the set "{even,odd}" forms a field (Definition 1.15) under these operations. Each of "even" and "odd" is an infinite set of integers. We need a framework for discussing arithmetic operations on sets. This framework permits us to construct the rational numbers, the real numbers, and other useful structures. We begin with a more general notion.

7.4. Definition. When S and T are sets, a *relation* between S and T is a subset of the Cartesian product $S \times T$. A *relation on S* is a subset of $S \times S$.

We usually define a relation R on a set S by stating a condition for pairs; the relation is the set of ordered pairs satisfying the condition.

7.5. Example. If $f \colon \mathbb{R} \to \mathbb{R}$, then the graph of f is a relation on $\mathbb{R}$. It is the set of ordered pairs $\{(x, y) \in \mathbb{R}^2 \colon y = f(x)\}$; each element of $\mathbb{R}$ is the first coordinate in exactly one such pair. The conditions "$|x| = |y|$" and "$x^2 + y^2 = 1$" also define relations on $\mathbb{R}$, but these relations are not the graphs of functions. ■

7.6. Definition. An *equivalence relation* is a relation R on a set S such that
1) $(x, x) \in R$ for all $x \in S$ (the *reflexive property*).
2) $(x, y) \in R$ if and only if $(y, x) \in R$ (the *symmetric property*).
3) $(x, y) \in R$ and $(y, z) \in R$ imply $(x, z) \in R$ (the *transitive property*).

7.7. Example. For every set S, the *equality relation* $R = \{(x, x) \colon x \in S\}$ is an equivalence relation on S. Echoing the notation for equality, we often write $x \sim y$ instead of $(x, y) \in R$ when R is an equivalence relation.

Let S be the set of students at a college. The condition "x and y have been in a class together" generally does not define an equivalence relation on S. It defines a relation that is reflexive and symmetric but not transitive. In other words, x and x have been in a class together, and if x and y have shared a class then y and x have shared a class, but $(x, y) \in R$ and $(y, z) \in R$ do not imply $(x, z) \in R$.

On the other hand, the condition "x and y were born in the same year" does define an equivalence relation on S. All three properties hold, because each person is born in only one year. ■

7.8. Remark. *Order relations.* The *divisibility relation* defined by $R = \{(m, n) \in \mathbb{N}^2 \colon m|n\}$ is not an equivalence relation. It is reflexive and transitive, but it is not symmetric. Indeed, it has the *antisymmetric* *property*: $(x, y) \in R$ and $(y, x) \in R$ together imply $x = y$. A relation that is reflexive, antisymmetric, and transitive is an *order relation*.

Another example of an order relation is the inclusion relation on the set S of subsets of a set X. (Recall that S is the "power set" of X - Definition 1.3). For all $A, B, C \in S$, we have $A \subseteq A$, $(A \subseteq B$ and $B \subseteq A)$ implies $A = B$, and $(A \subseteq B$ and $B \subseteq C)$ implies $A \subseteq C$. ∎

We will encounter many equivalence relations in this book. In this chapter, we focus on an equivalence relation associated with divisibility, called "congruence". The notions of congruence and modular arithmetic, introduced by Karl Friedrich Gauss (1777-1855), are so fundamental that we have special terminology and notation for them.

7.9. Definition. *Congruence.* Given a natural number n, the integers x and y are *congruent modulo n* if $x - y$ is divisible by n. We write this as $x \equiv y \bmod n$. The number n is the *modulus*.

7.10. Theorem. For every $n \in \mathbb{N}$, congruence modulo n is an equivalence relation on $\mathbb{Z}$.

Proof. Reflexive property: $x - x$ equals 0, which is divisible by n. Symmetric property: If $x \equiv y \bmod n$, then by definition $n | (x - y)$. Since $y - x = -(x - y)$, and since n divides $-m$ if and only if n divides m, we also have $n | (y - x)$, and hence $y \equiv x \bmod n$. Transitive property: If $n | (x - y)$ and $n | (y - z)$, then integers a, b exist such that $x - y = an$ and $y - z = bn$. Adding these equations yields $x - z = an + bn = (a + b)n$, so $n | (x - z)$. We conclude that the transitive property holds. ∎

7.11. Definition. Given an equivalence relation on S, the set of elements equivalent to $x \in S$ is the *equivalence class* containing x. The equivalence classes of the relation "congruence modulo n" are the *remainder classes* or *congruence classes* of integers modulo n. The set of congruence classes is written as $\mathbb{Z}_n$ or $\mathbb{Z}/n\mathbb{Z}$. When n is given, we use $\bar{a}$ to denote the congruence class containing a.

The equivalence classes of an equivalence relation on S form a partition of S; elements x and y belong to the same class if and only if (x, y) satisfies the relation. Conversely, if $A_1, \ldots, A_k$ is a partition of S, then the condition "x and y are in the same set in the partition" defines an equivalence relation on S (Exercise 7.2).

7.12. Remark. *Remainder classes.* By definition, a and b are congruent modulo n if and only if $a - b$ is divisible by n. By the Division Algorithm, there are unique integers k, r such that $a = kn + r$ and $0 \le r < n$; we have called r the "remainder" upon division by n. If $a = kn + r$ and

$b = ln + s$ with $r, s \in \{0, \ldots, n - 1\}$, then $n | (a - b)$ if and only if $r - s = 0$. Hence $a \equiv b \bmod n$ if and only if a and b have the same remainder upon division by n; this explains why the congruence classes are also called remainder classes. For $0 \le r < n$, the rth class in $\mathbb{Z}_n$ is $\{kn + r \colon k \in \mathbb{Z}\}$. ∎

The next lemma is the property of the congruence relation that allows us to define arithmetic with congruence classes.

7.13. Lemma. If $a \equiv r \bmod n$ and $b \equiv s \bmod n$, then $a + b \equiv r + s \bmod n$ and $a \cdot b \equiv r \cdot s \bmod n$.

Proof. The statements $a \equiv r \bmod n$ and $b \equiv s \bmod n$ mean there exist integers k, l such that $a = kn + r$ and $b = ln + s$. Adding these equations yields $a + b = (k + l)n + (r + s)$, which means $a + b \equiv r + s \bmod n$. Multiplying the equations yields $a \cdot b = kln^2 + (ks + lr)n + r \cdot s$, which means $a \cdot b \equiv r \cdot s \bmod n$. ∎

7.14. Example. Since $79 \equiv 4 \bmod 5$ and $23 \equiv 3 \bmod 5$, we can multiply the congruence classes to obtain $79 \cdot 23 \equiv 12 \bmod 5$. Since $12 \equiv 2 \bmod 5$, we can further reduce this to $79 \cdot 23 \equiv 2 \bmod 5$. ∎

Lemma 7.13 enables us to define arithmetic on congruence classes. The result of adding or multiplying two congruence classes will itself be a congruence class. Like arithmetic operations on real numbers, these operations take a pair of elements of a set as input and return an element of that set as output.

7.15. Definition. A *binary operation* on a set S is a function from $S \times S$ to S. On $\mathbb{Z}_n$, addition is the binary operation defined by letting the sum of the congruence classes $\bar{a}$ and $\bar{b}$ be the congruence class containing the integer $a + b$. Similarly, multiplication is the binary operation defined by letting the product of $\bar{a}$ and $\bar{b}$ be the congruence class containing the integer $a \cdot b$. In notation, we have $\bar{a} + \bar{b} = \overline{a + b}$ and $\bar{a} \cdot \bar{b} = \overline{a \cdot b}$.

In the formulas in Definition 7.15, the arithmetic operations between classes are the operations being defined; the operations on the right are previously known operations on integers. Lemma 7.13 guarantees that the binary operations we have specified are well-defined functions: when we choose integers a_1, a_2 in the class $\bar{a}$ and integers b_1, b_2 in the class $\bar{b}$, the congruence classes containing $a_1 + b_1$ and $a_2 + b_2$ are the same, and the congruence classes containing $a_1 \cdot b_1$ and $a_2 \cdot b_2$ are the same. There are many ways to perform the computation telling us the sum or product of two congruence classes, but the result of the computation is always the same congruence class.

For this reason, we seldom use the notation $\bar{a}$ for congruence classes. The expression $6+6\equiv 5\bmod 7$ is both a statement about integers and a statement about congruence classes. The validity of addition and multiplication with congruence classes is one reason we use the "equality-like" notation ($\equiv$) for the congruence relation.

7.16. Example. *Binary arithmetic.* We have already used arithmetic modulo 2. The congruence classes are "even" (0 mod 2) and "odd" (1 mod 2). The addition table modulo 2 states that the sum of two integers with the same parity is even, and the sum of two integers with opposite parity is odd. The multiplication table says that the product of two integers is odd if and only if they are both odd. ∎

+	0 1
0	0 1
1	1 0

*	0 1
0	0 0
1	0 1

7.17. Example. *Clock arithmetic.* Minutes on a clock behave like arithmetic mod 60. If it is now "quarter-past" and a 90 minute movie is starting, then it will end at "quarter-till". This is independent of the hour, just as the sum of two odd numbers is even no matter which odd numbers we use. ∎

7.18. Remark. *Modular computation.* Lemma 7.13 holds for all integers r and s; they need not lie between 0 and $n-1$. Thus when performing arithmetic operations in which we care only about the congruence class of the result modulo n, *we may at any time replace a number by a more convenient representative of its congruence class.* We may write the computation of Example 7.14 as $79 \cdot 23 \equiv 4 \cdot 3 \equiv 12 \equiv 2 \bmod 5$. Here the "mod 5" indicates that all four expressions belong to the same congruence class modulo 5. ∎

7.19. Example. *"Casting out nines": An integer is divisible by 9 if and only if the sum of its decimal digits is divisible by 9.* Since 10 is congruent to 1 modulo 9, every nonnegative power of 10 is also congruent to 1 modulo 9. Therefore $\Sigma_{n\geq 0}a_n 10^n \equiv \Sigma_{n\geq 0}a_n 1^n \equiv \Sigma_{n\geq 0}a_n \bmod 9$. This is called "casting out nines" and was used as a check by clerks adding columns of figures before adding machines were invented.

To check the computation of Σc_i, let s be the sum of the digits of the result, and for each i let b_i be the sum of the digits of c_i. If the addition is correct, then Σb_i must be congruent to s modulo 9. For example, suppose we add the numbers 123, 456, 789 and obtain 1268. The sums of the digits in the three numbers are 6, 15, 24, respectively, which sum to 45. The sum of the digits of 1268 is 17. Since 17 is not congruent to 45 modulo 9, we must have made a mistake. The correct sum is 1368. ∎

Addition of congruence classes behaves nicely: adding a multiple of n never changes the remainder modulo n, so the class $\bar{0}$ is an additive identity for arithmetic modulo n. Also, since $x - x$ belongs to this class, the class containing $-x$ is an additive inverse for the class containing x.

Multiplication does not always behave so well, though Lemma 7.13 implies that the class containing 1 is a multiplicative identity. The next lemma leads to many properties of modular arithmetic. We postpone developing them in order to give several applications of the lemma, including the solution of Problems 7.1 and 7.3.

7.20. Lemma. If a and n are relatively prime integers, then multiplication by $\bar{a}$ is a bijection from the set $S = \mathbb{Z}_n - \{0\}$ to itself.

Proof. Multiplying the numbers $1, 2, \ldots, n-1$ by the integer a yields the numbers $a, 2a, \ldots, (n-1)a$. We proved in Theorem 6.18 that when a and n are relatively prime integers, the numbers $a, 2a, \ldots, (n-1)a$ have different remainders modulo n. If $ka \equiv 0 \bmod n$, then n must divide k because a and n are relatively prime (this uses Proposition 6.13). Since n does not divide any of $1, \ldots, n-1$, we conclude that the numbers $a, 2a, \ldots, (n-1)a$ have $n-1$ different *nonzero* remainders modulo n. Hence they represent all elements of S. We conclude that multiplication by a permutes the nonzero congruence classes. ∎

7.21. Corollary. If a and n are relatively prime integers, then there is exactly one congruence class $\bar{b}$, called the *multiplicative inverse* of $\bar{a}$, such that $ab \equiv 1 \bmod n$.

APPLICATIONS

We use Corollary 7.21 first to present an *ad hoc* solution of Problem 7.1 and then to prove a theorem that provides another algorithm.

7.22. Solution. *Chinese Remainder Problem.* We seek a number x that is congruent to 1 mod 3, to 2 mod 5, and to 4 mod 7. Thus $x = 3n + 1$ for some integer n. Incorporating the second requirement, we have $3n + 1 \equiv 2 \bmod 5$, which becomes $3n \equiv 1 \bmod 5$. Since 3 and 5 are relatively prime, there is a unique congruence class modulo 5 as a solution; we have $n \equiv 2 \bmod 5$. Writing $n = 5m + 2$ yields $x = 3(5m + 2) = 15m + 7$.

Incorporating the third requirement, we have $15m + 7 \equiv 4 \bmod 7$. Since $15 \equiv 1 \bmod 7$ and $7 \equiv 0 \bmod 7$, we immediately obtain $m \equiv 4 \bmod 7$, so that $m = 7k + 4$ for some $k \in \mathbb{Z}$. Hence $x = 15(7k + 4) + 7 = 135k + 67$. The smallest positive number (of soldiers) is 67. ∎

This method can be combined with induction on the number of congruences to prove the next theorem. We present a short proof that yields another algorithm and avoids induction.

7.23. Theorem. (Chinese Remainder Theorem) If $\{n_i\}$ is a set of r natural numbers that are pairwise relatively prime, and $\{a_i\}$ are r arbitrary integers, then the system of congruences $x \equiv a_i \bmod n_i$ has a unique solution modulo $N = \Pi n_i$.

Proof. If x and x' are solutions, then they must be congruent modulo N. To see this, note that $x \equiv x' \equiv a_i \bmod n_i$ for each i, so $n_i | (x - x')$. Since the n_i's are relatively prime, this yields $N | (x - x')$, by Corollary 6.16.

Now we construct a solution. For each i, let $N_i = N/n_i$. Since n_i is relatively prime to the other moduli, we have $\gcd(N_i, n_i) = 1$. By Corollary 7.21, there is exactly one congruence class modulo n_i, call it $\bar{y}_i$, such that $N_i y_i \equiv 1 \bmod n_i$. Set $x = \Sigma_{j=1}^{r} a_j N_j y_j$. When we consider this equation modulo n_i, the terms with $j \neq i$ are congruent to 0, since $n_i | N_j$ for $j \neq i$. Only the term with $j = i$ remains, and from $N_i y_i \equiv 1 \bmod n_i$ we obtain $x \equiv a_i N_i y_i \equiv a_i \bmod n_i$. Hence this value of x satisfies all specified congruences. ∎

7.24. Example. Suppose we seek x such that $x \equiv 2 \bmod 5$, $x \equiv 4 \bmod 7$, and $x \equiv 3 \bmod 9$. This yields $N = 315$ and $N_1, N_2, N_3 = 63, 45, 35$.

i	a_i	n_i	N_i	$N_i \bmod n_i$	y_i
1	2	5	63	3	2
2	4	7	45	3	5
3	3	9	35	−1	−1

By the Chinese Remainder Theorem, we obtain a solution by setting x to be $2 \cdot 63 \cdot 2 + 4 \cdot 45 \cdot 5 + 3 \cdot 35 \cdot (-1) = 1047$. (Hand computation using this procedure should check the solution at this stage!) All numbers congruent to 1047 modulo 315 are solutions. The one with smallest absolute value is $1047 - 3 \cdot 315 = 102$. ∎

When the moduli are not pairwise relatively prime, there may be no solution, or it may be possible to modify the problem to use the Chinese Remainder Theorem anyway (Exercise 7.26).

The Solution of Problem 7.2 uses a different equivalence relation.

7.25. Solution. *The Newspaper Problem.* In Problem 7.2, the check for x dollars and y cents is paid instead as y dollars and x cents. Note that x and y are between 0 and 99. After subtracting 50 cents for the newspaper, the remaining money is twice the original value of the check.

We can encode this information as $100y + x - 50 = 2(100x + y)$, which simplifies to $98y - 199x = 50$. This is a diophantine equation, which we can solve by the method of Example 6.12. After some calculation, we obtain $(x, y) = (-1650 + 98j, -3350 + 199j)$ for $j \in \mathbb{Z}$. To enforce $0 < x < 100$, we take $j = 17$ and obtain $(x, y) = (16, 33)$. This answer checks, since $\$33.16 - \$0.50 = 2 \times \$16.33$.

A natural equivalence relation leads us to a uniform approach for all possible prices of the newspaper. We define an equivalence relation on integer pairs (m, n) representing m dollars plus n cents by saying that (a, b) and (c, d) are equivalent if they represent the same amount of money, which means $(a, b) = (c + n, d - 100n)$ for some $n \in \mathbb{Z}$. The problem states that $(y, x - 50)$ and $(2x, 2y)$ are equivalent, each representing the amount of money remaining after buying the newspaper. After setting $y = 2x + n$ and $x - 50 = 2y - 100n$, we eliminate y to obtain $3x + 50 = 98n$. Since $x \geq 0$, we have $n \geq 0$. Since x is an integer, $98n - 50$ must be divisible by 3; this holds if we choose $n = 1$. With $n = 1$ we obtain $x = 16$ and $y = 2x + 1 = 33$, as above.

When the newspaper costs k cents, we obtain $3x + k = 98n$. For $k = 75$, this requires n to be positive and divisible by 3, but $n \geq 3$ implies $x \geq (98 \cdot 3 - 75)/3 = 73$. Since $y = 2x + n$, this yields $y \geq 149$, which violates the conditions of the problem. Hence there is no solution when $k = 75$. Each choice of n up to 99 yields solutions for various k. For example, when $n = 99$, we have the solution $\$0.99$ for the original check, if the newspaper costs $\$97.02$. See also Exercise 7.29. ∎

FERMAT'S LITTLE THEOREM

When p is prime, Lemma 7.20 implies that multiplication by a nonzero integer a permutes the nonzero congruence classes modulo p and leaves 0 fixed. In other words, the function $f_a \colon \mathbb{Z}_p \to \mathbb{Z}_p$ defined by $f_a(x) = ax$ is a bijection (below we illustrate $f_5 \colon \mathbb{Z}_{13} \to \mathbb{Z}_{13}$).

Because every vertex is the head of one arrow and is the tail of one arrow, the functional digraph of a permutation is a collection of pairwise disjoint cycles. The element a itself lies on a cycle, so repeated multiplication by a (modulo p) must eventually return a to itself. On the step immediately before returning to a, the cycle reaches 1; otherwise, a would be at the head of arrows from 1 and another vertex.

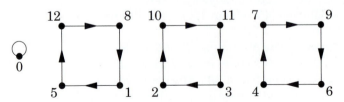

Our next theorem shows that the length of this cycle must divide p, so that $a^{p-1} \equiv 1 \bmod p$. Proved by Pierre de Fermat (1601-1665), the theorem has many generalizations and applications. It also has many proofs, applying various fundamental ideas. Exercise 7.37 asks for a proof using equivalence relations. Exercise 6.36 asks for a proof using induction and binomial coefficients. In Example 9.34, we present a proof that uses multinomial coefficients. The proof we give here, due to Leonhard Euler (1707-1783), is one of the most direct.[†]

7.26. Theorem. (Fermat's Little Theorem) If p is prime and a is not a multiple of p, then $a^{p-1} \equiv 1 \bmod p$.

Proof. Our analysis of the functional digraph for multiplication by a shows that some power of a between a^0 and a^{p-1} is congruent to 1 modulo p. By the Well-Ordering Property, we may choose k to be the smallest such integer. The theorem follows if we prove that k divides $p-1$, because $p-1 = mk$ implies that $a^{p-1} = (a^k)^m \equiv 1^m \equiv 1 \bmod p$.

To prove that k divides $p-1$, we partition the nonzero congruence classes modulo p into subsets of size k. Let $S_x = \{x, xa, xa^2, \ldots, xa^{k-1}\}$ be the set of all multiples of x by powers of a. There are k distinct powers of a modulo p. The elements of S_x are distinct, because $xa^i \equiv xa^j \bmod p$ with $i < j$ implies $a^{j-i} \equiv 1 \bmod p$, which contradicts the choice of k. (This cancellation of x uses the primality of p, which implies that multiplication by x permutes the nonzero classes.)

Hence $|S_x| = k$, and we need only show that the sets S_x partition the nonzero congruence classes. Two elements are in the same set if one is a power of a times the other (modulo p). Hence "being in the same set" is the relation R on $[p-1]$ defined by $(x, y) \in R$ if xy^{-1} is a power of a. Since $xx^{-1} \equiv 1 \equiv a^k \bmod p$, R is reflexive. If $xy^{-1} \equiv a^i \bmod p$, then $yx^{-1} \equiv a^{k-i} \bmod p$, so R is symmetric. If $xy^{-1} \equiv a^i \bmod p$ and $yz^{-1} \equiv a^j \bmod p$, then $xz^{-1} \equiv a^{i+j} \bmod p$, so R is transitive. Hence the sets S_x are the equivalence classes of an equivalence relation on $[p-1]$, and they partition $[p-1]$. We conclude that k divides $p-1$. ∎

7.27. Example. We have seen that $5^4 \equiv 1 \bmod 13$. The smallest powers of 4,3,2 congruent to 1 modulo 13 are 4^6, 3^3, and 2^{12}. In each case, we obtain $a^{12} \equiv 1 \bmod 13$. ∎

7.28. Example. *Computation with Fermat's Little Theorem.* We can rapidly compute remainders for large numbers involving powers. For example, we have $11^{902} = 11^{30 \cdot 30 + 2} = (11^{30})^{30} \cdot 11^2 \equiv 1^{30} \cdot 121 \bmod 31 \equiv -3 \bmod 31 \equiv 28 \bmod 31$. ∎

7.29. Corollary. If p is prime and $a \in \mathbb{Z}$, then $a^p \equiv a \bmod p$.

[†]For further reading about this theorem, see Andre Weil, *Number Theory: An Approach through History*, Birkhäuser (Boston 1984).

The contrapositive of this immediate corollary of Fermat's Little Theorem enables us to solve Problem 7.3 most of the time.

7.30. Solution. *Primality testing.* Suppose a and p are integers. The contrapositive of Corollary 7.29 states that if a^p is not congruent to a modulo p, then p is not a prime number. Thus, finding such a number a proves that p is not prime without knowing any factors of p.

For example, suppose we want to test whether 341 is prime. If we choose $a = 7$, we have an easy computation. Because $7^3 = 343 \equiv 2 \bmod 341$ and $2^{10} = 1024 \equiv 1 \bmod 341$, we can compute

$$7^{341} = 7^{3 \cdot 113 + 2} \equiv 2^{113} 7^2 \equiv 2^{110+3} 7^2 \equiv 8 \cdot 49 \equiv 392 \equiv 51 \bmod 341.$$

Since $51 \not\equiv 7 \bmod 341$, we conclude that 341 cannot be prime.

We can apply this test even without such a clever choice for a. Although it still takes some work, computing the congruence class of a^{341} never requires 341 multiplications. Using repeated squaring, we can compute the congruence classes of the numbers $\{a^{2^k}\}$. The binary representation of 341 tells us which of these to multiply together to compute a^{341}. For example, suppose $a = 3$. Repeated squaring yields

$$3^2 = 9 \qquad 3^8 = 81^2 = 6561 \equiv 82 \bmod 341$$
$$3^4 = 81 \qquad 3^{16} \equiv 82^2 \equiv 245 \bmod 341$$

and so on. The binary representation of 341 is 101010101. If we multiply together the congruence classes of 3^n for $n = 1, 4, 16, 64, 256$, we have the congruence class of 3^{341}. ∎

Modular multiplications are fast on computers. If n is not prime, then computing the congruence class of a^n for a few random choices of a is likely to prove that n is not prime, but this doesn't always work. There are some numbers such that $a^n \equiv a \bmod n$ for every a even though n is not prime. One such number is 561, which has prime factorization $3 \cdot 11 \cdot 17$. Such numbers are called "Carmichael numbers". Although we won't prove it here, we state a characterization. A natural number n is a Carmichael number if and only if 1) n has no repeated prime factors, and 2) whenever p divides n, also $p - 1$ divides $n - 1$. For example, 2,10,16 all divide 560.

CONGRUENCE AND GROUPS (optional)

In the remainder of this chapter, we present a bit more formal discussion of the arithmetic properties of $\mathbb{Z}_n$. We have proved that addition and multiplication modulo n are well-defined. This enables us to specify addition and multiplication tables for the elements of $\mathbb{Z}_n$. Below we illustrate these for $\mathbb{Z}_6$ and $\mathbb{Z}_7$.

+	0 1 2 3 4 5
0	0 1 2 3 4 5
1	1 2 3 4 5 0
2	2 3 4 5 0 1
3	3 4 5 0 1 2
4	4 5 0 1 2 3
5	5 0 1 2 3 4

*	0 1 2 3 4 5
0	0 0 0 0 0 0
1	0 1 2 3 4 5
2	0 2 4 0 2 4
3	0 3 0 3 0 3
4	0 4 2 0 4 2
5	0 5 4 3 2 1

+	0 1 2 3 4 5 6
0	0 1 2 3 4 5 6
1	1 2 3 4 5 6 0
2	2 3 4 5 6 0 1
3	3 4 5 6 0 1 2
4	4 5 6 0 1 2 3
5	5 6 0 1 2 3 4
6	6 0 1 2 3 4 5

*	0 1 2 3 4 5 6
0	0 0 0 0 0 0 0
1	0 1 2 3 4 5 6
2	0 2 4 6 1 3 5
3	0 3 6 2 5 1 4
4	0 4 1 5 2 6 3
5	0 5 3 1 6 4 2
6	0 6 5 4 3 2 1

Since adding a multiple of n does not change a congruence class modulo n, the class 0 is an *identity element* for the addition modulo n. Furthermore, $(n-i)+i \equiv n \equiv 0 \bmod n$, so the class $n-i$ acts as an *additive inverse* of i. As verified earlier, the sum of two congruence classes is a congruence class. Also, $(a+b)+c \equiv a+(b+c) \bmod n$. These properties make $\mathbb{Z}_p$ a "group" under addition mod p.

7.31. Definition. *Group.* A *group* is a set S together with a binary operation $\circ$ on $S^{\dagger}$ satisfying the following properties:
1) There is an element $e \in S$ such that for every $x \in S$, $x \circ e = x = e \circ x$ (e is called the *identity element*).
2) For every $x \in S$, there is an element $y \in S$ such that $x \circ y = e = y \circ x$ (y is called the *inverse* of x).
3) For every $x, y, z \in S$, $(x \circ y) \circ z = x \circ (y \circ z)$ (*associative property*).

The elements of a field (Definition 1.15) form a group under addition. The nonzero elements of a field form a group under multiplication. Whenever the binary operation in a group is written as $+$, we express the inverse of x as $-x$, write $y-x$ for $y+(-x)$, and name the identity element 0. We have done this for $\mathbb{Z}_n$ under addition mod n.

What about multiplication in $\mathbb{Z}_n - \{0\}$? We know that 1 is a multiplicative identity, but we soon run into trouble. When n is composite, that is $n = ab$, we have $a \cdot b \equiv 0 \bmod n$, so discarding 0 does not permit the remaining elements to form a group under multiplication (see the table for $\mathbb{Z}_6$).

When p is prime, $p|ab$ implies $p|a$ or $p|b$. Hence ab is not congruent to $0 \bmod p$ when a and b are not congruent to $0 \bmod p$. Thus multiplication is a binary operation on $\mathbb{Z}_p - \{0\}$. The associative property follows from the associative property of integer multiplication, since we can select any integers from these congruence classes to do the computation. Finally, we verify below that multiplicative inverses exist in $\mathbb{Z}_p - \{0\}$ whenever p is prime. The table above exhibits multiplicative inverses in $\mathbb{Z}_7 - \{0\}$; we have $6 \cdot 6 \equiv 1$, $5 \cdot 3 \equiv 1$, $4 \cdot 2 \equiv 1$, and $1 \cdot 1 \equiv 1$.

$\dagger$The definition of a binary operation includes the property that $x \circ y \in S$ for all $x, y \in S$; this is the property of *closure* under $\circ$.

7.32. Corollary. If p is prime, then $\mathbb{Z}_p - \{0\}$ is a group under multiplication.

Proof. We have verified all the needed properties except the existence of inverses. Consider $a \neq 0$. Since a and p are relatively prime, Corollary 7.21 implies that there is some nonzero class $\bar{b}$ such that $\bar{a}\bar{b} \equiv \bar{1}$. The class $\bar{b}$ is the desired $(\bar{a})^{-1}$. Note also that $\bar{b}\bar{a} \equiv \bar{1}$, since multiplication modulo p is commutative. ■

In our discussion, we have completed all the details of proving that $\mathbb{Z}_p$ is a field if (and only if) p is prime.

When can a number be its own multiplicative inverse?

7.33. Lemma. If p is prime and $a \in \mathbb{N}$, then $a^2 \equiv 1 \bmod p$ if and only if $a \equiv 1 \bmod p$ or $a \equiv -1 \bmod p$.

Proof. If $a^2 \equiv 1$, then p divides $a^2 - 1$, which equals $(a+1)(a-1)$. When a prime divides a product, it must divide one of the factors (Corollary 6.14). Hence p divides $a+1$ or $a-1$, yielding $a \equiv -1 \bmod p$ or $a \equiv 1 \bmod p$. Conversely, if a is in one of these classes, then p divides $(a+1)(a-1)$, and a^2 is in the same congruence class as 1. ■

7.34. Theorem. (Wilson's Theorem) $(p-1)! \equiv -1 \bmod p$ for p prime.

Proof. This holds for $p = 2$ because $1 \equiv -1 \bmod 2$. For $p > 2$, we use Lemmas 7.20 and 7.33. For each $1 \leq i \leq p-1$, there is exactly one i' in $[p-1]$ such that $ii' = 1$. The numbers 1 and $p-1$ are self-inverse (i.e., $p-1 \equiv -1 \bmod p$), and by the lemma the others form disjoint pairs of inverses. Hence $\Pi_{i=2}^{p-2} i \equiv 1 \bmod p$, and $\Pi_{i=1}^{p-1} i \equiv p-1 \equiv -1 \bmod p$. ■

Wilson's Theorem was only conjectured by John Wilson (1741-1793); Joseph Louis Lagrange (1736-1813) gave the first proof in 1770.

EXERCISES

7.1. Determine whether the following relations R are equivalence relations on the given set S.
    a) $S = \mathbb{N} - \{1\}$; $(x, y) \in R$ if and only if $\gcd(x, y) > 1$.
    b) $S = \mathbb{R}$; $(x, y) \in R$ if and only if there exists $n \in \mathbb{Z}$ such that $x = 2^n y$.

7.2. Suppose S is the disjoint union of sets $A_1, \ldots, A_k$. Let R be the relation consisting of pairs $(x, y) \in S \times S$ such that x, y belong to the same member of $\{A_1, \ldots, A_k\}$. Prove that R is an equivalence relation on S.

7.3. Let S be a finite set, let C be a fixed subset of S, and let $\mathbf{P}$ be the collection of all subsets of S. Let R be a relation defined on $\mathbf{P}$ by $(A, B) \in R$ if and only if $A \cap C = B \cap C$. Is R an equivalence relation? Give a proof or a counterexample.

7.4. Suppose $f: \mathbb{R} \to \mathbb{R}$. Let $O(f)$ be the set of functions g for which there exist positive constants $c, a \in \mathbb{R}$ such that $|g(x)| \le c|f(x)|$ for all $x > a$. Define a relation R on the set S of functions mapping $\mathbb{R}$ to $\mathbb{R}$ by putting $(g, h) \in R$ if and only if $g - h \in O(f)$. Prove that R is an equivalence relation on S.

7.5. Find the flaw in the following argument that the symmetric and transitive properties imply the reflexive property for a relation R on S: "Consider $x \in S$. If $(x, y) \in R$, then the symmetric property implies that $(y, x) \in R$. Now the transitive property applied to (x, y) and (y, x) implies that $(x, x) \in R$."

7.6. (!) Prove that every year (also leap years) has at least one Friday the 13th.

7.7. (–) Suppose k is an odd number. Prove that $k^2 - 1$ is divisible by 8.

7.8. Suppose p is an odd prime. Determine all solutions to $2n^2 + n \equiv 0 \bmod p$.

7.9. (!) Suppose $m, n, p \in \mathbb{Z}$ and 5 divides $m^2 + n^2 + p^2$. Prove that 5 divides at least one of $\{m, n, p\}$.

7.10. (–) Use modular arithmetic to prove that $k^n - 1$ is divisible by $k - 1$ for all $n, k \in \mathbb{N}$ with $k \ge 2$.

7.11. (!) *Primes and factorials.*
a) Compute the exponent of each prime in the factorization of $k!$.
b) Use the answer to (a) to prove that N is divisible by $k!$ if N is the product of k consecutive natural numbers.

7.12. (+) Prove that there are infinitely many primes of the form $4n + 3$ and infinitely many primes of the form $6n + 5$, where $n \in \mathbb{N}$. (Hint: Show first that the divisors of a number congruent to -1 mod 4 cannot all be congruent to 1 mod 4, and the divisors of a number congruent to -1 mod 6 cannot all be congruent to 1 or 3 mod 6. Comment: Dirichlet proved more generally that if a and b are relatively prime, then there are infinitely many primes of the form $an + b$, but this is beyond the techniques we have available.)

7.13. (–) Prove that if two natural numbers have the same number of copies of each digit in their decimal representations, then they differ by a multiple of 9.

7.14. (–) What is the congruence class of 10^n modulo 11? Use this to determine the remainder when 654321 is divided by 11.

7.15. (!) The base 10 representation of an integer is *palindromic* if the digits read the same when written forward or backward. Prove that every palindromic integer with an even number of digits is divisible by 11. More generally, prove that every integer whose base k representation is palindromic and has even length is divisible by $k + 1$.

7.16. Define $f: \mathbb{Z}_n \to \mathbb{Z}_n$ by $f(x) = x^2$. For which $n \in \mathbb{N}$ is f injective?

7.17. (–) Prove that the first six powers of 10 belong to distinct congruence classes modulo 7. (Comment: Gauss asked whether the powers of 10 yield $n - 1$ distinct congruence classes modulo n for infinitely many n; this remains unanswered. The moduli 5 and 13 fail, even though they are primes.)

7.18. Suppose n is a 6-digit decimal integer such that the set of digits in the decimal expansion is $\{1,2,3,4,5,6\}$ and for each i the i-digit number formed by the leftmost i digits is divisible by i. The integer 123456, for example, does not work, because 1234 is not divisible by 4. Determine all possible values of n.

7.19. (+) Suppose n is an integer whose base 10 representation is a permutation of the digits 0 through 9. Suppose also that the first i digits of n, read from left to right, form an integer that is divisible by i, for $1 \le i \le 10$. Determine all possible values of n. (Hint: Various values of i impose constraints or combinations of constraints that restrict the possibilities; for example, the 10th digit must be 0, and then the fifth digit must be 5. Division is not needed.)

7.20. (!) *Test for divisibility by 7.*
   a) Let $a_k \cdots a_0$ be the base 10 representation of n. We can determine whether n is divisible by 7 by treating n as $\Sigma a_i 10^i$ and reducing the powers of 10 modulo 7; we have discussed this approach to divisibility by 9. Apply this to check whether 7 divides 535801.
   b) Given a positive integer n, let $f(n)$ be the integer formed by subtracting twice the last base 10 digit of n from the number formed by the remaining digits of n. For example, if $n = 154$, then $f(n) = 15 - 8 = 7$. Prove that $7|n$ if and only if $7|f(n)$. Apply this to check whether 7 divides 535801. (Hint: To prove that $7|n$ if and only if $7|f(n)$, prove first that $7|n$ if and only if $7|[10f(n)]$.)

7.21. *Test for divisibility by n* (generalization of Exercise 7.20). Let n be a positive integer. Let $f(n)$ be the integer formed by subtracting j times the last base 10 digit of n from the number formed by the remaining digits of n (example above has $j = 2$). Prove that if s is not divisible by 2 or 5 and $10j \equiv -1 \bmod s$, then n is divisible by s if and only if $f(n)$ is divisible by n. Describe the resulting tests for divisibility by 17 and by 19, and illustrate how they work on 323, which equals $17 \cdot 19$.

7.22. (!) *Primes and threes.*
   a) Prove that the sum of the digits in the base 10 expansion of a natural number n is a multiple of 3 if and only if n is a multiple of 3.
   b) Prove that $6|x$ when $x + 1$ and $x - 1$ are prime, with one exception.
   c) Suppose $x + 1$ and $x - 1$ are prime. Form a new number by concatenating the digits of one with the digits of the other. Thus $\{11, 13\}$ can become 1113 or 1311. Prove that the resulting number is not prime, with one exception.

7.23. Suppose that $n \in \mathbb{N}$, $a, b \in \mathbb{Z}$, and $d = \gcd(a, n)$. Consider arithmetic modulo n. Prove that there is no congruence class $\bar{x}$ that solves the congruence equation $\bar{a}\bar{x} = \bar{b}$ unless d divides b, in which case there are d solutions.

7.24. (!) 1500 soldiers arrive in training camp. A few soldiers desert the camp. The drill sergeants divide the remaining soldiers into groups of five and discover that there is 1 left over. When they divide them into groups of 7, there are 3 left over, and when they divide them into groups of 11, there are again 3 left over. Determine the number of deserters.

7.25. Find all integers that are congruent to 1 mod 7, 3 mod 8, and 5 mod 9. Which of these integers has the smallest *absolute* value?

7.26. Suppose $x \equiv 3 \bmod 6$, $x \equiv 4 \bmod 7$, and $x \equiv 5 \bmod 8$. Explain why the Chinese Remainder Theorem does not apply to compute x. Transform the problem to an equivalent problem where the Chinese Remainder Theorem can be used. Compute the smallest positive solution for x. Give a precise (and concise) reason why there is no smaller positive number that works.

7.27. Derive a description of all integers congruent to $x \bmod a$, $y \bmod b$, and $z \bmod c$, given that n is such an integer.

7.28. Let m be a positive integer. Give an example of a polynomial f with integer coefficients and leading coefficient 1 such that $f(x) \equiv 0 \bmod m$ for all $x \in \mathbb{Z}$. (Comment: Compare this with Corollary 4.19.)

7.29. (+) Analyze the Newspaper Problem in full (Solution 7.25). In particular, determine the prices k for which the problem has a solution.

7.30. (–) Use Fermat's Little Theorem to find a number between 0 and 12 that is congruent to 2^{100} modulo 13.

7.31. (–) In the proof of Fermat's Little Theorem, we partitioned $[p-1]$ into equivalence classes of size k, where k is the smallest positive integer such that $a^k \equiv 1 \bmod p$.

a) For each nonzero element $a \in \mathbb{Z}_{13}$, compute the smallest k such that $a^k \equiv 1 \bmod 13$. (Comment: In a multiplicative group, the smallest k such that a^k is the identity is called the *order* of a.)

b) List the partitions of $[12]$ that arise in the proof of Fermat's Little Theorem when $p = 13$ (no proof required).

7.32. Fermat's Little Theorem implies that p divides $2^p - 2$ if p is prime. Fermat conjectured that the converse is also true, meaning that p divides $2^p - 2$ only if p is prime, but he was wrong. Euler provided the counterexample of $p = 341$. Prove that 341 is not prime and that 341 divides $2^{341} - 2$, verifying Euler's counterexample. (Hint: use Fermat's Little Theorem.)

7.33. (+) A *cyclic shift* of a p-tuple x is a p-tuple obtained by adding a constant (modulo p) to the indices of the elements of x; shifting x by $p+i$ positions produces the same p-tuple as shifting x by i positions. Let R be the relation on $[a]^p$ (the set of p-tuples with entries in $\{1, \ldots, a\}$) defined by putting $(x, y) \in R$ if the p-tuple y can be obtained from x by a cyclic shift.

a) Prove that R is an equivalence relation on $[a]^p$.

b) Use part (a) and Lemma 7.20 to prove that p divides $a^p - a$. (Hint: partition a set of size $a^p - a$ into subsets of size p.)

c) Use part (a) to prove Fermat's Little Theorem.

7.34. Wilson's Theorem states that $(p-1)! \equiv -1 \bmod p$ if p is prime.

a) Prove the converse: if $(p-1)! \equiv -1 \bmod p$, then p is prime.

b) Use Wilson's Theorem to prove that if p is an odd prime, then $2(p-3)! \equiv -1 \bmod p$.

7.35. (!) Theorem 6.18 states that if a and p are relatively prime, then $\{a, 2a, \ldots, (p-1)a\}$ have distinct remainders modulo p. Use this to give another proof of Fermat's Little Theorem.

7.36. We say that k is a square modulo n if $k \equiv j^2 \bmod n$ for some j. Suppose that $n = m^2 + 1$ for some $m \in \mathbb{N}$. Prove that if k is a square modulo n, then $-k$ is also a square mod n.

7.37. Prove that the set of all permutations of $[n]$, viewed as a set of functions from $[n]$ to $[n]$, forms a group under the operation of composition. Use the properties of functions.

7.38. Prove that the polynomials of degree k with coefficients in $\mathbb{Z}_p$ form a group under addition modulo p.

7.39. *Uniqueness of inverses.* Suppose a finite set S is a group under the binary operation $\circ$; in particular, every $x \in S$ has an inverse under $\circ$. Let $f_y : S \to S$ be the function defined by $f_y(x) = y \circ x$. Prove that f_y is surjective. Use this to conclude that f_y is injective. Use this to conclude that y has exactly one inverse.

Chapter 8

The Rational Numbers

One cannot generally group a given whole number of objects into a specified number of equal parts. We introduce rational numbers to study ratios of integers. The precise definition of the set $\mathbb{Q}$ of rational numbers uses the set of integers and the concept of equivalence relation developed in Chapter 7. We present applications to such things as Pythagorean triples, probability, and games. We also discuss the philosophical issue of the existence of real numbers that are not rational.

8.1. Problem. *The Billiard Problem.* Suppose a square billiard table has corners at $\{(0,0),(1,0),(1,1),(0,1)\}$. A ball leaves the origin along a line with slope s. If the ball reaches a corner, it stops (or falls off the table). Whenever it hits the side of the table not at a corner, it continues to travel on the table, but the slope of the line is multiplied by -1. Does this process terminate? When $s = 3/5$, the answer is "yes". ∎

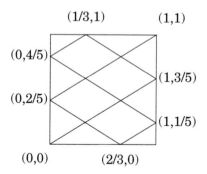

8.2. Problem. *Pythagorean triples.* What are the integer solutions to $a^2 + b^2 = c^2$? The positive solutions measure sides of right triangles. ∎

8.3. Problem. *Iterated averaging.* Starting with $\{0, 1\}$, what numbers can be found by iteratively averaging two numbers already found? ∎

CONSTRUCTING THE RATIONALS

The division algorithm emphasizes the remainder when an integer m is divided by a nonzero integer n; here we focus on the ratio. The definition of a rational number requires care, since many fractions represent the same ratio. Let $F = \mathbb{Z} \times (\mathbb{Z} - \{0\})$. Each element of F is an ordered pair of integers, which we think of as the numerator and (nonzero) denominator of a fraction. To express the idea that different pairs represent the same ratio, we define an equivalence relation on F.

8.4. Lemma. Let $F = \mathbb{Z} \times (\mathbb{Z} - \{0\})$, and let $\sim$ be a relation on F defined by $(a, b) \sim (c, d)$ if and only if $ad = bc$. The relation $\sim$ is an equivalence relation on F.

Proof. We must verify the reflexive, symmetric, and transitive properties. *Reflexive property:* $(a, b) \sim (a, b)$, since $ab = ba$. *Symmetric property:* If $(a, b) \sim (c, d)$, then by definition $ad = bc$, which is equivalent to $cb = da$, which by definition means $(c, d) \sim (a, b)$. *Transitive property:* Suppose $(a, b) \sim (c, d)$ and $(c, d) \sim (e, f)$, which means $ad = bc$ and $cf = de$. To prove that $(a, b) \sim (e, f)$, we want to show that $af = be$. Multiplying the two known equations yields $adcf = bcde$. Since $d \neq 0$, we can cancel d to obtain $acf = bce$. If $c \neq 0$, then we can also cancel c to obtain $af = be$. If $c = 0$, then $ad = bc$ and $cf = de$ imply that a and e are 0 as well, and again we have $af = be$. ∎

8.5. Definition. The set of *rational numbers*, written as $\mathbb{Q}$, is the set of equivalence classes of $F = \mathbb{Z} \times (\mathbb{Z} - \{0\})$ under the relation $\sim$ defined above. We write $\frac{m}{n}$ or m/n to denote the rational number that is the equivalence class containing the pair (m, n). We write $\frac{a}{b} = \frac{c}{d}$ to mean that (a, b) and (c, d) belong to the same equivalence class.

To use the rationals as numbers, we must first define addition, multiplication, the additive and multiplicative identities, and the positive set for $\mathbb{Q}$. Definition 8.6 specifies these in terms of the known operations on integers. As in Chapter 1, we can then define subtraction by $x - y = x + (-y)$ and define order by $x < y$ when $y - x$ is positive.

8.6. Definition. We define the rational numbers 0 and 1 to be $\frac{0}{1}$ and $\frac{1}{1}$, respectively. The *sum* and *product* of $\frac{a}{b}, \frac{c}{d} \in \mathbb{Q}$ are

$$\frac{a}{b} + \frac{c}{d} = \frac{ad + bc}{bd} \qquad \text{and} \qquad \frac{a}{b} \cdot \frac{c}{d} = \frac{ac}{bd}.$$

The rational number $\frac{a}{b}$ is *positive* if $ab > 0$.

8.7. Theorem. With Definition 8.6, the set $\mathbb{Q}$ of rational numbers forms an ordered field.

Proof. The proof has two main parts; both rely on integer arithmetic. We first verify that the operations in $\mathbb{Q}$ are well-defined. We must also verify that they make $\mathbb{Q}$ into an ordered field. We prove the order axioms here and leave the field axioms to Exercise 8.6. When proving that the operations are well-defined (independent of the representatives chosen from the classes), we select $\frac{a}{b} = \frac{a'}{b'}$ and $\frac{c}{d} = \frac{c'}{d'}$.

Addition is well-defined: $\frac{a}{b} + \frac{c}{d} = \frac{a'}{b'} + \frac{c'}{d'}$. It suffices to prove that $\frac{ad+bc}{bd} = \frac{a'd'+b'c'}{b'd'}$, by the definition of rational addition. By the definition of the equivalence relation, the condition for equality is $(ad + bc)b'd' = bd(a'd' + b'c')$. Using the properties of integer arithmetic, this condition becomes $(ab' - ba')dd' = bb'(cd' - dc')$. This equality follows from $ab' - ba' = cd' - dc' = 0$, which in turn holds because $(a, b) \sim (a', b')$ and $(c, d) \sim (c', d')$.

Multiplication is well-defined: $\frac{a}{b} \cdot \frac{c}{d} = \frac{a'}{b'} \cdot \frac{c'}{d'}$. It suffices to prove that $\frac{ac}{bd} = \frac{a'c'}{b'd'}$, by the definition of rational multiplication. The equality is equivalent to $acb'd' = bda'c'$, which follows from $ab' = ba'$ and $cd' = dc'$.

The positive set is well-defined: $\frac{a}{b} > 0$ if and only if $\frac{a'}{b'} > 0$. This holds since $ab' = ba'$ requires that ab and $a'b'$ have the same sign.

The positive set is closed under addition. From $\frac{a}{b} > 0$ and $\frac{c}{d} > 0$, we have $ab > 0$ and $cd > 0$. This yields $(ad + bc)bd > 0$ for each choice of signs for a, b, c, d, and hence $\frac{a}{b} + \frac{c}{d} > 0$.

The positive set is closed under multiplication. Again $\frac{a}{b} > 0$ and $\frac{c}{d} > 0$ imply $ab > 0$ and $cd > 0$. Now the number of positive elements in $\{a, b, c, d\}$ is even; hence $acbd > 0$, which yields $\frac{a}{b} \cdot \frac{c}{d} > 0$.

The trichotomy property holds. For each nonzero class $\frac{a}{b}$, we have $-\frac{a}{b} = \frac{-a}{b}$. The closure properties of the positive set imply that it contains exactly one of $\frac{a}{b}$ and $-\frac{a}{b}$. ∎

The function f defined by $f(m) = m/1$ is an injection of $\mathbb{Z}$ into $\mathbb{Q}$ that preserves all the arithmetic properties of the integers, so we may treat the rational numbers $\{m/1: m \in \mathbb{Z}\}$ as being the integers.

8.8. Example. *Geometric interpretation of rational numbers.* We have defined the rationals as equivalence classes of pairs of integers (with the second coordinate nonzero). For each such pair (m, n), define the line $L(m, n)$ through the point $(0, 0)$ by $L(m, n) = \{(x, y) \in \mathbb{R} \times \mathbb{R}: mx = ny\}$.

Pairs are mapped to the same line if and only if they are equivalent pairs. Furthermore, if $b \neq 0$, then the integer point $(b, a) \in \mathbb{R} \times \mathbb{R}$ lies on the line $L(a, b)$. Hence we have established a bijection between the rational numbers and the lines through the origin (other than the vertical line) that pass through integer points. The inverse of this bijection assigns to each line the rational number that is the slope of the line. ■

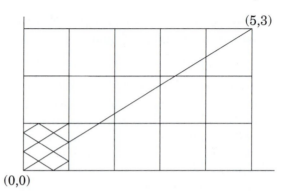

$(5,3)$

$(0,0)$

8.9. Solution. *The Billiard Problem.* Our ball starts at the origin and bounces off the walls of the unit square; let L be the line with slope s along which it starts. Vertical direction changes after each unit of vertical travel, but the magnitude of the vertical rate of travel remains the same. The same statement holds for horizontal motion. Thus "reaching a corner" means simultaneously having traveled integer amounts m vertically and n horizontally. This occurs if and only if L contains the integer point (m, n); the path followed by the ball is then a folding of the segment from $(0, 0)$ to (m, n). The line L contains such a point if and only if s is rational. (See Exercise 8.7 for a stronger statement.) ■

The set of rational numbers is countable (Exercise 8.8), but a bijection from $\mathbb{Q}$ to $\mathbb{N}$ cannot list the rational numbers in the usual order "$\leq$". Whenever x, y are distinct rational numbers, we can find another rational number between them, such as their average. There is no "next" rational number under the usual ordering "$\leq$", and thus we cannot use induction with "$\leq$" to prove statements about rational numbers.

In Chapter 6, we proved that integers have unique prime factorizations. We thus can write the numerator and denominator of a fraction as products of prime powers. If they have a common factor, we can cancel it to obtain another representative of the same rational number. This leads to a canonical representative, like the canonical representative between 0 and $p - 1$ for a congruence class modulo p.

8.10. Definition. A fraction a/b is *in lowest terms* if a and b have no common factors and $b > 0$.

8.11. Remark. A fraction is in lowest terms if and only if its denominator is the smallest positive number among the denominators of all representatives of the same rational number.

In modular arithmetic, any convenient member of a congruence class can be used for computation; the canonical member is not always best. This observation also holds for rational numbers; when adding them, for example, it is convenient to use fractions with a common denominator instead of keeping them in lowest terms.

Selecting the representative of a rational number in lowest terms is an example of the technique of *extremality*. Selecting a member of a set that is extremal in some respect provides extra leverage. For example, the representative of a rational number in lowest terms is the one with smallest positive denominator. Another extremal description of this representative comes from the geometric interpretation of rational numbers as lines: if $x = a/b$ in lowest terms, then (b, a) is the integer point with positive first coordinate that is closest to the origin on the line associated with x.

IRRATIONAL NUMBERS

Next we argue that there are numbers that are not rational. Due to its importance, we give several proofs that there is no rational number whose square is 2.

8.12. Example. *Irrationality of* $\sqrt{2}$. We claim that the equation $x^2 = 2$ has no rational number x as a solution. Otherwise, we may assume $x = a/b$ is a rational solution, with (a, b) written in lowest terms. We have $a^2 = 2b^2$, so a^2 is even. Since the square of any odd integer is odd, a must be even. But now a^2 is divisible by 4, and hence $2b^2$ is divisible by 4, so b^2 is even. We conclude that b also is even. Now a, b are both divisible by 2, contradicting the choice of a/b in lowest terms. ∎

This argument generalizes to prove that no prime number has a rational square root (Exercise 8.9a), but stronger results are available. A slightly different argument prohibits rational square roots for all natural numbers (except the squares of integers), again by extremality.

8.13. Theorem. The positive integer k has no rational square root if k is not the square of an integer.

Proof. By contradiction; suppose m/n is a rational expression for $\sqrt{k}$, chosen so that n is positive and minimal. If m/n is not an integer, then there is an integer q such that $m/n - 1 < q < m/n$. This is equivalent to

$0 < m - nq < n$. Since $m - nq \neq 0$, we can write

$$\frac{m}{n} = \frac{m(m - nq)}{n(m - nq)} = \frac{m^2 - mnq}{n(m - nq)} = \frac{n^2k - mnq}{n(m - nq)} = \frac{nk - mq}{m - nq}.$$

Since $0 < m - nq < n$, we have found a representation of m/n with smaller positive denominator, which contradicts the choice of n. Therefore, if the square root of k is rational, it must be an integer. ∎

The proof above does not use the prime factorization of integers and is generalized in Exercise 8.11. For a proof using prime factorization, see Exercise 8.9. The argument in Example 8.12 for the irrationality of $\sqrt{2}$ generalizes to describe rational zeros of polynomials with integer coefficients.

8.14. Theorem. (Rational Zeros Theorem) Suppose $c_0, \ldots, c_n$ are integers with $n \geq 1$ and $c_0, c_n \neq 0$, and let $f(x)$ be the polynomial $f(x) = \Sigma_{i=0}^{n} c_i x^i$. If $r = p/q$ in lowest terms is a rational solution to the equation $f(x) = 0$, then p must divide c_0 and q must divide c_n.

Proof. If $f(r) = 0$, we can multiply both sides of $f(r) = 0$ by q^n to obtain $\Sigma_{i=0}^{n} c_i p^i q^{n-i} = 0$. If we move the term $c_n p^n$ to the other side, we have

$$-c_n p^n = \Sigma_{i=0}^{n-1} c_i p^i q^{n-i} = q\Sigma_{i=0}^{n-1} c_i p^i q^{n-1-i}.$$

Since q divides one side of this, it must also divide the other side. But q and p are relatively prime, since p/q was chosen in lowest terms, and hence q must divide c_n. Similarly, if we move the term $c_0 q^n$ to the other side, we have

$$-c_0 q^n = \Sigma_{i=1}^{n} c_i p^i q^{n-i} = p\Sigma_{i=1}^{n} c_i p^{i-1} q^{n-i}.$$

Now p divides one side and hence the other, which implies that p divides c_0, since p, q are relatively prime. ∎

8.15. Example. *No rational solutions.* If the equation $x^3 - 6 = 0$ has a rational solution r, written as p/q in lowest terms, then q must divide 1 and p must divide 6. The only possibilities are $r = \pm 1, \pm 2, \pm 3, \pm 6$, none of which work. Hence the cube root of 6 is irrational. ∎

8.16. Example. *Solutions to quadratics.* The quadratic formula gives $(-b \pm \sqrt{b^2 - 4ac})/2a$ as the solutions to $ax^2 + bx + c = 0$. Even when a, b, c are integers, the solutions may be irrational. For example, $(1 + \sqrt{5})/2$ is a solution to the equation $x^2 - x - 1 = 0$. This number is not rational, because the Rational Zeros Theorem implies that the only possible rational solutions to this equation are ± 1, which do not satisfy the equation. Nevertheless, products of irrational numbers may be rational. For example, $\dfrac{\sqrt{5}+1}{2} \cdot \dfrac{\sqrt{5}-1}{2} = 1$. ∎

PYTHAGOREAN TRIPLES

Why should we care that there is no rational solution to $x^2 = 2$? We believe there is a number $\sqrt{2}$ that is the ratio of two physical quantities. The length of the diagonal of a square with side-length 1 is a quantity that we believe is $\sqrt{2}$; it satisfies $x^2 = 2$. In elementary geometry, we construct right angles with straightedge and compass. We can thus construct a right triangle whose short sides have unit length. The length of the third side is $\sqrt{2}$, by the Pythagorean Theorem.

The ancients believed that all numbers were rational. It is said that therefore the person who discovered irrational numbers was murdered (by drowning). Are irrational numbers "crazy"? The Billiard Problem and decimal expansions (see Chapter 13) show that irrational numbers may exhibit complicated behavior, but the name "irrational number" does not arise from "crazy". The relationship between the psychological meaning of "irrational" (lacking reason) and the mathematical meaning of "irrational" (not a ratio of integers) is that "ratio" and "reason" come from the same Greek root. The Pythagoreans allowed only rational numbers in their reasoning.

8.17. Theorem. (Pythagorean Theorem) If a, b, c are the lengths of the sides of a right triangle, with c the length of the side opposite the right angle, then $a^2 + b^2 = c^2$.

Proof. (Sketch) We assume the notions of right angle, triangle, rectangle, and area. We assume that the area of a rectangle is the product of the lengths of two neighboring sides, the area of a region is the sum of the areas of the regions formed by cutting it by line segments, and the area of congruent regions is the same. These notions imply that the area of a right triangle is half the product of the short sides, because the diagonal of a rectangle cuts it into two pieces of equal area. By considerations of symmetry, the inner and outer quadrilaterals in the illustration are squares. Our notions of area tell us that the area of the large square equals the area of the small square plus the sum of the areas of four congruent triangles. These two formulas for the area are $(a + b)^2$ and $c^2 + 2ab$, and equating them yields $a^2 + b^2 = c^2$. ∎

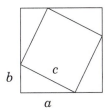

We can now describe all the integer solutions to $a^2 + b^2 = c^2$; such triples (a, b, c) are called *Pythagorean triples*.

8.18. Theorem. The Pythagorean triples are the integer multiples of triples of the form $(2rs, r^2 - s^2, r^2 + s^2)$ or $(r^2 - s^2, 2rs, r^2 + s^2)$, where r, s are integers.

Proof. Since $(2rs)^2 + (r^2 - s^2)^2 = (r^2 + s^2)^2$, these are indeed Pythagorean triples. If we multiply each element of such a triple by n, then we multiply the equality by n^2 on both sides, so all integer multiples of these triples also belong to the set. To prove that we have a characterization, we must prove that every Pythagorean triple can be described in this way. Suppose a, b, c are integers with $a^2 + b^2 = c^2$ and $b, c \neq 0$. We can normalize this by dividing by c^2 to obtain $x^2 + y^2 = 1$, where $x = a/c$ and $y = b/c \neq 0$.

The point (x, y) has rational coordinates and lies on the unit circle in the Cartesian plane. It is helpful to describe the circle using parametric equations, where the parameter is the slope of the line from $(-1, 0)$ to (x, y). This slope is $t = y/(1 + x)$, which is rational because x and y are rational.

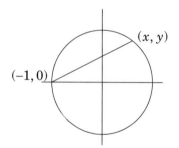

Substituting $y = t(x + 1)$ into $x^2 + y^2 = 1$ yields $x^2 + t^2(x + 1)^2 = 1$, which we rewrite as $x^2 + \dfrac{2t^2}{1 + t^2} x + \dfrac{t^2 - 1}{1 + t^2} = 0$. The first coordinate of a point on both the line and the circle must satisfy this quadratic equation. If the roots are α and β, then

$$x^2 + \frac{2t^2}{1 + t^2} x + \frac{t^2 - 1}{1 + t^2} = (x - \alpha)(x - \beta).$$

In particular, $\alpha \cdot \beta = (t^2 - 1)/(1 + t^2)$. Since $x = -1$ (from the point $(-1, 0)$) is one root, the other root must be $(1 - t^2)/(1 + t^2)$. Since t is rational, this value is also rational. Let $t = s/r$ in lowest terms. We obtain the solution $x = \dfrac{1 - t^2}{1 + t^2} = \dfrac{1 - s^2/r^2}{1 + s^2/r^2} = \dfrac{r^2 - s^2}{r^2 + s^2}$ and $y = t(x + 1) = \dfrac{s}{r} \dfrac{2r^2}{r^2 + s^2} = \dfrac{2rs}{r^2 + s^2}$. Multiplying by $r^2 + s^2$ gives us the integer triple $(r^2 - s^2, 2rs, r^2 + s^2)$, where the ratios equal the original ratios a/c and b/c.

Because the ratios are the same, the original triple (a, b, c) is a multiple of this triple by a rational number z. If z is an integer, then we have the desired conclusion. Otherwise $r^2 - s^2, 2rs, r^2 + s^2$ have a common factor $n \neq \pm 1$ (from the denominator of z in lowest terms). Hence n also divides the sum and difference of $r^2 - s^2$ and $r^2 + s^2$, which are $2r^2$ and $2s^2$. Since r and s have no common factors, this requires $n = \pm 2$. Hence the terms of the triple are all even, and r, s must both be odd. Now let $R = (r + s)/2$ and $S = (r - s)/2$. We compute $2RS = (r^2 - s^2)/2$, $R^2 - S^2 = rs$, and $R^2 + S^2 = (r^2 + s^2)/2$. Hence $(a, b, c) = (2RS, R^2 - S^2, R^2 + S^2)$, and again we have expressed (a, b, c) in the desired form. ∎

The famous *Fermat's Last Theorem* is the statement that $x^n + y^n = z^n$ has no solution in integers if $n \geq 3$. Fermat wrote this in the margin of a book in the 17th century, claiming to have a marvelous proof that would not fit in the margin, but he died without presenting a proof to anyone. Mathematicians labored for 350 years to find a proof. Andrew Wiles succeeded in 1994.

FRACTIONS AND PROBABILITY

Fractions arise in many contexts. When performing many experiments, we may consider the fraction of successful experiments. A batting average in baseball, for example, is the ratio of hits (successes) to at-bats (experiments), usually written as a three-place decimal. In Solution 5.27 and Exercise 5.25, we counted the poker hands of various types. When we view all possible hands as equally likely, the ratio of the number of ways to obtain a particular type of hand to the total number of hands is the probability of obtaining that type. Fractions and probabilities may lead to surprising results when analyzing data.

8.19. Example. *Simpson's Paradox.* Many people believe that if competitor A performs better than competitor B in every category, then B cannot be rated higher than A. We present a counterexample using airline performance.[†] The phenomenon is called *Simpson's Paradox.*

In 1987, airlines in the United States had to report the percentage of their flights that arrived on time at each of the nation's 30 busiest airports. Alaska Airlines served only five of these airports and performed better than America West at every one of them, but overall America West had a higher on-time average at these airports.

The explanation is that on-time performance is largely controlled by weather conditions. Alaska Airlines serves primarily Seattle, with habitual bad weather; America West serves sunny Phoenix. Although Alaska always did better under comparable conditions, the overall

---

[†]A. Barnett, How numbers can trick you, *Technology Review* (1994), 38-45.

Destination	Alaska % on time	Alaska # arrivals	America West % on time	America West # arrivals
Los Angeles	88.9	559	85.6	811
Phoenix	94.8	233	92.1	5255
San Diego	91.4	232	85.5	448
San Francisco	83.1	605	71.3	449
Seattle	85.8	2146	76.7	262
Total	86.7	3775	89.1	7225

statistic for America West is dominated by service under easy conditions, while Alaska is judged primarily by the airport where weather
conditions prevent good performance. ∎

When the same experiment is performed many times, we expect
the fraction of successes to approach a fixed value called the probability
of success. Alternatively, consider a finite set A whose elements are
equally likely to occur. If B is a subset of A, then we call B an "event"
and define the *probability* of the event B to be $|B|/|A|$. This is the proba-
bility that a "randomly chosen element" of A will belong to B. In this
context, a probability is a rational number between 0 and 1. More general situations allow the probability of an event to be an arbitrary real
number in the interval $[0, 1]$.

Suppose an experiment has two possible outcomes, with probabilities x and $1 - x$. When the payoffs for these outcomes are a and b, we
say that the "expectation" (expected payoff) is $ax + b(1 - x)$. This is the
average payoff we expect if we perform the experiment many times. We
will return to probability and expectation in Chapter 9.

8.20. Application. *The Odd/Even Finger Game.* Players A and B try
to outwit each other in the following game. On each play of the game,
each player shows 1 or 2 fingers. The payoffs from A to B are listed in
the matrix below for the four possible outcomes.

	A shows 1	A shows 2
B shows 1	−2	+3
B shows 2	+3	−4

This seems like a fair game, but it favors B. If B always shows 1 or
always shows 2, then A can use this information to win. Hence B
should show 1 sometimes and 2 sometimes, perhaps showing 1 with
probability x and 2 with probability $1 - x$. For example, a player can
implement $x = 2/3$ by rolling a die and showing 1 if the roll is at most 4
and showing 2 otherwise. Player A may know the strategy x, but A
does not know what B will show on a particular play.

Knowing the strategy x, A can compute the expectation of the outcome that results from playing either column and play only the column

with the smaller expectation. Hence the expectation that x guarantees for B is the minimum of the expectations in the two columns. These expections are $-2x + 3(1-x)$ and $3x - 4(1-x)$, which simplify to $3 - 5x$ and $7x - 4$; B wants to maximize the minimum of these. Since one decreases with x and the other increases, the minimum is maximized when they are equal; $3 - 5x = 7x - 4$ yields $x = 7/12$. By choosing $x = 7/12$, B guarantees an average payoff per game of at least $1/12$.

Player A can keep B from winning any more than $1/12$, on the average. Player A can play column 1 with probability y, limiting the expected payoff to B to at most $\min \max \{ -2y + 3(1-y), 3y - 4(1-y) \}$. The minimum occurs at $y = 7/12$, where the two values are equal.

Alternatively, can treat the players symmetrically, with B using strategy x and A using strategy y. The expected payoff (to B) is $-2xy + 3y(1-x) + 3x(1-y) - 4(1-y)(1-x)$, equal to $7x - 4 + y(7 - 12x)$. If B chooses $x = 7/12$, then the expected payoff is $7(7/12) - 4 = 1/12$, which is independent of y. Thus B can ensure a positive expectation no matter how A chooses y. We can also rewrite the expression as $7y - 4 + x(7 - 12y)$. By choosing $y = 7/12$, A makes this independent of x and ensures that the average payoff to B is at most $1/12$. ∎

FURTHER PROPERTIES OF RATIONALS (optional)

We have seen examples of proofs about rational numbers using reduction to lowest terms and using the closure of the rational numbers under arithmetic operations. The next proof uses a different technique. We may prove a statement for $\mathbb{Q}$ in the same way that we built the numbers; first prove it for the natural numbers, then for the integers, then for the rational numbers.

8.21. Theorem. Suppose $f : \mathbb{Q} \to \mathbb{Q}$ satisfies $f(x + y) = f(x) + f(y)$ for all $x, y \in \mathbb{Q}$. Then $f(wx) = wf(x)$ for all $w, x \in \mathbb{Q}$.

Proof. First suppose $w = 1$; here the statement is a triviality. This provides the basis step for a proof by induction in the case $w \in \mathbb{N}$. For the induction step, suppose it is true when $w = n$. Then

$$f((n + 1)x) = f(nx + x) = f(nx) + f(x) = nf(x) + f(x) = (n + 1)f(x),$$

where we have used the distributive law, the defining property of f, the induction hypothesis, and the distributive law again. To prove the claim for $w = 0$, we need only show that $f(0) = 0$, which follows from $f(0) = f(0 + 0) = f(0) + f(0) = 2f(0)$. For $w = -1$, we use $0 = f(0) = f(x - x) = f(x) + f(-x)$, which implies $f(-x) = -f(x)$. Now we can prove the claim for $w \in \mathbb{Z}$: $f((-n)x) = f((-1)nx) = -f(nx) = -nf(x)$ for $n \in \mathbb{N}$.

Next suppose w is the reciprocal of the integer n. We have $f(x) = f(n(x/n)) = nf(x/n)$, and therefore $f(x/n) = (1/n)f(x)$. Note that at each stage we proved the statement for arbitrary $x \in \mathbb{Q}$, so these steps are justified. Now that we have the statement for all integers and for reciprocals of natural numbers, we can write $w \in \mathbb{Q}$ as a/b in lowest terms and conclude $f((a/b)x) = af((1/b)x) = (a/b)f(x)$. ∎

The statement of Theorem 8.21 is false when $\mathbb{Q}$ is replaced by $\mathbb{R}$; the conclusion then requires the additional hypothesis that f be continuous (continuity is discussed in Chapter 15).

8.22. Definition. A rational number is a *dyadic rational* if it can be expressed as a fraction whose denominator is a power of 2.

8.23. Solution. *Iterated averages and dyadic rationals.* We solve Problem 8.3: which numbers can be generated from $\{0, 1\}$ by iteratively taking the average (arithmetic mean) of two numbers already in the set?

Since the average of two rational numbers is rational, only rational numbers in the interval $[0, 1]$ can arise. In addition, the only numbers that can arise are dyadic rationals, since 0 and 1 are dyadic rationals, and the average of two dyadic rationals is also a dyadic rational.

To complete the solution, we prove that every dyadic rational in the interval $[0, 1]$ is generated. Except for 0 itself, each such rational is expressed in lowest terms as $(2j + 1)/2^k$ for some nonnegative integers j, k. We prove by induction on k that $(2j + 1)/2^k$ is achievable. For $k = 0$, the only such number is 1 itself. For the induction step, suppose $k > 0$, and consider the number $x = (2j + 1)/2^k$ in the interval $(0, 1)$. The number x is the average of $(2j)/2^k$ and $(2j + 2)/2^k$, which equal $j/2^{k-1}$ and $(j + 1)/2^{k-1}$ and lie in $[0, 1]$. Since one of $\{j, j + 1\}$ is even, one of these numbers is not in lowest terms (the numerator is an odd number times a power of 2 with positive exponent). After canceling these factors of 2, we have expressed x as the average of two dyadic rationals with smaller exponents in the denominator. By the induction hypothesis, each of these is achievable, so x also is achievable. ∎

In Chapter 13, we consider decimal and binary expansions of real numbers in the interval $[0, 1]$. We will see that the dyadic rationals are precisely the real numbers whose binary expansions terminate.

EXERCISES

8.1. Let a/m and b/n be rational numbers in lowest terms. Prove that $(an + bm)/(mn)$ is in lowest terms if and only if m and n are relatively prime.

8.2. Prove that a fraction is in lowest terms if and only if its denominator is the smallest positive number among the denominators of all representatives of the same rational number.

8.3. (!) Determine all (x, y) such that $1/x + 1/y = 1/(x + y)$.

8.4. (–) Suppose that a, b, c, d are positive integers with $a/b < c/d$. Prove that $a/b < (a + c)/(b + d) < c/d$. Give a simple interpretation of this in terms of batting averages. Give another interpretation of this using slopes of lines.

8.5. Suppose a, b, c, d are positive integers with $a \leq c \leq d$ and $c/d \leq a/b$. Prove that $b - a \leq d - c$. Prove that this conclusion does not always hold if $a \leq d < c$ and $c/d \leq a/b$.

8.6. Complete the proof that $\mathbb{Q}$ is an ordered field, by verifying the field axioms for rational addition and multiplication. The desired statements should be reduced to statements about integers, and then properties of integer arithmetic should be used to prove them. Division is not allowed, but nonzero integers can be canceled from both sides of an equality. The required properties are:

a) Addition in $\mathbb{Q}$ satisfies the identity, commutative, associative, and inverse properties, with 0 being the identity element and $(-a)/b$ being the additive inverse of a/b.

b) Multiplication in $\mathbb{Q} - \{0\}$ satisfies the identity, commutative, associative, and inverse properties, with 1 being the identity element and d/c being the multiplicative inverse of c/d.

c) The distributive law holds for rational arithmetic.

8.7. (!) In the Billiard Problem, for each corner of the square determine the condition on the slope s so that the process ends there.

8.8. (!) Prove that the set of rational numbers is countable.

8.9. *Square roots of integers.*

a) Generalize the first proof given for the irrationality of $\sqrt{2}$ to prove that the square root of every prime number is irrational (i.e., do not use the Rational Zeros Theorem).

b) Use part (a) and the unique factorization of integers to prove that the square root of an integer is irrational unless it is an integer.

8.10. Use the Rational Zeros Theorem to prove that the kth root of an integer is not a rational number unless it is an integer.

8.11. (+) Let p be a polynomial with integer coefficients and leading coefficient 1. Without using the Rational Zeros Theorem, prove that if $p(t) = 0$ for some $t \in \mathbb{Q}$, then $t \in \mathbb{Z}$. (Hint: if $t \notin \mathbb{Z}$, write $t = m/n$ with $n > 1$ and minimal. Let $q = t - \lfloor m/n \rfloor$, and use the numerators of $\{q^k\}$ to obtain a decreasing sequence of positive integers.)

8.12. Use the parametrization of Pythagorean triples to prove that every integer greater than two is a member of a Pythagorean triple not containing 0. (Hint: give a construction when n is even and another construction when n is odd. The fact that $(k + 1)^2 - k^2 = 2k + 1$ may be useful.)

8.13. Each of three containers has two marbles; one contains two white marbles, one contains two black marbles, and one contains one white and one black. A container is selected at random (each equally likely), and one of the two marbles inside is selected at random (each equally likely). Given that the selected ball is black, what is the probability that the other ball in its container is black?

8.14. Assume that a child is equally likely to be born male or female. If we are told that a family has two children, and at least one of the children is male, what is the probability that the other child is male?

8.15. In baseball, "batting average" is defined as the fraction "Hits/(At-bats)". Consider two players A and B. Suppose their performance in day games and night games is as follows:

	Day A	Day B	Night A	Night B
Hits	a	c	w	y
At-bats	b	d	x	z

Find values for a, b, c, d, w, x, y, z so that A has a higher batting average than B in both day games and night games but B has a higher batting average overall.

8.16. (!) Consider universities H and Y, each having 100 professors. Construct an example where, in each of the categories "assistant professors", "associate professors", and "full professors", the fraction who are women is higher at H than at Y, and yet Y has more women professors than H.

8.17. (+) A man has a watch with indistinguishable hands. An act of violence between midnight and the following noon simultaneously kills him and stops his watch. Can we always determine the time of death from this information if:
    a) the watch has hour, minute and second hands?
    b) the watch has only hour and minute hands?

8.18. (!) Let $\mathbb{Q}^* = \mathbb{Q} - \{0\}$. Suppose that $f: \mathbb{Q}^* \to \mathbb{Q}^*$ and that f satisfies $f(x + y) = f(x)f(y)/[f(x) + f(y)]$ whenever $x, y \in \mathbb{Q}^*$. Suppose $c = f(1)$. Compute $f(x)$ in terms of c for every $x \in \mathbb{Q}^*$. (Hint: consider the function $g = 1/f$.)

8.19. *The Finger Game (Example 8.20).*
    a) Which values of x in the interval $[0, 1]$ guarantee a positive expectation for B no matter what A does?
    b) We have seen that when each player shows one finger with probability 7/12, B expects to win an average of 1/12 dollars per game. With these strategies, what *fraction* of the games does B expect to win?

8.20. Suppose the payoffs from A (column player) to B (row player) in a matrix game where each has two strategies are $\left(\begin{smallmatrix} a & b \\ c & d \end{smallmatrix}\right)$. Determine the value of the game in all cases, meaning the maximum amount that B can guarantee receiving by playing the first row with probability x and the second with probability $1 - x$.

PART III

DISCRETE MATHEMATICS

Chapter 9

Combinatorial Reasoning

Why do we count? We can measure the performance of an algorithm by counting the operations it may use. We compare the sizes of sets and determine the likelihood of events by counting. In this chapter we discuss methods of counting, ideas of probability, and several applications. In particular, we explore the binomial coefficients more fully and develop the multinomial coefficients. These concepts enable us to solve problems such as the following.

9.1. Problem. *Summation of Integer Powers.* Given a correct formula for $\sum_{i=0}^{n} i^k$, we can prove it by induction. Lacking a formula, how can we discover it? ∎

9.2. Problem. *Bertrand's Ballot Problem.* Suppose that candidates A and B in an election receive a and b votes, respectively, with $a \geq b$, and that the votes are counted in random order. What is the probability that candidate A never trails? ∎

9.3. Problem. *Bernoulli trials.* Repeated performances of an experiment with a fixed probability of success are called *Bernoulli trials*, after Jakob Bernoulli (1654-1705). When we perform n trials of an experiment that has probability p of success, and the outcome of one trial cannot affect the outcome of any other trial, we expect to have about np successes. How can we make this intuition precise? ∎

9.4. Problem. *"Hitting for the cycle".* How often does a baseball player get a single, a double, a triple, and a home run in the same game? In Solution 9.38, we solve a special case for a particular batter. ∎

MORE ON THE BINOMIAL COEFFICIENTS

We begin with another model for counting problems.

9.5. Example. *Lattice paths.* In addition to the alternative models for selection problems described in Chapter 5, we consider a model using lattice paths in the plane. We start from the origin and take steps of unit length rightward or upward. When we take n steps, we have a *lattice path* of length n. The final location is determined by how many steps we take to the right; if there are k steps to the right, we reach the point $(k, n - k)$. Furthermore, the actual path is determined by which steps are taken to the right. This establishes a bijection between the lattice paths reaching $(k, n - k)$ and the binary n-tuples with k ones, so the number of lattice paths to $(k, n - k)$ is $\binom{n}{k}$. The illustration below shows the number of paths to specified points. ■

1					
1	5				
1	4	10			
1	3	6	10		
1	2	3	4	5	
1	1	1	1	1	1

9.6. Proposition. For nonnegative integers a, b, the number of lattice paths from the origin to the point (a, b) is $\binom{a + b}{a}$.

Proof. See the discussion in Example 9.5. ■

This lattice path or "block-walking" argument suggests an inductive formula for the binomial coefficients. It permits inductive proofs of identities when a combinatorial proof doesn't come to mind. This inductive formula is sometimes called Pascal's Formula in honor of Blaise Pascal (1623-1662). The triangular array of numbers with all the $\binom{n}{k}$'s in the nth row is called Pascal's Triangle, though it was known to Chinese mathematicians much earlier.

$$
\begin{array}{ccccccccccc}
&&&&& 1 &&&&& \\
&&&& 1 && 1 &&&& \\
&&& 1 && 2 && 1 &&& \\
&& 1 && 3 && 3 && 1 && \\
& 1 && 4 && 6 && 4 && 1 & \\
1 && 5 && 10 && 10 && 5 && 1
\end{array}
$$

9.7. Lemma. (Pascal's Formula) If $n \geq 1$, then $\binom{n}{k} = \binom{n-1}{k} + \binom{n-1}{k-1}$.

Proof. We give three proofs. All use the same idea, phrased in different models.

Proof 1. By Proposition 9.6, the number of lattice paths reaching $(k, n - k)$ is $\binom{n}{k}$. Each path arrives at $(k, n - k)$ from exactly one of the points $(k, n - k - 1)$ and $(k - 1, n - k)$. By Proposition 9.6 again, there are $\binom{n-1}{k}$ paths of the first type and $\binom{n-1}{k-1}$ paths of the second type.

Proof 2. Using the subset model, we count the k-sets in $[n]$. There are $\binom{n-1}{k}$ such sets not containing n and $\binom{n-1}{k-1}$ such sets containing n.

Proof 3. $(1 + x)^n = (1 + x)(1 + x)^{n-1}$. Using the binomial theorem, we expand both $(1 + x)^n$ and $(1 + x)^{n-1}$ to obtain

$$\Sigma_{k=0}^{n} \binom{n}{k} x^k = (1 + x) \Sigma_{k=0}^{n-1} \binom{n-1}{k} x^k = \Sigma_{k=0}^{n-1} \binom{n-1}{k} x^k + \Sigma_{k=0}^{n-1} \binom{n-1}{k} x^{k+1}.$$

Changing the index in the last summation yields $\Sigma_{k=1}^{n} \binom{n-1}{k-1} x^k$. Since $\binom{n-1}{n}$ and $\binom{n-1}{-1}$ equal 0, we can add 0 to each summation to obtain $\Sigma_{k=0}^{n} \binom{n}{k} x^k = \Sigma_{k=0}^{n} [\binom{n-1}{k} + \binom{n-1}{k-1}] x^k$. By Corollary 4.19, the corresponding coefficients must be equal. ∎

Pascal's Formula allows us to prove the formula for $\binom{n}{k}$ by induction on n (Exercise 9.1). The proof requires knowing that $\binom{0}{0} = 1$ and that $\binom{n}{k} = 0$ when $k < 0$ or $k > n$. We prove two more identities involving binomial coefficients and list others in the exercises.

9.8. Lemma. $k\binom{n}{k} = n\binom{n-1}{k-1}$.

Proof. Each side counts the k-person committees with a designated chairperson that can be formed from a set of n people. On the left, we select the committee and then select the chair from it; on the right, we select the chair first and then fill out the rest of the committee. ∎

Combinatorial proofs of summation formulas often consist of defining a set whose size is the total and partitioning that set into subsets whose sizes are the terms in the sum; this again is "counting two ways".

9.9. Theorem. (The Summation Identity). $\Sigma_{i=0}^{n} \binom{i}{k} = \binom{n+1}{k+1}$.

Proof. The right side counts the binary $n + 1$-tuples with $k + 1$ ones. We can partition this set into disjoint subsets according to which position holds the rightmost 1. The number of ways to form the sequence so the rightmost 1 is in position $i + 1$ is $\binom{i}{k}$. ∎

$$\underbrace{\qquad\qquad}_{k \ \text{ones}} \quad 1 \ 0 \ 0 \ 0$$

$$\underset{1 \ \text{positions}}{\uparrow} \quad \underset{i \ \ i+1}{\uparrow \ \uparrow} \qquad \underset{n+1}{\uparrow}$$

The block-walking version of this proof counts the paths to $(k, n-k)$ according to the height at which they take the last step to the right. Exercise 9.2 requests a proof by induction.

9.10. Solution. *Summation of integer powers.* Formulas for sums of powers (Problem 9.1) are easy to verify by induction but difficult to guess. The Summation Identity provides a method that automatically generates the answer and the proof. Notice that $i = \binom{i}{1}$. Therefore, the Summation Identity proves the summation formula for the first n natural numbers by $\sum_{i=0}^{n} i = \sum_{i=0}^{n} \binom{i}{1} = \binom{n+1}{2} = n(n+1)/2$. End of proof! For the squares, we must rewrite i^2 using binomial coefficients. Since $i^2 = 2\binom{i}{2} + i = 2\binom{i}{2} + \binom{i}{1}$, we have

$$\sum_{i=0}^{n} i^2 = 2\sum_{i=0}^{n}\binom{i}{2} + \sum_{i=0}^{n}\binom{i}{1} = 2\binom{n+1}{3} + \binom{n+1}{2} = \frac{n(n+1)(2n+1)}{6},$$

where for the last step we extract the common factor $n(n+1)$ from the formulas for $\binom{n+1}{3}$ and $\binom{n+1}{2}$. Using this approach, we can sum $f(0), \ldots, f(n)$ for any polynomial function f. ∎

This method eliminates the guesswork but not the "grunt-work" to obtain the exact formula, as we must express i^k in terms of $\{\binom{i}{j}: 0 \le j \le k\}$ to apply the Summation Identity. Since i^k equals $k!\binom{i}{k}$ plus a polynomial of degree less than k, the Summation Identity yields a proof by induction that the formula is always a polynomial function of n having degree $k+1$. Using a bit more care, we derive in Theorem 9.12 the first two terms of this polynomial.

9.11. Remark. *Binomial coefficients as polynomials.* When viewed as a function of n, the binomial coefficient $\binom{n}{k} = \frac{1}{k!} n(n-1)\cdots(n-k+1)$ is a polynomial of degree k. The coefficient of n^k is $\frac{1}{k!}$, and the coefficient of n^{k-1} is $\frac{1}{k!}\sum_{j=1}^{k-1}(-j) = \frac{-1}{k!}\binom{k}{2} = \frac{-1}{2(k-2)!}$. The kth power of an enormous number is much larger than its $k-1$st power (see Chapter 15). When considering large inputs, we may be content with knowing the leading term of a polynomial, or perhaps the first two terms.

In the next theorem, we use $O(n^k)$ to indicate the set of polynomials of degree at most k. The "Big Oh" notation applies more generally to describe growth conditions for arbitrary functions, as defined in Exercises 3.19 and 7.4. When f is a polynomial, we therefore write $f(x) = 2x^k + O(x^{k-1})$ to mean that f belongs to the set of polynomials of degree k with leading coefficient 2. This is a common abuse of notation. It allows us to write $k!\binom{n}{k} = n^k - \binom{k}{2}n^{k-1} + O(n^{k-2})$. Also, when we subtract a polynomial of lower degree from a polynomial f we do not change the leading term. Hence if f has degree k and g has degree $k-1$, we can write $f(n) - g(n) = f(n) + O(n^{k-1})$. ∎

9.12. Theorem. For fixed $k \in \mathbb{N}$, the value of $\Sigma_{i=0}^{n} i^k$ is a polynomial function of n with leading term $\frac{1}{k+1} n^{k+1}$ and next term $\frac{1}{2} n^k$.

Proof. As observed in Remark 9.11,

$$k!\binom{i}{k} = i(i-1)\cdots(i-k+1) = i^k - \binom{k}{2}i^{k-1} + g(i)$$

for some polynomial g of degree at most $k-2$. Solving for i^k yields $i^k = k!\binom{i}{k} + \binom{k}{2}i^{k-1} - g(i)$.

Now we apply induction on k. For $k = 1$, the formula $\Sigma_{i=0}^{n} i = \frac{1}{2}n^2 + \frac{1}{2}n$ agrees with the claim. For $k > 1$, we have $\Sigma_{i=0}^{n} i^k = k!\Sigma_{i=0}^{n}\binom{i}{k} + \binom{k}{2}\Sigma_{i=0}^{n} i^{k-1} - \Sigma_{i=0}^{n} g(i)$. By the induction hypothesis, the term of degree j in $g(i)$ contributes a polynomial of degree $j+1$ to $\Sigma_{i=0}^{n} g(i)$. Summing these contributions yields $\Sigma_{i=0}^{n} g(i) = O(n^{k-1})$. Also, the induction hypothesis yields $\binom{k}{2}\Sigma_{i=0}^{n} i^{k-1} = \binom{k}{2}\frac{1}{k}n^k + O(n^{k-1})$, and the Summation Identity yields $k!\Sigma_{i=0}^{n}\binom{i}{k} = k!\binom{n+1}{k+1}$.

These three formulas yield $\Sigma_{i=0}^{n} i^k = k!\binom{n+1}{k+1} + \frac{k-1}{2}n^k + O(n^{k-1})$. We now use Lemma 9.8 to replace $\binom{n+1}{k+1}$ with $\frac{n+1}{k+1}\binom{n}{k}$, and next we substitute for $\binom{n}{k}$ the polynomial expression in Remark 9.11. We obtain

$$
\begin{aligned}
\Sigma_{i=0}^{n} i^k &= k!\frac{n+1}{k+1}\binom{n}{k} + \frac{k-1}{2}n^k + O(n^{k-1}) \\
&= k!\frac{1}{k+1}(n+1)\frac{1}{k!}[n^k - \binom{k}{2}n^{k-1}] + \frac{k-1}{2}n^k + O(n^{k-1}) \\
&= \frac{1}{k+1}n^{k+1} + [\frac{1}{k+1}(1 - \binom{k}{2}) + \frac{k-1}{2}]n^k + O(n^{k-1}).
\end{aligned}
$$

To complete the induction step, we simplify the coefficient of n^k:

$$\frac{1}{k+1}[1 - \binom{k}{2}] + \frac{k-1}{2} = \frac{2 - k(k-1) + (k+1)(k-1)}{2(k+1)} = \frac{2+k-1}{2(k+1)} = \frac{1}{2} \quad\blacksquare$$

ANALYSIS OF GAUSSIAN ELIMINATION (optional)

Consider the problem of solving a system of n linear equations in n unknowns. The *substitution method* or *Gaussian elimination* uses successive elimination of variables. Using one of the equations, we express one variable x_i in terms of the others. We substitute this expression for x_i in the other equations to "eliminate" x_i from the problem. This leaves a system of $n-1$ equations in $n-1$ unknowns. We solve it (if possible) for the other variables (using this procedure inductively), and then we use the expression for x_i to compute x_i. (If there is no solution of the smaller system, then there is no solution of the original system.)

As a function of n, we ask how many steps this takes. Knowledge of the leading term often suffices when studying algorithmic efficiency.

9.13. Definition. A *system of linear equations* in n variables $x_1, \ldots, x_n$ is a set of equations of the form $a_{i,1}x_1 + \cdots + a_{i,n}x_n = b_i$, where all the *coefficients* $a_{i,j}$ and *constants* b_i are real numbers. The system is *homogeneous* if the constants b_i all equal 0. A *multiplicative operation* is a multiplication or a division. An *additive operation* is an addition or a subtraction.

9.14. Theorem. Using at most n^2 multiplicative operations and $n^2 - n$ additive operations, solving a system of n linear equations in n variables can be reduced to solving a system of $n - 1$ linear equations in $n - 1$ variables. Overall, the system can be solved using at most $2n^3/3 + O(n^2)$ arithmetic operations.

Proof. If all the coefficients are 0, then there is no solution (unless the system is homogeneous, and then all values are solutions). Hence we may assume there is an equation in which the coefficient on some variable is nonzero. By relabeling, we may assume that this is the nth equation and the variable x_n. To express x_n in terms of the other variables, we divide this equation by the coefficient $a_{n,n}$. We obtain $x_n = b' - c_1 x_1 - \cdots - c_{n-1}x_{n-1}$, with $b' = b_n/a_{n,n}$ and $c_i = a_{n,i}/a_{n,n}$. This takes n divisions and $n - 1$ subtractions.

To substitute this into the ith equation, we change the coefficient of x_j from $a_{i,j}$ to $a_{i,j} - c_j a_{i,n}$. We also change the right side to $b_i - b'a_{i,n}$. For each such equation, we perform n multiplications and n subtractions. Thus we eliminate x_n and create the smaller system by using n^2 multiplicative operations and $(n - 1)n$ additive operations. Overall, the elimination phase takes $\sum_{i=1}^{n} i^2 + \sum_{i=1}^{n} i(i - 1) = 2n^3/3 + O(n^2)$ arithmetic operations.

Computing x_n using its expression in terms of the other variables takes $n - 1$ multiplications and $n - 1$ additions. Thus computing the solutions after the elimination phase takes $O(n^2)$ operations. The bulk of the work occurs in the elimination phase. When we sum the two contributions, the total number of operations remains $2n^3/3 + O(n^2)$. ∎

If the coefficients of the system are rational and the system has a unique solution, then the solution is rational; this follows by induction on the size of the system.

PROBABILITY AND EXPECTATION

Questions of probability can lead to counting problems difficult both to state and to solve. It has been remarked that probability theory is the area of mathematics in which an expert is most likely to blunder. One explanation may be the ambiguity of the language used in questions of chance and probability. In particular, the expression "at random" may have more than one interpretation.

9.15. Example. Consider the question "The Smiths have two children, and at least one is a boy; what is the probability that both are boys?" The correct answer depends on the procedure by which the information "At least one is a boy" is obtained. We assume that when we list the older child first, the four possibilities Boy-Boy, Boy-Girl, Girl-Boy, Girl-Girl are equally likely. Hence 1/3 of the families having at least one boy have two boys. On the other hand, the speaker may have encountered only the older child, noted that it was a boy, and said "at least one is a boy". If the information arose in this way, then the answer is 1/2. ∎

9.16. Example. *Bertrand's Paradox.* Joseph Louis Francois Bertrand (1822-1900) observed another ambiguity. Choose a chord of the unit circle at random; what is the probability that its length exceeds $\sqrt{3}$ (the length of the side of an inscribed equilateral triangle)? The answer depends on the meaning of "at random". We could place a "spinner" at the center and spin it twice to select two points on the circumference to be the endpoints of the chord. In this model, the probability that the chord length exceeds $\sqrt{3}$ is 1/3.

Alternatively, we could choose the midpoint of the chord by throwing a dart at the circle that is equally likely to land in regions of equal area. The midpoint of a chord uniquely determines the chord, and in this model the probability that the length exceeds $\sqrt{3}$ is 1/4. Other reasonable models yield other values for the probability (Exercise 9.21). ∎

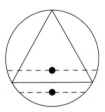

These examples illustrate that the study of probability requires precise language and clearly defined models for experiments.

9.17. Solution. *Bertrand's Ballot Problem.* Bertrand also posed and solved the Ballot Problem. Candidates A and B receive a and b votes, respectively, and we assume that $a \geq b$. We ask for the probability that A never trails, given that the votes are counted in random order. This could mean that the ballot box contains $a + b$ slips of paper, which can be removed from the box in $(a + b)!$ equally likely orders. Alternatively, it could mean that the lists of who received the ith vote are equally likely. A list such as *ABAABAB* with final score (a, b) is determined by the positions of the A's, which can be chosen in $\binom{a+b}{a}$ ways, so there are $\binom{a+b}{a}$ lists with final score (a, b).

In this problem, the two models give the same answer. Consider the model with $(a + b)!$ outcomes. Changing the order of paper slips for one candidate does not change whether the candidate ever trails. Hence the $a!b!$ orderings of the paper slips that correspond to the list *ABAABAB* give the same answer. Also, the events for distinct lists with the same final score are equally likely. Under either model, then, it suffices to count the lists in which A never trails and divide by $\binom{a+b}{a}$.

We use *election* to mean a list of A's and B's. An election is *good* if A never trails; otherwise it is *bad*. To count the good elections with final score (a, b), we count the bad elections with final score (a, b) and subtract them from the total. An election is bad if there is a k such that the score reaches $(k, k + 1)$. The minimal such k is the first time A trails. Modify the election after this time by changing every A to a B and every B to an A. Now A gets $b - k - 1$ additional votes and B gets $a - k$ additional votes, so the final score of the new election is $(b - 1, a + 1)$.

Every election with final score $(b - 1, a + 1)$ is won by B, since $a \geq b$. Therefore, in such an election there is a least k when the score is $(k, k + 1)$. Switching the votes after this point as done previously generates an election with final score (a, b). The second map is the inverse of the first, and this establishes a bijection between the set of bad elections with final score (a, b) and the set of all elections with final score $(b - 1, a + 1)$. Hence there are $\binom{a+b}{a+1}$ bad elections. To obtain the probability of the good elections, we compute

$$\frac{\binom{a+b}{a} - \binom{a+b}{a+1}}{\binom{a+b}{a}} = 1 - \frac{b}{a+1} = \frac{a-b+1}{a+1}. \quad \blacksquare$$

9.18. Remark. *Lattice paths and Catalan numbers.* The switching argument in Solution 9.17 is due to Antoine Désiré André (1840-1917). Graphing the successive vote totals as points in the plane yields a lattice path to (a, b). The path never steps above the diagonal if and only if

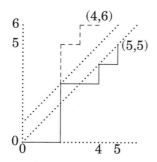

candidate A never trails in the election. We can translate the switching argument into the language of lattice paths; the bijection maps the bad elections into lattice paths reaching $(b - 1, a + 1)$, via a reflection of the portion after $(k, k + 1)$ through the line with equation $y = x + 1$.

In the special case where $a = b = n$, the number of good paths is $\frac{1}{n+1}\binom{2n}{n}$. These numbers are known as the *Catalan numbers*; we shall see that they provide the solution to many counting problems (Exercises 9.24-9.26, Problem 12.4, Exercises 12.38-12.41). ∎

After we have defined an experiment and associated a probability with each outcome of the experiment, we can proceed with a precise mathematical development. At this stage, we restrict our development to experiments with finitely many outcomes, although some of the exercises allow infinitely many outcomes.

9.19. Definition. A *finite sample space* is a finite set S and a function $P: S \to [0, 1]$ such that $\Sigma_{a \in S} P(a) = 1$. We think of S as the set of possible outcomes of an experiment, with $P(a)$ denoting the probability of the outcome a. An *event* A is a subset of S. We write $\text{Prob}(A) = \Sigma_{a \in A} P(a)$ and call this the *probability* of the event A.

Two events are *mutually exclusive* if they are disjoint subsets of S; in this case $\text{Prob}(A \cup B) = \text{Prob}(A) + \text{Prob}(B)$. Events A and B are *independent* if $\text{Prob}(A \cap B) = \text{Prob}(A) \cdot \text{Prob}(B)$. The *conditional probability* $\text{Prob}(A|B)$, read as "the probability of A given B", equals $\text{Prob}(A \cap B)/\text{Prob}(B)$, defined whenever $\text{Prob}(B) \neq 0$.

When the outcomes in the sample space are equally likely, $\text{Prob} A = |A|/|S|$ for each event A. The rules of sum and product then apply to the probabilities of mutually exclusive events and independent events. Even when sample points are not equally likely, the probability of mutually exclusive events is the sum of their probabilities, and the probability of independent events is the product of their probabilities. When A and B are independent,

$$\text{Prob}(A|B) = \text{Prob}(A \cap B)/\text{Prob}(B) = \frac{\text{Prob}(A)\text{Prob}(B)}{\text{Prob}(B)} \text{Prob}(A).$$

This justifies the term "independent"; the probability that A occurs is unaffected by knowing whether B occurs.

Conditional probability explains the confusion in Example 9.15. By stating that "at least one [child] is a boy", we are using conditional probability. The answer depends on whether the given event is "the first child is a boy" or "the children are not both girls".

The next example underscores the distinction between "the fraction of students in each year that are math majors" and "the fraction of math majors that are in each year". We use it to motivate a general formula about conditional probability.

9.20. Example. Suppose all college students are freshman, sophomores, juniors, or seniors. If we know the fractions of students in each year that are math majors, can we determine the fraction of math majors that are senior? The answer is yes if we also know the number of students in each year.

Suppose the fractions of students that are math majors are 1/3, 1/4, 1/5, 1/6 in the succeeding years of school. Suppose also that the populations in the four years are 1500, 1400, 1250, 1200, respectively. We can now compute the populations inside the table below. There are 1300 math majors altogether, and thus the fraction of math majors that are seniors is 2/13. ∎

	Fr	So	Jr	Sr
Math	500	350	250	200
Other	1000	1050	1000	1000
Total	1500	1400	1250	1200

9.21. Remark. *Bayes' Formula.* The method in Example 9.20 generalizes to a common situation involving conditional probability. Consider events A and $\{B_i\}$, where $\{B_i\}$ are mutually exclusive events with probabilities that sum to 1. We are interested in the conditional probability $P(B_i|A)$, but instead we are given $\text{Prob}(A|B_i)$. We can put Example 9.20 in this framework by viewing A as the event of being a math major and B_i as the event of being in the ith year. If $\text{Prob}(A|B_i) = a_i$ and $\text{Prob}(B_i) = b_i$, then we can compute

$$\text{Prob}(B_i|A) = \frac{\text{Prob}(B_i \cap A)}{\text{Prob}(A)} = \frac{\text{Prob}(A \cap B_i)}{\Sigma_j \text{Prob}(A \cap B_j)} = \frac{\text{Prob}(A|B_i)\text{Prob}(B_i)}{\Sigma_j \text{Prob}(A|B_j)\text{Prob}(B_j)} = \frac{a_i b_i}{\Sigma a_j b_j} \qquad ∎$$

We often associate a numerical value with each outcome of an experiment. The value of the outcome is then a "random variable". We can also study the "expected" value of this variable. Intuitively, this corresponds to averaging the values over many trials. When the outcomes have different probabilities, we take a weighted average, using the probabilities as weights.

9.22. Definition. Given a sample space S, a *random variable* is a function $X: S \to \mathbb{R}$; it assigns to each element of S a numerical value. We write $X = k$ to denote the event that the value of X is k. The *expectation* or *expected value* of X, written $E(X)$, is $\Sigma_{a \in S} X(a)\text{Prob}(a)$. By grouping together the sample points for each event $X = k$, we can also write $E(X) = \Sigma_k k \cdot \text{Prob}(X = k)$.

9.23. Example. *Average grade.* We select a student at random from a class of n students, with each student having probability $1/n$ of being chosen. Suppose that the numbers of students receiving grades A, B, C, D, E are a, b, c, d, e, respectively, where $a + b + c + d + e = n$. Let X be the random variable whose value is the numerical value of the chosen student's letter grade. The average grade of the students is $(4a + 3b + 2c + 1d + 0e)/n$, which is the expectation $E(X)$. ∎

9.24. Solution. *The binomial distribution.* Suppose we flip a coin n times, where the probability of heads is p on each toss. The sample space is the Cartesian product $\{H, T\}^n$, the set of lists of length n formed from $\{H, T\}$. In Bernoulli trials, no flip affects the outcome of any other flip; the events H_i and T_j that correspond to a head on the ith flip and a tail on the jth flip are independent if $i \neq j$. Thus the probability of a particular list with k heads and $n - k$ tails is $p^k(1 - p)^{n-k}$.

Let X be the number of heads; this is a random variable with the possible values $0, \ldots, n$. The probability of the event $X = k$ is $\binom{n}{k}p^k(1 - p)^{n-k}$, since there are $\binom{n}{k}$ arrangements with k heads, each having probability $p^k(1 - p)^{n-k}$. To compute the expectation, we have

$$E(X) = \Sigma_{k=0}^n k\binom{n}{k}p^k(1 - p)^{n-k} = \Sigma_{k=1}^n n\binom{n-1}{k-1}p^k(1 - p)^{n-k}$$

$$= np\Sigma_{k=1}^n \binom{n-1}{k-1}p^{k-1}(1 - p)^{(n-1)-(k-1)} = np[p + (1 - p)]^{n-1} = np$$

Here we dropped the term for $k = 0$ (it equals 0), used Lemma 9.8 to extract the factor of n, and applied the Binomial Theorem. ∎

There is a simpler way to compute $E(X)$ here. It uses a fundamental and intuitive property of expected value. For example, the expected total number of newspapers sold daily at all the newstands in New York is the sum of the expected number sold at each newsstand. We can compute the total sales for a year in two ways. We can sum the sales for each day, or we can sum the sales for each newsstand. We then divide by the number of days to obtain the expectation.

9.25. Proposition. (Linearity of expectation). If $X_1, \ldots, X_n$ are random variables defined on a sample space S, and X is the random variable ΣX_i, then $E(X) = \Sigma E(X_i)$.

Proof. Using the definition of expectation, the distributive law, and interchanging the order of summation,

$$E(X) = \Sigma_{a \in S} X(a)\text{Prob}(a) = \Sigma_{a \in S}[\Sigma_{i=1}^n X_i(a)]\text{Prob}(a)$$

$$= \Sigma_{i=1}^n [\Sigma_{a \in S} X_i(a)\text{Prob}(a)] = \Sigma_{i=1}^n E(X_i) \quad ∎$$

9.26. Solution. *The binomial distribution, revisited.* When we perform n independent trials with success probability p, we can define a variable X_i for the ith trial, with $X_i = 1$ if this trial is a success, and $X_i = 0$

if it is not. Letting X be the total number of successes, we have $X = \Sigma X_i$. Each X_i has expectation $E(X_i) = 1 \cdot p + 0 \cdot (1 - p) = p$. By the linearity of the expectation, $E(X) = \Sigma E(X_i) = np$. ∎

The conclusion in Solution 9.26 does not require independence for the various events H_i of getting heads on the ith trial. This simpler computation thus gives a stronger result than the computation in Solution 9.24, because there the formula for the probabilities of the sample points depends on independence of the trials.

The random variables X_i in Solution 9.26 are called *indicator variables* because their value (0 or 1) indicates whether a particular event happens. Their use often simplifies the computation of expectation.

9.27. Application. Suppose A, B, and n other people line up in random order. What is the expected number of people between A and B?

For each i, let $X_i = 1$ if the ith person stands between A and B, and let $X_i = 0$ otherwise. The expected number of people between A and B is then $E(X)$, where $X = \Sigma X_i$. Since $E(X_i) = 1/3$ for each i, we have $E(X) = \Sigma E(X_i) = n/3$. ∎

MULTINOMIAL COEFFICIENTS

Many counting problems that involve two options generalize naturally to questions about m options, where $m \in \mathbb{N}$. In the Bertrand Ballot problem, there are $\binom{a+b}{b}$ elections that reach the final score (a, b). How does this generalize when we have m candidates and the final score is $(a_1, \ldots, a_m)$? We have seen that the binomial coefficient $\binom{n}{k}$ counts n-tuples consisting of k 1's and $n - k$ 0's (or k A's and $n - k$ B's). We have also seen that it is the coefficient of $x^k y^{n-k}$ in the expansion of $(x + y)^k$. We generalize these questions to m candidates or m types of letters or polynomials with m variables.

9.28. Definition. Suppose $k_1, \ldots, k_m$ are nonnegative integers summing to n. The *multinomial coefficient*, written $\binom{n}{k_1, \ldots, k_m}$, is the number of ways to arrange n objects of m types in a row, where there are k_i objects of type i.

The binomial coefficient counts arrangements of two types of objects. Suppose we have three types of objects. If our objects are a, b, b, c, c, then there are six arrangements with the a in each of five possible locations, because we choose the positions for the two b's from the remaining four positions to determine the arrangement. Hence $\binom{5}{1, 2, 2} = 30$. Generalizing this, we can choose positions for the first type of object, then positions for the second type, and so on, to obtain the formula $\binom{n}{k_1,\ldots,k_m} = \binom{n}{k_1}\binom{n-k_1}{k_2}\binom{n-k_1-k_2}{k_3}\cdots$. In Theorem 9.29, we provide a more direct argument for a simplification of this formula.

9.29. Theorem. If $k_1,\ldots,k_m$ are nonnegative integers summing to n, then $\binom{n}{k_1,\ldots,k_m} = \dfrac{n!}{k_1!\cdots k_m!}$.

Proof. Let M be the number of arrangements consisting of k_i letters of the ith type, for each i. We can turn such an arrangement into an arrangement of distinct objects by putting labels (e.g., subscripts) on the k_i letters of type i, for each i. For each i in a particular arrangement, we can assign the labels in $k_i!$ ways. Hence in total we have formed $M\Pi_{i=1}^m k_i!$ arrangements of n distinct letters. Since we have made the letters distinct, this must equal $n!$, the total number of arrangements of n distinct letters. Hence $M = n!/\Pi_{i=1}^m k_i!$. ∎

9.30. Example. Roll a balanced six-sided die 21 times. What is the probability of rolling exactly one 1, exactly two twos, and so on, up to exactly six sixes? Answer: $\binom{21}{1,2,3,4,5,6}(1/6)^{21} = 0.0000935969$. ∎

The name "multinomial coefficient" arises from the expansion of polynomials with several variables.

9.31. Corollary. The number $\binom{n}{k_1,\ldots,k_m}$ is the coefficient of $x_1^{k_1}\cdots x_m^{k_m}$ in the expansion of $(x_1+\cdots+x_m)^n$.

Proof. The monomial $x_1^{k_1}\cdots\cdot x_m^{k_m}$ arises once in the expansion of $(x_1+\cdots+x_m)^n$ for each way to arrange a set consisting of k_i copies of x_i for each i. Each such arrangement corresponds to a term in the expansion of the product. The jth position in the arrangement corresponds to the term chosen from the jth factor in $(x_1+\cdots+x_m)\cdots(x_1+\cdots+x_m)$. ∎

9.32. Example. *Trinomial expansion.*
$$(x+y+z)^3 = x^3 + y^3 + z^3 + 3x^2y + 3x^2z + 3y^2z + 3y^2x + 3z^2x + 3z^2y + 6xyz.$$ ∎

9.33. Corollary. If p is prime and $\Sigma_{i=1}^m k_i = p$ with $0 \le k_i < p$, then p divides $\binom{p}{k_1,\ldots,k_m}$.

Proof. Because the multinomial coefficient is the size of a finite set, Theorem 9.29 implies that $M = p!/\Pi_{i=1}^m k_i!$ is an integer. Writing this as $p! = M\Pi_{i=1}^m k_i!$, we observe that the left side is divisible by p. The factorials on the right side do not have p as a factor. Since p is prime, this implies that p divides M. ∎

Corollary 9.33 yields a remarkably short proof of Fermat's Little Theorem, due to Gottfried Wilhelm Leibniz (1646-1716). Exercise 6.36 requests a related proof using the binomial theorem.

9.34. Example. *Fermat's Little Theorem.* To prove that $a^{p-1} \equiv 1 \bmod p$ when p is prime and a is an integer not divisible by p, we prove that $a^p \equiv a \bmod p$. Modular computation allows us to assume that a is positive. Expressing a as $\Sigma_{i=1}^a 1$, we consider the expansion of $(1 + \cdots + 1)^p$, which we treat as $(x_1 + \cdots + x_a)^p$ with each x_i equal to 1. By Corollary 9.31, the coefficient of $x_1^{k_1} \cdots x_m^{k_m}$ is $\binom{p}{k_1, \ldots, k_m}$. For each term x_i^p, in which the exponents on the variables other than x_i are 0, the coefficient equals 1; there are a of these. By Corollary 9.33, all the other coefficients are divisible by p. Hence $a^p = (1 + \cdots + 1)^p \equiv a \bmod p$. ∎

9.35. Remark. *Combinatorial proofs.* 1) Writing $m \in \mathbb{N}$ as $\Sigma_{i=1}^m 1$ leads to combinatorial interpretations. Consider $m^2 = m(m-1) + m$. Algebraically, we apply the distributive law. Geometrically, we partition an m-by-m square; deleting an m-by-1 strip leaves an m-by-$(m-1)$ rectangle. Combinatorially, we count the contributions to $m^2 = (\Sigma_{i=1}^m 1)^2$ in two ways. For easier record-keeping, we consider $(\Sigma_{i=1}^m x_i)^2$ and later set each $x_i = 1$. Squaring the sum yields m terms of the form x_i^2 and $\binom{m}{2}$ terms of the form $2x_i x_j$; thus $m^2 = 2\binom{m}{2} + m$.

2) In order to prove that a ratio is an integer, it suffices to prove that it counts some finite set. For example, when we partition a set S into sets of size k, the ratio $|S|/k$ counts the sets in the partition. We used this to prove Fermat's Little Theorem (Theorem 7.26). Similarly, the argument in Theorem 5.26 to compute $\binom{n}{k}$ can be interpreted as a combinatorial argument that the ratio $n!/[k!(n-k)!]$ is an integer.

The set of n-tuples formed from a set B of size k has size k^n. Thus, when we seek a combinatorial proof of an algebraic statement involving k^n, it is natural to use the set B^k. Example 9.34 illustrates this. ∎

9.36. Example. $k-1$ *divides* $k^n - 1$. Modular arithmetic provides the shortest proof of this; since $k \equiv 1 \bmod (k-1)$, also $k^n \equiv 1 \bmod (k-1)$. We have also considered an inductive proof and an algebraic proof using the geometric sum $(k^n - 1 = (k-1)\Sigma_{i=0}^{n-1} k^i)$. Now we present a combinatorial proof (see also Exercise 9.18).

There are k^n n-tuples with entries in $B = \{0, \ldots, k-1\}$, one of which is all 0; let $S = B^k - \{(0, 0, \ldots, 0)\}$. Each n-tuple in S has a leftmost nonzero value. Let A_i consist of the n-tuples in S whose leftmost nonzero value is i. If $|A_i| = |A_j|$ when $i \neq j$, then we have partitioned S into $k-1$ sets of equal size, which proves that $k-1$ divides $|S| = k^n - 1$. To prove $|A_i| = |A_j|$, we define a bijection from A_i to A_j by changing the leftmost nonzero element of each n-tuple in A_i from i to j. Below the sequences for $(k, n) = (5, 2)$ are listed by these classes, grouping the corresponding elements under this bijection in the same column. ∎

$$
\begin{array}{ccccccc}
A_1 & 01 & 10 & 11 & 12 & 13 & 14 \\
A_2 & 02 & 20 & 21 & 22 & 23 & 24 \\
A_3 & 03 & 30 & 31 & 32 & 33 & 34 \\
A_4 & 04 & 40 & 41 & 42 & 43 & 44
\end{array}
$$

9.37. Proposition. (The multinomial distribution) Suppose an experiment has m possible outcomes, with p_j being the probability of the jth outcome and $\Sigma_{j=1}^m p_j = 1$. If we perform n independent trials, then the probability that for each j the jth outcome occurs exactly k_j times is $\binom{n}{k_1, \ldots, k_m} p_1^{k_1} \cdots p_m^{k_m}$.

Proof. There are m^n possible lists of outcomes for n successive trials. Since the trials are independent, the probability of each particular list in which the jth outcome occurs precisely k_j times is $\Pi_{j=1}^m p_j^{k_j}$. The number of lists of this type is the number of ways to arrange these outcomes (k_j of type j for all j) in a row, which equals the multinomial coefficient $\binom{n}{k_1, \ldots, k_m}$. The different lists are mutually exclusive events, so the probability is the number of lists of this type times the probability of each one. ∎

9.38. Solution. *Hitting for the cycle.* Our baseball player bats randomly, meaning each at-bat is an independent trial. The result is a single with probability .15, a double with probability .06, a triple with probability .02, a home run with probability .07, and otherwise an out. This describes a good hitter, whose batting average is .300 and "slugging average" (expected number of bases per at-bat) is .610. "Hitting for the cycle" means getting at least one hit of each of the four types in a single game. What is the probability that this player hits for the cycle if he bats exactly 5 times in a game?

There are 5 ways this can occur. There can be one of each hit and one out, or two of one type of hit and one each of the three others. We use the multinomial distribution to compute each probability below:

one of each hit, one out: $5!(.15)(.06)(.02)(.07)(.70) = 0.0010584$.
two singles: $(5!/2)(.15)^2(.06)(.02)(.07) = 0.0001134$.
two doubles: $(5!/2)(.15)(.06)^2(.02)(.07) = 0.00004536$.
two triples: $(5!/2)(.15)(.06)(.02)^2(.07) = 0.00001512$.
two home runs: $(5!/2)(.15)(.06)(.02)(.07)^2 = 0.00005292$.

These events are mutually exclusive, so we add the probabilities, obtaining the answer 0.0012852, which is about 1 in 800. This overestimates the probability that this player hits for the cycle in a given game, due to many factors we have not considered. ∎

GENERATING FUNCTIONS (Optional)

Many counting problems can be solved with a technique combining combinatorics and algebra. In proving that $(1 + x)^r = \Sigma_{n=0}^r \binom{r}{n} x^n$, we argue combinatorially that the coefficient of x^n must be $\binom{r}{n}$. This assigns no value to x; we view x as a formal symbol, not as a number.

More generally, consider a sequence $\langle a \rangle = (a_0, a_1, a_2, \cdots)$. We think of a_n as the solution to a counting problem with n as a parameter. We associate the expression $a_0 x^0 + a_1 x^1 + a_2 x^2 + \cdots$ with the sequence $\langle a \rangle$. In this expression, we do not treat x^n as a number; it is merely a place-holder for the term a_n in the sequence. In Chapter 14 we consider the issues of convergence that arise when x is treated as a number; then the expression is called an "infinite series". Here we treat x as a formal variable, and convergence of infinite series is irrelevant.

9.39. Definition. A *formal power series* is an expression of the form $\Sigma_{n=0}^\infty a_n x^n$ in which x is treated as a formal variable, not as a number. The formal power series $\Sigma_{n=0}^\infty a_n x^n$ is the *generating function* for the sequence $\langle a \rangle$. Given two formal power series $\Sigma_{n=0}^\infty a_n x^n$ and $\Sigma_{n=0}^\infty b_n x^n$, their *sum* is the formal power series in which the coefficient of x^n is $a_n + b_n$, and their *product* is the formal power series in which the coefficient of x^n is $\Sigma_{j=0}^n a_j b_{n-j}$.

Generating functions arise in the theory of probability. When a_n is the probability that a nonnegative integer-valued random variable X has the value n, the generating function $\Sigma a_n x^n$ is called the *probability generating function* of X (see Exercise 16.50).

By the definition, two formal power series are equal if and only if they have the same sequence of coefficients. The definitions of sum and product of formal power series agree with our experience in multiplying polynomials. These definitions allow us to build generating functions to solve counting problems; we want the coefficient of x^n to be is the number of solutions when the value of the parameter equals n. Consider subsets of an r-element set. When $r = 1$, there is one way to choose the element and one way not to choose it, so the generating function is $1 + x$. To build the generating function for a fixed r, we take r such factors, because we make such a choice for each element. The contributions to the coefficient of x^n correspond to the subsets of size n. This principle applies more generally.

9.40. Lemma. Consider sets A, B, C with $C = A \times B$. For $n \in \mathbb{N} \cup \{0\}$, let a_n, b_n, c_n be the number of elements of "size" n in A, B, C, respectively. If the size of each $\gamma = (\alpha, \beta) \in C$ with $\alpha \in A$ and $\beta \in B$ is the sum of the sizes of α and β, then the generating function for $\langle c \rangle$ is the product of the generating functions for $\langle a \rangle$ and $\langle b \rangle$.

Proof. Choose $\gamma = (\alpha, \beta)$ with $\alpha \in A$ and $\beta \in B$. If z has size n, then for some k the sizes of α and β are k and $n - k$. Any element of A having size k can be paired with any element of B having size $n - k$, since the coordinates of elements in a Cartesian product are chosen independently. The subsets using a particular size from A are disjoint, so we sum over the possible values of k to obtain $c_n = \Sigma_{k=0}^n a_k b_{n-k}$. Hence the generating functions satisfy the definition of product. ∎

9.41. Application. *Selections with repetition.* Suppose we have r types of objects, and we want to select objects, with no restriction on the number of objects chosen of each type. How many we select of one type is independent of how many we select of any other type. Hence the overall set of selections is an r-fold Cartesian product of the sets of selections involving each type individually. The generating function is the formal power series in which the coefficient of x^n is the number of ways to do this and obtain a total of n elements. Using Lemma 9.40, this generating function is the product of r generating functions for the situation involving one type of object. When we have only one type of object, there is one way to choose n copies of it, so the generating function in this case is $\Sigma_{n=0}^\infty x^n$. We conclude that the generating function for selections of n objects from r types of objects is $(\Sigma_{n=0}^\infty x^n)^r$.

   We solved this problem in Theorem 5.31 by bijective arguments. We know that the number of ways to select n objects is $\binom{n+r-1}{r-1}$, and by definition this is the coefficient of x^n in the generating function. Hence we have given a combinatorial proof of an algebraic identity about formal power series: $(\Sigma_{n=0}^\infty x^n)^r = \Sigma_{n=0}^\infty \binom{n+r-1}{r-1} x^n$. ∎

   We can view a polynomial as a formal power series with finitely many nonzero coefficients. We proved in Corollary 4.19 that polynomials are equal as functions if and only if they have the same coefficients. For formal power series this is the *definition* of equality, because we do not treat the formal variable as a number. The definition of multiplication for formal power series agrees with the definition of multiplication for polynomials when the power series have finitely many terms. The formal power series 1 (the expression $1x^0 + 0x^1 + 0x^2 + \cdots$) is a multiplicative identity. Hence we may study multiplicative inverses.

9.42. Theorem. For $r \in \mathbb{N}$, the formal power series expansion of the generating function $(1 - x)^{-r}$ is $\Sigma_{n=0}^\infty \binom{n+r-1}{r-1} x^n$.

Proof. When we multiply the two formal power series $1 - x$ and $\Sigma_{n=0}^\infty x^n$, we obtain 1. Hence we write $\Sigma_{n=0}^\infty x^n = (1-x)^{-1}$ as formal power series. When we raise the series to the rth power, we obtain $\Sigma_{n=0}^\infty \binom{n+r-1}{r-1} x^n$ by Application 9.41. ∎

The special case $r = 1$ is the formal sum of the geometric series: $(1-x)^{-1} = \sum_{n=0}^{\infty} x^n$. Exercises 9.40-9.47 involve generating functions for combinatorial problems. In Chapter 15, we consider convergence questions for power series; this allows us to treat (convergent) power series as functions, thus justifying the name "generating *function*".

EXERCISES

9.1. Use Pascal's Formula to prove by induction on n that $\binom{n}{k} = \dfrac{n!}{k!(n-k)!}$.

9.2. Prove by induction that $\sum_{i=0}^{n} \binom{i}{k} = \binom{n+1}{k+1}$ for integers $k, n \geq 0$.

9.3. Use Pascal's Formula to prove the binomial theorem by induction on n.

9.4. Count the ways to choose distinct subsets $A_0, A_1, \ldots, A_n$ of $[n]$ such that $A_0 \subset A_1 \subset \cdots \subset A_n$. What happens if repetitions are allowed in the list of $n+1$ sets?

9.5. By grouping a set of n^2 dots in a different way, give a combinatorial proof that $n^2 = 2\binom{n}{2} + n$.

9.6. By counting a set in two ways, prove that $\binom{k}{l}\binom{n}{k} = \binom{n}{l}\binom{n-l}{k-l}$.

9.7. *Summing the cubes.*
  a) Prove directly that $m^3 = 6\binom{m}{3} + 6\binom{m}{2} + m$.
  b) Use part (a) to prove that $\sum_{i=1}^{n} i^3 = (\dfrac{n(n+1)}{2})^2$ (without using induction).
  c) Prove part (a) by counting a set in two ways. (Hint: consider the expansion of $(\sum_{i=1}^{m} x_i)^3$.)

9.8. Give three proofs that $n^3 + 5n$ is divisible by 6 for every $n \in \mathbb{N}$.
  a) Use induction.
  b) Use modular arithmetic.
  c) Use an expression for $n^3 + 5n$ in terms of binomial coefficients.

9.9. Let S_n be the hexagonal arrangement consisting of n rings of dots, as illustrated below for $n \in \{1, 2, 3\}$. Let a_n be the number of dots in S_n. Compute a_n. Compute $\sum_{k=1}^{n} a_k$. (Hint: use the Summation Identity.)

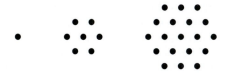

9.10. By counting a set in two ways, prove that $\sum_{i=0}^{k} \binom{m}{i}\binom{n}{k-i} = \binom{m+n}{k}$.

9.11. By counting a set in two ways, prove that $\sum_{i=-m}^{n} \binom{m+i}{r}\binom{n-i}{s} = \binom{m+n+1}{r+s+1}$.

9.12. By counting a set (of selections with repetition) in two ways, prove that
$$\sum_{i=0}^{k} \binom{m+k-i-1}{k-i}\binom{n+i-1}{i} = \binom{m+n+k-1}{k}.$$

9.13. By counting a set in two ways, prove that $\Sigma_{A \subseteq [n]} \Sigma_{B \subseteq [n]} |A \cap B| = n4^{n-1}$. (Hint: consider the set of triples (x, A, B) such that $A, B \subseteq [n]$ and $x \in A \cap B$; count this set in two ways.)

9.14. By counting an appropriate geometric arrangement of points, prove that $2\Sigma_{i=1}^{q-1} \lfloor ip/q \rfloor = (p-1)(q-1)$ if p and q are relatively prime.

9.15. Evaluate $\Sigma_{S \subseteq [n]} \Pi_{i \in S} 1/i$.

9.16. We wish to form a necklace by placing distinguishable beads (numbered 1 through n) on a circular string. Each such arrangement can be rotated or flipped; two necklaces are indistinguishable if one can be rotated or flipped so that it looks like the other. Prove that indistinguishability is an equivalence relation. Count the equivalence classes with n beads. (The beads are distinguished by their labels, so there is no problem of periodicity.)

9.17. Suppose n is prime, and there are k types of feathers available, each in unlimited supply. We wish to place one feather at each corner of a k-cornered hat. Each such arrangement can rotate; unlike necklaces, hats cannot be worn upside-down. Two arrangements of feathers are indistinguishable if one can be rotated to look like the other. Prove that indistinguishability is an equivalence relation. Count the equivalence classes of hats that can be formed using a total of n feathers from the k types.

9.18. Consider a track meet with k^n contestants. In each round, the remaining contestants are placed in groups of size k to run. The winner in each group advances to the next round.

   a) Use this to give another combinatorial proof that $k-1$ divides $k^n - 1$.

   b) How many races are run in the entire competition?

9.19. Consider a dial having a pointer that is equally likely to point to each of n regions numbered $1, 2, \ldots, n$. When we spin the dial three times, what is the probability that the sum of the selected numbers is n?

9.20. Suppose we roll two dice, one red and one green. Under each assumption below, what is the probability that the roll is double-sixes?

   a) The red die shows a six.

   b) At least one of the dice shows a six. Does the method of obtaining this information affect the answer?

9.21. (+) *Bertrand's Paradox.* In Example 9.16 (generating a random chord of the unit circle), let p be the probability that the length of the chord exceeds $\sqrt{3}$.

   a) Suppose the endpoints of the chord are generated by two random spins on the circumference of the circle. Prove that $p = 1/3$. (Assume that spinner points to an arc with probability proportional to the length of the arc.)

   b) Suppose the midpoint of the chord is generated by throwing a dart at the circle. Prove that $p = 1/4$. (Assume that the probability the dart lands in a region is proportional to the area of the region.)

   c) Devise a model for generating the chord that yields $p = 1/2$.

9.22. In Bertrand's Ballot Problem (Problem 9.2), suppose the outcome is (a, b), with $a > b$, and the votes are counted in random order. What is the probability that A is always ahead of B? What is the probability that the score is tied at some point during the election after the beginning?

9.23. (+) Suppose m 0's and n 1's are placed in some order around a circle. A position is *good* if every arc of the circle starting at that position and moving clockwise contains more 0's than 1's.

a) Prove that no matter how the elements are arranged, there are exactly $m - n$ good positions. (Comment: This is a special case of the Cycle Lemma of Dvoretzky and Motzkin [1947].)

b) Apply part (a) to solve Bertrand's Ballot Problem.

9.24. Let A be the set of lattice paths from $(0,0)$ to (n,n) that do not move above the line given by $y = x$. Let B be the set of nondecreasing functions $f: [n] \to [n]$ such that $f(i) \le i$ for all i. Establish a bijection from A to B.

9.25. A *ballot list* of length $2n$ is a binary $2n$-tuple $(b_1, \ldots, b_{2n})$ such that for each i, the number of 1's in $\{b_1, \ldots, b_i\}$ is at least as large as the number of 0's. In the language of Solution 9.17, ballot lists are equivalent to "good elections" with total score (n, n). Establish bijections from the set of ballot lists of length $2n$ to each of the sets below.

a) $2n + 1$-tuples of nonnegative integers in which consecutive entries differ by 1 and $a_1 = a_{2n+1} = 0$.

b) Arrangements of $2n$ people in 2 rows of length n so that heights are increasing in each row and column.

9.26. (+) Place $2n$ points on the boundary of circle. Establish a bijection to prove that the number of ways to pair up the points by drawing noncrossing chords equals the number of ballot lists of length $2n$.

9.27. The fraction of the games that a tennis player wins against each of her four opponents is $.6, .5, .45, .4$, respectively. Suppose that she plays 30 matches against each of the first two and 20 matches against each of the last two. Given that she wins a particular match, what is the conditional probability that it is against the ith opponent, for $i \in \{1, 2, 3, 4\}$.

9.28. Players A and B alternate flipping a coin, with A flipping first. The first player to obtain "heads" wins. Suppose the probability of heads is p on each flip. Determine x, the probability that A wins, as a function of p. Evaluate the formula in the special case of a fair coin, $p = .5$. (Hint: obtain a linear equation that x must satisfy.)

9.29. Consider a dial having a pointer that is equally likely to point to each of n regions numbered $1, 2, \ldots, n$ in cyclic order. When the selection is k, the gambler receives 2^k dollars.

a) What is the expected payoff per spin of the dial?

b) Suppose that the gambler has the following option. After the pointer comes to rest, the gambler can accept that payoff or flip a coin to change it. If the coin shows heads, the pointer moves one spot counterclockwise; if tails, it moves one spot clockwise. When should the gambler flip the coin? What is the expected payoff under the optimal strategy?

9.30. Suppose n envelopes contain the amounts $a_1, \ldots, a_n$ in dollars, where $a_1 \le a_2 \cdots \le a_n$. A gambler is presented two successive envelopes. He does not know which they are, but he knows that for $1 \le i \le n-1$ the probability is p_i that the envelopes contain a_i and a_{i+1} dollars. He chooses one of these two envelopes at random and opens it. He can then either keep that amount or switch to the other envelope. Suppose he sees a_k dollars in the open envelope. In terms of the data of the problem, determine whether he should switch.

9.31. Suppose X is a random variable that takes values only in $[n]$. Prove that $E(X) = \sum_{k=1}^{n} \text{Prob}(X \ge k)$.

9.32. (!) Consider an experiment in which a coin is flipped until heads first appears. Suppose the probability of heads is p on each flip. Prove that the expectation of the number of flips in the experiment is $1/p$. (Hint: Obtain a linear equation that the expectation must satisfy.)

9.33. (!) *The Coupon Collector Problem.* A restaurant gives one of n types of coupons with each meal, with each of the n types having equal probability. A diner receives a free meal after collecting one coupon of each type. Prove that the expectation of the number of meals a diner must purchase to get all the coupons is $n \sum_{i=1}^{n} 1/i$. (Hint: apply Exercise 9.31 and the linearity of expectation.)

9.34. Consider a set of $2n$ people.
  a) How many ways are there to partition $2n$ people into n pairs?
  b) Suppose that the set consists of n men and n women. If this set is partitioned at random into n pairs, what is the expected number of male-female couples in the resulting partition?

9.35. Suppose n distinct pairs of socks are put into the laundry, each sock having one mate. Afterwards, socks are drawn one by one, and whenever a pair is obtained it is folded. Determine the expected number of pairs of socks obtained among the first k socks drawn. (Hint: use the linearity of expectation.)

9.36. (!) Find one polynomial p such that $p(n) = 3^n$ for $n = 0, 1, 2, 3, 4$. (Hint: express 3^n as $(1+1+1)^n$ and use Theorem 9.29, letting $k_3 = n - k_1 - k_2$.)

9.37. Consider an experiment in which all the monomials in k variables with total degree n are equally likely to occur (0 is allowed as an exponent).
  a) Determine the probability that all k variables have positive exponent in the chosen monomial.
  b) For $(n, k) = (10, 4)$, determine the probability that the exponents are different. (Here 0 is allowed as an exponent.)

9.38. Suppose n and k are natural numbers. Prove that there is exactly one choice of integers $m_1, \ldots, m_k$ such that $0 \le m_1 < m_2 < \cdots < m_k$ and
$$n = \binom{m_1}{1} + \binom{m_2}{2} + \cdots + \binom{m_k}{k}.$$
(Hint: observe that $\binom{m}{k} = \sum_{i=1}^{k} \binom{m-i}{k+1-i}$. Comment: This is called the "k-nomial" expansion of n, by analogy with the q-ary expansion.)

9.39. (+) The goal of this problem is to determine which polynomials p with rational coefficients have the property that $p(n) \in \mathbf{Z}$ if $n \in \mathbf{Z}$. Let I be the set of polynomials with this property. Recall that the sum $p + q$ of two functions p, q

on a set S is the function h such that $h(x) = p(x) + q(x)$. Similarly, the scalar multiple $n \cdot p$ is the function h such that $h(x) = n \cdot p(x)$.

a) Show that if $p, q \in I$ and $n \in \mathbb{Z}$, then $p + q \in I$ and $n \cdot p \in I$.

b) Show that $p_j \in I$, where $p_j(x) = \binom{x}{j}$, and that $\Sigma_{j=0}^k n_j \binom{x}{j} \in I$ for $\{n_j\} \subseteq \mathbb{Z}$.

c) Suppose f is a polynomial of degree k with rational coefficients. Prove that f can be expressed as $f(x) = \Sigma_{j=0}^k b_j \binom{x}{j}$, where the b_j's are rational. (Hint: one way to prove this uses induction on the degree of the polynomial.)

d) Prove that $f \in I$ if and only if $f(x) = \Sigma_{j=0}^k b_j \binom{x}{j}$, where the b_j's are integers. (Hint: evaluate f at the integers in the set $\{0, \ldots, k\}$. Note that $\binom{0}{0} = 1$, by our convention that $0! = 1$.)

9.40. Use generating functions to obtain a formula for the number of jars of 12 American coins (five types) having between 2 and 6 coins of each type. (Hint: let a_n be the number of ways to do this selecting n coins instead of 13, and obtain the generating function for the sequence $\langle a \rangle$.

9.41. Use generating functions to prove that $\Sigma_{k=0}^n \binom{n}{k}^2 = \binom{2n}{n}$.

9.42. Use generating functions to solve Exercises 9.10-9.12.

9.43. Suppose that $b_n = \Sigma_{k=0}^n a_k$ and that $A(x)$ is the generating function for the sequence $\langle a \rangle$. Obtain the generating function for $\langle b \rangle$ in terms of $A(x)$.

9.44. Let a_n be the number of ways to express n as a sum of distinct natural numbers. Let b_n be the number of ways to express n as a sum of odd natural numbers written in nonincreasing order. Derive expressions for the generating functions of the sequences $\langle a \rangle$ and $\langle b \rangle$.

9.45. (+) Establish a bijection to prove that the numbers a_n and b_n defined in Exercise 9.42 are equal. (Hint: Proposition 4.23 states that every natural number has a unique expression as an odd number times a power of 2.)

9.46. The partitions of n are the nonincreasing lists of positive integers with sum n (see Exercise 5.34). Given $k \in \mathbb{N}$, let a_n be the number of partitions of n using integers of size at most k. Derive an expression for the generating function of $\langle a \rangle$.

9.47. Let a_n be the number of ways to select $r \in \mathbb{N}$, roll a six-sided die r times, and obtain a sum of n. Derive an expression for the generating function of $\langle a \rangle$.

Chapter 10

Two Principles of Counting

In this chapter, we study two proof techniques in discrete mathematics, the Pigeonhole Principle and the Inclusion-Exclusion Principle. We consider these together because both are fairly easy to prove but have elegant applications that may require some cleverness to discover. They also can make it possible to avoid lengthy analysis by cases. The Inclusion-Exclusion Principle is used to solve counting problems. The Pigeonhole Principle is a principle of counting in the sense that it considers cardinalities of sets, but its applications are to existence problems and extremal problems rather than to enumerative problems.

THE PIGEONHOLE PRINCIPLE

"Of three ordinary people, two must have the same sex."[†] The Pigeonhole Principle is also called the "Dirichlet drawer principle" in honor of Peter Gustav Lejeune-Dirichlet (1805-1859). It implies that extracting $n + 1$ shoes from a closet containing n pairs of shoes must produce a matched pair of shoes; they cannot all come from different pairs. We proved a version of the Pigeonhole Principle as early as Chapter 2: in any set of real numbers, some number must be at least as large as the average. We have made arguments already that could be phrased using the Pigeonhole Principle (see Exercise 5.11 and Theorem 6.18). The principle itself is elementary; the subtlety arises in the applications.

10.1. Theorem. (Pigeonhole Principle) Placing more than kn objects into n classes puts more than k objects into some class.

[†]This observation is attributed to Prof. D.J. Kleitman of the Massachusetts Institute of Technology.

Proof. We prove the contrapositive. If no class has more than k objects, then the total number of objects is at most kn. This uses the property, proved by induction in Proposition 4.9, that the n inequalities $m_i \leq k$ can be summed to obtain the inequality $\Sigma_{i=1}^{n} m_i \leq kn$. ∎

To apply the Pigeonhole Principle, we must determine what should play the role of the objects and what should play the role of the classes. Sometimes the Pigeonhole Principle pops up in proof by contradiction.

10.2. Example. *Existence of multiplicative inverses modulo p.* If a and p are relatively prime, then there exists some $b \in \{1, \ldots, p-1\}$ such that $ab \equiv 1 \bmod p$. Otherwise, $a, 2a, \ldots, (p-1)a$ fall into the $p-2$ nonzero congruence classes other than $\bar{1}$. By the Pigeonhole Principle, two fall in the same class. If ia and ja fall in the same class, then $ia \equiv ja \bmod p$ yields $p|(i-j)a$. Since a and p are relatively prime, this implies $p|(i-j)$ (by Proposition 6.13), which implies $i = j$ (since they are less than p), which is a contradiction. ∎

10.3. Example. *A society of friends.* Suppose that "being friends" is a symmetric relation. We prove that in any set S of people with $|S| \geq 2$, there must be two people that have the same number of friends in S. If $|S| = n$, then each person in S has between 0 and $n-1$ friends in S. We cannot have a person with 0 friends and a person with $n-1$ friends, however, because a person with $n-1$ friends is a friend of everyone else. Hence at most $n-1$ distinct numbers of friends arise among the n people, and some pair must have the same total. ∎

10.4. Example. *Midpoints between integer points.* Given five points in the plane with integer coordinates ("integer points"), the midpoint of the segment joining some pair of them also has integer coordinates. To prove this, we must understand when a midpoint has integer coordinates. The midpoint of the segment from (a, b) to (c, d) is $(\frac{a+c}{2}, \frac{b+d}{2})$. If this is an integer point, then a and c have the same parity (both odd or both even), and similarly b and d have the same parity. This suggests putting the integer points into four classes by the parity of their x and y coordinates: (odd,odd), (odd,even), (even,odd), and (even,even). With five points, we must have two in the same class, and then the segment joining them has an integer midpoint. With four points, we can have one in each class and avoid having an integer midpoint. ∎

10.5. Example. *Forcing divisible pairs.* If S is a set of $n+1$ numbers in $[2n]$, then S contains a pair of numbers such that one divides the other. This result is best possible in the following sense: the n numbers $\{n+1, n+2, \ldots, 2n\}$ do not contain such a pair. To apply the pigeonhole principle, we partition $[2n]$ into n classes such that for every pair of numbers in the same class, one divides the other. Recall that every natural number has a unique representation as an odd number times a power of two. For fixed k, the set $\{(2k-1)2^{j-1}: j \geq 1\}$ has the desired property; the smaller of any pair in this set divides the larger. Since there are only n odd numbers less than $2n$, we get the right number of classes. Explicitly, the kth class consists of those numbers in $\{2^{j-1}(2k-1): j \in \mathbb{N}\}$ that are at most $2n$. ∎

     The examples above could have been phrased as extremal problems: What is the largest number of integer points in the plane such that no segment joining two of them has an integer midpoint? What is the largest size of a subset of $[2n]$ such that no element divides another? The pigeonhole principle establishes a bound, and a construction shows that the bound is best possible. In such a problem, it does not suffice to present the construction and show that one cannot add one more element to it, because this does not prevent the existence of larger configurations constructed in other ways. For example, to build a large set that avoids divisible pairs in Example 10.5, it would be reasonable to choose primes. When $n = 5$, we pick the set $\{2, 3, 5, 7\}$ in this way, at which point we cannot add any more elements from $[10]$ without creating a divisible pair. That does not prove that the largest such set has size 4, and indeed $\{6, 7, 8, 9, 10\}$ is a larger example. To solve an extremal problem, our proof must show that all possible examples satisfy the bound.

10.6. Example. *Longest monotone sublist.* Consider a list of $n^2 + 1$ distinct numbers. A subset of the positions forms a *monotone sublist* if the numbers in those positions form an increasing list or a decreasing list when taken in order. For example, in the list 3,2,1,6,5,4,9,8,7,10, the numbers 3,6,9,10 form an increasing sublist of length 4. Erdős and Szekeres proved in 1935 that every list of $n^2 + 1$ distinct numbers contains a monotone sublist of length at least $n + 1$. Suppose $a_1, \ldots, a_{n^2+1}$ is the list. For each k, let x_k be the maximum length of an increasing sublist ending with a_k, and let y_k be the maximum length of a decreasing sublist ending with a_k. For the example above, the values of these parameters are:

k	1	2	3	4	5	6	7	8	9	10
a_k	3	2	1	6	5	4	9	8	7	10
x_k	1	1	1	2	2	2	3	3	3	4
y_k	1	2	3	1	2	3	1	2	3	1

If there is no monotone sublist of length $n + 1$, then x_k and y_k never exceed n, and there are only n^2 possible pairs (x_k, y_k). Since there are $n^2 + 1$ values of k, the pigeonhole principle implies that two of the pairs are the same. Suppose $(x_i, y_i) = (x_j, y_j)$ with $i < j$. If $a_i < a_j$, then $x_j > x_i$; if $a_i > a_j$, then $y_j > y_i$. This contradiction implies that a number exceeding n must appear in one of the pairs. Since there is a list of n^2 distinct numbers having no monotone sublist of length $n + 1$ (Exercise 10.17), the result is best possible. ∎

10.7. Example. *A domino tiling problem.* A six by six checkerboard with 36 squares can be covered exactly by 18 dominoes consisting of two squares each; this is a *tiling* of the checkerboard by dominoes. We prove that every such tiling can be cut between some pair of adjacent rows or adjacent columns without cutting any dominoes. In the picture below, the tiling can be cut along the middle horizontal line.

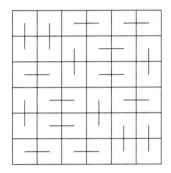

Consider an arbitrary tiling. Every domino cuts one line between two adjacent rows or between two adjacent columns. There are 18 dominoes and 10 possible lines to be cut, so the average number of cuts per line is 1.8. Since every set of numbers contains a number that is at most the average, this implies that some line is cut at most once. This is not strong enough to prove the claim, as it leaves the possibility that every line is cut at least once.

To complete the proof, we observe that every line is cut by an even number of dominoes; this implies that a line cut by at most one domino is not cut at all. The observation is easy: having an odd number of dominoes crossing a line would leave an odd number of squares on each side to be paired up by dominoes that don't cross the line, but each set of dominoes covers an even number of squares. ∎

10.8. Example. *The Chess Player Problem.* A chess player wants to practice for a championship match over a period of 11 weeks. She wants to play at least one game per day but at most 132 games altogether. No matter how she schedules the games, there must be a period of consecutive days on which she plays a total of *exactly* 22 games.

We can study the total played on consecutive days by considering partial sums. Let a_i be the total number of games played on days 1 through i, and set $a_0 = 0$. Then $a_j - a_i$ is the total number of games played on days $i+1$ through j. We seek an i and a j such that $a_i + 22 = a_j$. This suggests considering both $\{a_j\colon 1 \le j \le 77\}$ and $\{a_i + 22\colon 0 \le i \le 76\}$. Since there is at least one game each day, the numbers in $\{a_j\}$ are distinct, as are the numbers in $\{a_i + 22\}$. Hence a duplication among these 154 numbers implies the desired result. Since $a_{77} \le 132$, and $a_{76} + 22 \le 153$, we have 154 numbers in [153], and some number must repeat. Because $a_{76} + 23$ could be as large as 154, this argument does not work to force a period of consecutive days with exactly k games if $k \ge 23$. ∎

We often use the pigeonhole principle to prove existence of some configuration. Earlier we proved existence statements by building an example of the desired object. The pigeonhole principle provides non-constructive proofs of existence statements; it can also be an effective way to avoid case analysis.

Our examples suggest several remarks about the use of the pigeonhole principle. The classes may have different sizes. Partial sums may help with problems involving order or sums. An example showing that a claim is best possible can suggest appropriate classes and objects to use in applying the pigeonhole principle to prove the claim. Finally, the pigeonhole principle can be combined with proof by contradiction or other techniques.

THE INCLUSION-EXCLUSION PRINCIPLE

The rules of sum and product used to solve elementary counting problems are not helpful for counting problems involving forbidden conditions, because they lead to lengthy analysis by cases. In contrast, the Inclusion-Exclusion Principle leads quickly to formulas that solve such problems. The principle is based on the containment relation on the collection of subsets of a finite set.

10.9. Problem. *Derangements.* Suppose a professor collects homework papers from n students and redistributes them at random for peer grading. In this case, "at random" means that each of the $n!$ permutations is equally likely. A permutation in which no student receives his or her own paper is a *derangement*. What is the probability that a random permutation is a derangement? ∎

10.10. Problem. *Dice-Rolling.* Suppose we roll a six-sided die until each of the numbers one through five have appeared at least once. What is the probability that we succeed in the first n rolls? ∎

10.11. Problem. *Euler totient.* Given a positive integer m, let $\phi(m)$ be the number of elements of $[m]$ that are relatively prime to m. The function $\phi \colon \mathbb{N} \to \mathbb{N}$ is the *Euler totient function*; Exercise 10.30 explores some of its properties. How can we compute $\phi(m)$? ∎

We begin by discussing the Euler totient for numbers with few prime factors. Recall that m and r are relatively prime if and only if $\gcd(m, r) = 1$. If m is a power of a prime p, then all numbers in $[m]$ are relatively prime to m except the multiples of p. There are m/p such multiples, so $\phi(m) = m - m/p$.

Next suppose that m has precisely two prime factors, p and q. We eliminate all m/p multiples of p and all m/q multiples of q in $[m]$, but this means that we have eliminated all the multiples of pq twice. We add these back in to correct the count and obtain $\phi(m) = m - m/p - m/q + m/pq$.

In general, m will have prime factors $p_1, \ldots, p_n$. Initially we include all of $[m]$. Excluding the multiples of each prime factor discards more than once every element divisible by more than one of the prime factors. When we then include the sets divisible by two prime factors, we will have included too often the elements divisible by more than two of them. Eventually the process of including and excluding will count each element the proper number of times.

Before presenting the solution formula for the totient function, we develop the general setting for the Inclusion-Exclusion Principle. In general, suppose we have a universe U of objects, and we want to count the objects that appear in none of the n subsets $A_1, \ldots, A_n$. In the totient problem, the universe is $[m]$, and the set A_i is the set of multiples of the ith prime factor. In the derangements problem, the universe U is the collection of all permutations of $[n]$. In order to count those with no fixed point, we will let $A_i \subseteq U$ be the set of permutations fixing i, and then the derangements are precisely the permutations that appear in none of these sets.

This yields a general model for solving counting problems with forbidden conditions. Let $N_{\varnothing}$ denote the number of elements of U that appear in none of the n specified subsets $A_1, \ldots, A_n$ of U. If $n = 1$, then $N_{\varnothing}$ counts the elements outside A_1, so $N_{\varnothing} = |U| - |A_1|$.

For $n = 2$, consider the Venn diagram below. We don't want to count elements in A_1 or A_2, so we subtract those from the total. This subtracts the elements of $A_1 \cap A_2$ twice, so we add those in again, obtaining $N_{\varnothing} = |U| - |A_1| - |A_2| + |A_1 \cap A_2|$. The Venn diagram makes it apparent that every element outside $A_1 \cup A_2$ makes a net contribution of 1, and every element inside $A_1 \cup A_2$ makes a net contribution of 0. (If we want to count the elements belonging to *at least* one of the sets, the formula is $|A_1| + |A_2| - |A_1 \cap A_2|$.)

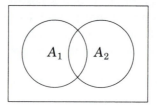

Before deriving the general formula, we also discuss the case $n = 3$ explicitly. The reader may use the Venn diagram below to keep track of the Including and Excluding as we describe it.

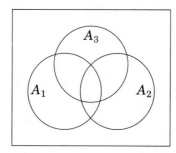

Again we start with all the elements, $|U|$. We don't want to include elements belonging to any of $\{A_i\}$, so we subtract $|A_1| + |A_2| + |A_3|$ from $|U|$. Any element belonging to more than one of the sets has been taken away more than once, so we add $|A_1 \cap A_2| + |A_2 \cap A_3| + |A_1 \cap A_3|$ to correct this. Now an element in none of the sets contributes 1 to the count, an element in exactly one set contributes $1 - 1 = 0$, and an element in exactly two sets contributes $1 - 1 - 1 + 1 = 0$, but an element in all three sets contributes $1 - 1 - 1 - 1 + 1 + 1 + 1 = 1$. We subtract $|A_1 \cap A_2 \cap A_3|$ as a final correction. Thus the inclusion-exclusion formula for $N_\emptyset$ when there are three forbidden sets is

$$|U| - (|A_1| + |A_2| + |A_3|) + (|A_1 \cap A_2| + |A_2 \cap A_3| + |A_1 \cap A_3|) - |A_1 \cap A_2 \cap A_3|.$$

In general, for each subset S of the indices $1, \ldots, n$, we weight $|\cap_{i \in S} A_i|$ negatively if $|S|$ is odd and positively if $|S|$ is even. The count arising when $S = \emptyset$ is $|U|$, because each element is in every one of no sets. Just as a sum over no terms is the additive identity 0 and a product over no factors is the multiplicative identity 1, so an intersection over no sets is the "intersective identity" U.

10.12. Theorem. (Inclusion-Exclusion Principle) Given a universe U of items and subsets $A_1, \ldots, A_n$ of the items, the number $N_\emptyset$ of items belonging to none of the subsets is given by

$$N_\emptyset = \Sigma_{S \subseteq [n]} (-1)^{|S|} |\cap_{i \in S} A_i|.$$

Proof. We need only show that each item belonging to none of the sets contributes 1 to the total and that all other items contribute 0. An item in none of the sets appears only in the term for $S = \varnothing$, so its contribution is 1. Suppose the sets among $\{A_i\}$ that contain the item x are precisely $\{A_i : i \in T\}$, where T is a non-empty subset of the index set $[n]$. The item x is counted in the term for each subset of T. It contributes +1 for each $S \subseteq T$ of even size and -1 for each $S \subseteq T$ of odd size. Hence the total contribution for x is $\Sigma_{S \subseteq T}(-1)^{|S|} = \Sigma_{k=0}^{|T|}(-1)^k\binom{|T|}{k}$.

There are several ways to show that this sum is 0. We can treat it as a special case of $\Sigma_{k=0}^{t}\binom{t}{k}y^k$, with y set to -1. By the binomial theorem, the sum is $(1+y)^t$, and when $y = -1$ it equals 0 since $t > 0$.

We give also a direct combinatorial proof by establishing a bijection between the subsets of odd size and the subsets of even size. Define a function f on the subsets of odd size as follows. If the element 1 appears in S, delete it. If 1 does not appear, add it. The result is always a subset of even size. Furthermore, f is invertible; given the even subsets, applying the same procedure inverts f. Hence f pairs the subsets of even size with those of odd size, and there are the same number of subsets having each parity. ∎

The Inclusion-Exclusion Principle is useful when we can model our problem as counting the elements outside some sets $A_1, \ldots, A_n$ and when the quantities $|\cap_{i \in S} A_i|$ are easy to compute.

10.13. Solution. *Euler totient.* Suppose m has n distinct prime factors $p_1, \ldots, p_n$. Within the universe $U = [m]$, we define the set A_i to be the multiples of p_i. The numbers relatively prime to m are the elements in none of $A_1, \ldots, A_n$. To apply the Inclusion-Exclusion Formula, we need the sizes of intersections of these sets. For the intersection of the sets indexed by the elements of $S \subseteq [n]$, we have $|\cap_{i \in S} A_i| = m/\Pi_{i \in S} p_i$. By the Inclusion-Exclusion Principle, we thus have

$$\phi(m) = N_\varnothing = \Sigma_{S \subseteq [n]}(-1)^{|S|}|\cap_{i \in S} A_i| = \Sigma_{S \subseteq [n]}(-1)^{|S|}m/\Pi_{i \in S} A_i.$$

For example, $60 = 2^2 \cdot 3 \cdot 5$, so we compute

$$\phi(60) = 60 - \frac{60}{2} - \frac{60}{3} - \frac{60}{5} + \frac{60}{6} + \frac{60}{10} + \frac{60}{15} - \frac{60}{30} = 16. \quad \blacksquare$$

10.14. Solution. *Derangements.* We can model Problem 10.9 by writing the numbers $1, \ldots, n$ (for the papers) in the positions $1, \ldots, n$ (for the students). We want to count the permutations of $[n]$ such that no i is in position i. An instance of i in position i is a fixed point; the derangements are the permutations with no fixed points.

Within the universe U of permutations of $[n]$, let A_i be the set of permutations that leave i fixed. Because derangements have no fixed points, $D_n = N(\varnothing)$. Consider a set $S \subseteq [n]$ with $|S| = k$. A permutation lies in all sets indexed by S if and only if it fixes $\{i : i \in S\}$. It can permute the other elements arbitrarily (including fixing them), so $|\cap_{i \in S} A_i| = (n - k)!$. There are $\binom{n}{k}$ choices of S with size k, and we weight these contributions by $(-1)^{|S|}$, so the formula is

$$D_n = \Sigma_{k=0}^n (-1)^k \binom{n}{k}(n - k)! = n! \Sigma_{k=0}^n (-1)^k / k!.$$

Dividing by $n!$ yields $\Sigma_{k=0}^n (-1)^k / k!$ for the probability of a derangement. This alternating sum converges rapidly to $1/e$, where $e = 2.71828\ldots$. Surprisingly, the probability is almost independent of n and tends to a nonzero limit as n grows. ∎

In the derangements computation, the size of $\cap_{i \in S} A_i$ depends only on $|S|$. This allows us to combine the terms for all sets S with $|S| = k$. The factor $\binom{n}{k}$ appears, multiplied by the size of each $\cap_{i \in S} A_i$ with $|S| = k$. We obtain a summation with $n + 1$ terms instead of a summation with 2^n terms. This simplification occurs often.

There are n^k functions from a k-set A to an n-set B. Using combinatorial arguments, we counted the injective ones. These correspond to listing k distinct elements of B in order and assigning them to $a_1, \ldots, a_k$, and there are $n!/(n - k)!$ ways to do this. We use the Inclusion-Exclusion Principle to count the surjective functions.

10.15. Example. *Surjective functions.* How many functions from A to B are surjective? Let A_i be the set of functions that omit the ith element of $B = \{b_1, \ldots, b_n\}$. Given a set $S \subseteq [n]$ of indices, $\cap_{i \in S} A_i$ is the set of functions that omit the corresponding $|S|$ elements of B. There are $(n - |S|)^k$ of these functions, since we can map A onto the remaining elements without restriction (possibly missing more elements). When we gather together all $\binom{n}{j}$ terms with $|S| = j$, for each j, we obtain $\Sigma_{j=0}^n (-1)^j \binom{n}{j}(n - j)^k$ for the number of surjective functions. ∎

10.16. Solution. *Dice-rolling.* The Inclusion-Exclusion Principle applies to events in a finite probability space as well as to sets in a finite universe. When we normalize so that the total count of the universe is 1, we can interpret counting probability outside certain events as counting elements outside certain sets.

Suppose we roll a fair six-sided die n times and want to know the probability that each of the values 1,2,3,4,5 appears during the experiment. If A_i is the event that i does not appear, we want the probability Prob($\varnothing$) outside all these events. The probability that we do not see one fixed value is $(5/6)^n$. The probability that k such events from $\{A_i\}$ occur, meaning that k values fail to occur, is $[(6 - k)/6]^n$. This holds for each of the $\binom{n}{k}$ choices of k values, so the inclusion-exclusion formula yields

$$\text{Prob}(\varnothing) = 1 - 5(\tfrac{5}{6})^n + 10(\tfrac{4}{6})^n - 10(\tfrac{3}{6})^n + 5(\tfrac{2}{6})^n - (\tfrac{1}{6})^n$$

For $n = 5, 10, 15, 20$, the probabilities are .015, .356, .698, .873, respectively. The probability first exceeds .5 for $n = 12$. ∎

EXERCISES

The first 23 problems are related to the Pigeonhole Principle, the others to Inclusion-Exclusion. The answers to most problems using the Inclusion-Exclusion Principle must be left as summations.

10.1. (–) Suppose that during a major league baseball season there are 140,000 at-bats and 35,000 hits. Which of the following must be true?
a) There is some player who hits exactly .250.
b) There is some player who hits at least .250.
c) There is some player who hits at most .250.

10.2. Each year, the Grievance Committee consists of three professors. How many professors must there be in the department to avoid having the same committee in a period of eleven years?

10.3. Suppose that S is a subset of $\{1, 2, \ldots, 3n\}$ and that S contains $2n + 1$ elements. Prove that S must contain three consecutive numbers. Show that this is best possible by exhibiting a set of size $2n$ for which the conclusion is false.

10.4. Let S be a set of $n + 1$ numbers in $[2n]$. Prove that S contains a pair of relatively prime numbers. Show that this is best possible by exhibiting a set of size n for which the conclusion is false.

10.5. (!) Prove that every set of seven distinct integers contains a pair whose sum or difference is a multiple of 10.

10.6. Suppose that the numbers 1 through 10 appear in some order around a circle. Prove that some set of three consecutive numbers sums to at least 17.

10.7. (!) The numbers 1 through 12 have fallen off the face of a clock and have been replaced in some random order. Prove that some set of three consecutive numbers has sum at least 20. Prove that some set of five consecutive numbers has sum at least 33. For three consecutive numbers, use more detailed analysis to determine whether it is possible for all the sums to be in $\{19, 20\}$.

10.8. (!) Prove that every set of five points in the square of area 1 has two points separated by distance at most $\sqrt{2}/2$. Prove that this is best possible by exhibiting five points with no pair less than $\sqrt{2}/2$ apart. (Warning: studying perturbations of the set found for the second part does not solve the first part.)

10.9. *Pigeonhole generalization.* Suppose $p_1, \ldots, p_k$ are natural numbers. Determine the minimum n such that for every way of distributing n objects into classes $1, \ldots, k$, there is some i such that class i receives at least p_i objects.

10.10. On a field 400 yards long, ten people each mark off football fields of length 100 yards. Prove that some point belongs to at least four of the fields.

10.11. (!) The *fractional part* of x is the amount by which it exceeds $\lfloor x \rfloor$. Suppose $x \in \mathbb{R}$ and $n \in \mathbb{N}$, and let $S = \{x, 2x, \ldots, (n-1)x\}$.
    a) Prove that if some pair of numbers in S have fractional parts that differ by at most $1/n$, then some number in S is within $1/n$ of an integer.
    b) Use part (a) to prove that some number in S is within $1/n$ of an integer.

10.12. Let S be a set of n integers. Prove that S has a non-empty subset whose sum is divisible by n. Show that this is best possible by exhibiting a set of $n-1$ integers that has no non-empty subset whose sum is divisible by n.

10.13. (+) Consider a collection S of $n+1$ positive integers summing to k. Call S "full" if for every $i \in k$, S has a subset with sum i. Prove that if $k \le 2n+1$, then S must be full. Show that this is best possible by exhibiting a set S of $n+1$ numbers summing to $2n+2$ that is not full.

10.14. Six students from class show up for a special help session. Prove that among the six there must be three who all know each other or three who all don't know each other.

10.15. Use congruence classes to determine the maximum size of a subset of [99] that has no two numbers differing by 3.

10.16. (!) Given $n, k \in \mathbb{N}$, use congruence classes to determine the maximum size of a subset of $[n]$ that has no two numbers differing by k.

10.17. (!) Prove that the Erdős-Szekeres result is best possible by constructing for each n (with proof) a list of n^2 distinct numbers having no monotone sublist of length $n+1$.

10.18. Consider an exam with three true/false questions, in which every student answers each question.
    a) How many students are needed to guarantee that no matter how they answer the questions, some two students agree on every question?
    b) How many students are needed to guarantee that no matter how they answer the questions, some two students agree on at least two questions? (Warning: parts (a) and (b) each require a proof for the upper bound and an example for the lower bound.)

10.19. (!) *The Key Problem.* A private club has 90 rooms and 100 members. Keys must be given to members such that each set of 90 members can be assigned to 90 distinct rooms whose doors they can open. Each key opens one door. The management wants to minimize the total number of keys. Prove that the minimum number of keys is 990. (Hint: Consider the scheme where 90 of the members have one key, and the remaining 10 members have keys to all 90 rooms. Prove that this works and that no scheme with fewer keys works.)

10.20. (+) Suppose the chess player of Example 10.7 plays on d consecutive days for a total of at most b games. Suppose we want to know whether there must be a total of exactly k games over some period of consecutive days, regardless of the schedule. Determine a formula $f(d, b)$ such that the argument of Example 10.7 works to prove the answer is "Yes" if $k \leq f(d, b)$.

10.21. (+) In Example 10.7, the chess player plays at most 132 days over 77 days. The argument there guarantees existence of a period of consecutive days with a total of exactly k games if $k \leq 22$. Using the pigeonhole principle and congruence classes modulo k, prove that there are also periods with exactly k games for $k \in \{23, 24, 25\}$. Construct a 77-day schedule of games such that no period of consecutive days has a total of exactly 26 games.

10.22. Suppose $m \geq 2n$, and let S be a set of m points on a circle with no two diametrically opposite. Say that $x \in S$ is "free" if fewer than n points of $S - x$ lie in the semicircle clockwise from x. Prove that S has at most n free points. (Hint: reduce the problem to the case $m = 2n$.)

10.23. Consider an n by n grid of dots at positions $\{(i, j) : 1 \leq i \leq n, 1 \leq j \leq n\}$ in the plane. Suppose each dot is colored black or white. How large must n be such that for every way to color the dots, there is a rectangle whose four corners all have the same color? (Warning: The answer must consist of a proof for the upper bound and an example for the lower bound.)

10.24. How many ways are there to place 10 distinct people within three distinct rooms? How many ways are there to place 10 distinct people within three distinct rooms so that every room receives at least one person?

10.25. How many decimal n-tuples contain at least one each of $\{1, 2, 3\}$?

10.26. (!) Say that an integer is "full" if its base 10 representation contains at least one of each digit $0, 1, \ldots, 9$. For this problem, a representation with fewer digits is considered a representation with m digits by adding leading 0's. Derive a summation formula for the number of full m-digit integers.

10.27. A bridge hand consists of 13 cards from a standard deck of 52 cards. What is the probability that a bridge hand has at least one card in each suit? What is the probability that it has no cards (a void) in at least one suit?

10.28. How many natural numbers less than 252 are relatively prime to 252?

10.29. How many natural numbers less than 200 have no divisor in $\{6, 10, 15\}$?

10.30. (!) Let $\phi(m)$ denote the Euler totient function (the number of elements of $[m]$ that are relatively prime to m). If p, q are distinct prime numbers, prove that $\phi(pq) = \phi(p)\phi(q)$. For $m \in \mathbb{N}$ in general, use the prime factorization of m to compute $\phi(m)$ as a product of numbers.

10.31. Suppose $A_1, \ldots, A_n$ are sets in a universe U. Let $T \subseteq [n]$ be a collection of indices, and let $N(T)$ be the number of elements of U that belong to the sets indexed by T but to no others among $A_1, \ldots, A_n$. By defining a new universe, prove the following generalization of the inclusion-exclusion formula:

$$N(T) = \Sigma_{T \subseteq S \subseteq [n]} (-1)^{|S|-|T|} |\cap_{i \in S} A_i|.$$

10.32. How many permutations of $[n]$ have no odd number as a fixed point?

10.33. (!) A math department has n professors and $2n$ courses, each professor teaching two courses each semester. How many ways are there to assign the courses in the fall semester? How many ways are there to assign the courses in the spring semester such that no professor teaches the same pair of courses in the spring as in the fall? If all the assignments are equally likely, what is the probability of this event?

10.34. Let D_n count the permutations of $[n]$ with no fixed points. Let E_n^k count the permutations of $[n]$ with exactly k fixed points, for $0 \le k \le n$.
   a) Derive a formula for E_n^k in terms of $\{D_j : 0 \le j \le n\}$.
   b) Derive a formula for $n!$ in terms of $\{D_j : 0 \le j \le n\}$.

10.35. (!) Given the five types of coins (pennies, nickels, dimes, quarters, half-dollars), how many ways can one select n coins so that no coin is selected more than 4 times? (Hint: use inclusion-exclusion and selections with repetition.)

10.36. Given two each of n types of letters, how many distinguishable permutations are there such that no two consecutive letters are the same?

10.37. (!) Consider a set of n boys and n girls. Use inclusion-exclusion to derive formulas for the number of ways to pair up the $2n$ people as lab partners and satisfy the following criteria. (No simple closed formulas are available.)
   a) For each i, the ith tallest boy is not matched to the ith tallest girl (same-sex pairs are allowed).
   b) Same condition as (a), but also each pair has one person of each sex.

10.38. How many ways are there to seat the people in n married couples around a merry-go-round so that no person sits next to his or her spouse? (Rotations of the seating arrangement are not distinguishable from each other.)

10.39. Use inclusion-exclusion to prove that $\sum_{k=0}^{n}(-1)^k\binom{n}{k} = 0$ if $n > 0$. What happens if $n = 0$?

10.40. Use inclusion-exclusion to prove that $\sum_{k=0}^{n}(-1)^k\binom{n}{k}2^{n-k} = 1$. (Do not use the binomial theorem.)

10.41. Use inclusion-exclusion and selections with repetition to prove that

$$\sum_{k=0}^{n}(-1)^k\binom{n}{k}\binom{n-k+r-1}{r} = \binom{r-1}{n-1}.$$

Chapter 11

Graph Theory

The "graphs" of graph theory differ from the graphs of functions. Informally, a graph consists of "vertices" and "edges" that form connections between them. For example, we can think of people as vertices and join two people by an edge if they have met. Graph theory helps answer questions about acquaintance, chemical bonding, electrical networks, transportation networks, binary vectors, etc. The techniques of include induction, parity, extremality, counting two ways, the pigeonhole principle, and inclusion-exclusion (even the Dart Board Problem).

We begin by stating several problems we will solve using graphs.

11.1. Problem. *The Königsberg Bridge Problem.* Some say that graph theory was born in the city of Königsberg in 1736. Located on the Pregel river, the parts of the city were linked by seven bridges as shown on the left below. The citizens wondered whether they could leave home, cross every bridge exactly once, and return home. This reduces to traversing the figure on the right, with heavy dots representing land masses and curves representing bridges. ■

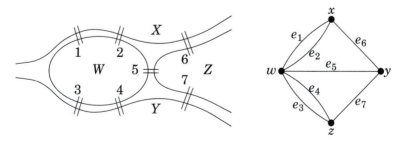

11.2. Problem. *The Marriage Problem.* Suppose there are n girls and n boys at a party, and each girl likes some subset of the boys. Under what conditions is it possible to match up the girls with the boys so that each girl is matched to a boy whom she likes? ■

11.3. Problem. *The Platonic Solid Problem.* A Platonic solid has congruent regular polygons as faces and has the same number of edges meeting at each corner. The tetrahedron, cube, and octahedron appear below. The dodecahedron and icosahedron are the only other Platonic solids. Why are these five the only ones? ∎

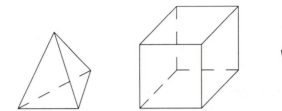

11.4. Problem. *The Art Gallery Problem.* A modern art gallery has the shape of a simple polygon in the plane, meaning a closed curve consisting of segments that meet only at successive vertices. What is the maximum number of stationary guards that may be needed to watch an art gallery with n corners? ∎

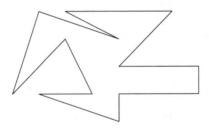

THE KÖNIGSBERG BRIDGE PROBLEM

To model the Königsberg Bridge Problem, we represent the land masses W, X, Y, Z by a set of elements $\{w, x, y, z\}$ called "vertices". We represent the seven bridges by a set of elements $\{e_1, e_2, e_3, e_4, e_5, e_6, e_7\}$ called "edges". We encode the information about which land masses lie at the ends of each bridge by associating with each edge $e_i \in E$ a pair of vertices. The informal notion of bridges joining two land masses permits us to answer the specific question of Problem 11.1 even before we make a formal definition of "graph".

11.5. Solution. *The Königsberg Bridge Problem.* The Swiss mathematician Leonhard Euler (1707-1783) observed in 1736 that the desired traversal of the bridges of Königsberg did not exist. Such a traversal would have to pass through a land mass on the way from each bridge to the next. Each time we visit a land mass, we enter it along one bridge and exit along another bridge. If we start and end in the same place, then we can also pair up the first departure from that land mass with the last entrance to it. Thus the desired traversal requires that the number of bridges at each land mass is even. This condition fails in the Königsberg example, so there is no such traversal. ∎

Suppose that in Problem 11.1 we add bridge 8 from W to Y and bridge 9 from X to Z. There will then be an even number of bridges at each land mass and many ways to traverse all the bridges; one example is the order 1,2,3,4,5,6,9,7,8.

We will prove that the requirement of an even number of bridges at each land mass, together with the ability to reach each bridge from every other, is also sufficient for traversability. In order to prove this and further results, we define "graph" precisely.

11.6. Definition. A *graph* G is a triple consisting of a *vertex set* $V(G)$, an *edge set* $E(G)$, and a function h_G that assigns to each edge $e \in E(G)$ an unordered pair of vertices.[†] When $h_G(e) = \{u, v\}$, we say that u and v are the *endpoints* of e and that e is *incident* to u and v. A graph G is *simple* if the function $h_G(e)$ is injective. In this case, we write $e = uv$ instead of $h_G(e) = \{u, v\}$.

11.7. Example. *The Königsberg graph.* The graph G we obtain from the Problem 11.1 has vertex set $\{w, x, y, z\}$ and edge set $\{e_i : 1 \le i \le 7\}$, with the endpoints of e_i for $1 \le i \le 7$ being $\{x, w\}$, $\{x, w\}$, $\{z, w\}$, $\{w, z\}$, $\{y, w\}$, $\{x, y\}$, $\{y, z\}$, respectively. This graph is not simple; $h_G(e_1) = h_G(e_2)$ and $h_G(e_3) = h_G(e_4)$. ∎

The terms "vertex" and "edge" come from geometry. As in Problem 11.1, we visualize graphs by drawing them in the plane. To each vertex we assign a point; to each edge we assign a curve that joins the points assigned to its vertices. We take this as an informal aid for visualization; in Definition 11.60 we define drawings more precisely.

11.8. Definition. The *degree* of a vertex $x \in V(G)$, written $d(x)$, is the number of edges in G incident to x. A *subgraph* of a graph G is a graph H such that $V(H) \subseteq V(G)$ and $E(H) \subseteq E(G)$; we also require $h_H(e) = h_G(e)$ for $e \in E(H)$. When H is a subgraph of G, we write $H \subseteq G$ and say "G contains H".

[†] In this chapter we consider only *finite* graphs, meaning that both the vertex set and the edge set are finite. Our model of graphs does not permit directed edges or loops (edges with equal endpoints). Definition 3.46 (functional digraph) does allow these possibilities.

11.9. Example. In the graph G of Problem 11.1, the degrees of w, x, y, z are $5, 3, 3, 3$, respectively. We can make all the vertex degrees even by adding two edges sharing no endpoints. We can also do this by adding three appropriate edges with one common endpoint, as illustrated on the left below. The resulting graph contains G.

On the right below, we illustrate a subgraph of G that has vertex degrees $2,2,1,0$ and is a simple graph. With appropriate names for the vertices, this drawing illustrates the Handshake Problem (Solution 4.20) for two couples, with handshakes as edges. ∎

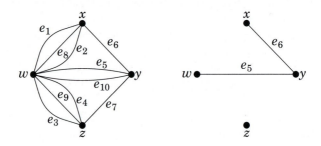

The solution of the Königsberg Bridge Problem uses vertex degrees and a precise notion of "traversal". We must travel each bridge only once, but we may visit land masses more than once.

11.10. Definition. A *trail* (of *length* k) in a graph G is a list $v_0, e_1, v_1,$ $e_2, \ldots, e_k, v_k$ that alternates between vertices and edges, such that 1) $h_G(e_i) = v_{i-1}v_i$ for all i, and 2) $e_1, \ldots, e_k$ are distinct elements of $E(G)$. A u, v-*trail* is a trail with first vertex u and last vertex v; these are its *endpoints*. A trail of positive length is *closed* if its endpoints are equal. A *circuit* is a closed trail. A trail in a graph is a *maximal* trail if it is not possible to insert vertices and edges in the list to obtain a longer trail.

11.11. Example. In the graph of Problem 11.1, $w, e_1, x, e_6, y, e_5, w, e_2, x$ is a trail of length 4 and $w, e_1, x, e_6, y, e_5, w, e_1, x$ is not a trail. Also $w, e_1, x, e_2, w, e_5, y, e_7, z, e_4, w$ is a closed trail (a circuit), and appending e_3, z to the end of it yields a maximal trail. ∎

We have shown that even vertex degrees are necessary for the existence of a circuit containing all edges of a graph. Even more obvious is the condition that each edge be reachable from every other edge. We say that an edge is *reachable* from another edge if there is a trail containing both. Euler remarked that these conditions are also sufficient, although no proof was published until 1871. In his honor we define a graph to be *Eulerian* if it has a single circuit containing all its edges, and we call such a circuit an *Eulerian circuit.*[†]

We first prove a lemma about maximal trails.

11.12. Lemma. If every vertex of a finite graph G has even degree, then every maximal trail of positive length in G is closed.

Proof. A trail contributes degree two when it passes through a vertex. Thus a non-closed trail uses an odd number of edges at each endpoint. If the endpoint has even degree, then a non-closed trail can be extended. We have proved the contrapositive of the claim. ∎

11.13. Theorem. A finite graph G is Eulerian if and only if each vertex has even degree and each edge is reachable from every other.

Proof. We have argued that the conditions are necessary; we prove that they are also sufficient.

Suppose that G satisfies the conditions; we may assume that G has at least one edge. Let T be a maximal trail of positive length in G; by Lemma 11.12, T is closed. If T does not include all of $E(G)$, let G' be the subgraph obtained from G by deleting $E(T)$. Since every edge of G is reachable from every other, there is a trail in G that starts with an edge of T and contains an edge of G'; let e be the first edge of G' on this trail, and let v be the vertex it follows.

Since T has even degree at every vertex, every vertex also has even degree in G'. Let T' be a maximal trail in G' beginning from v along e. By Lemma 11.12, T' is closed and ends at v. Hence we may insert T' (after deleting v from the ends) to obtain a trail properly containing T. This contradicts the maximality of T, so we conclude that T already contains all edges of G. ∎

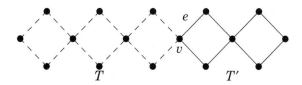

The proof of Theorem 11.13 uses extremality; we chose a maximal trail. Choosing an extremal example is a basic technique of proof. Induction proofs often amount to showing that a statement has no smallest counterexample. Our proofs of Fermat's Little Theorem (Theorem 7.26) and the Ballot Problem (Solution 9.17) illustrate other uses of extremality.

[†]The name "Euler" is pronounced as "oiler", because it is a Germanic name like "Freud", not a Greek name like "Euclid".

We close this section with some observations about vertex degrees. The vertex degrees and the number of edges satisfy a simple equation proved by a counting argument.

11.14. Theorem. (Degree-sum Formula) If G is a graph with m edges, then $m = \frac{1}{2}\Sigma_{v \in V(G)}d(v)$.

Proof. Summing the degrees counts each edge twice, since each edge has two endpoints and contributes to the degree of each endpoint. ∎

11.15. Example. *The d-dimensional cube Q_d.* The cube Q_d is a simple graph with 2^d vertices that are the d-tuples of 0's and 1's. Two vertices of Q_d form an edge if and only if they differ in exactly one coordinate. Since each coordinate of a binary d-tuple can be changed in exactly one way, each vertex has degree d. By the Degree-sum Formula, Q_d thus has $d2^{d-1}$ edges. We show Q_2 and Q_3 below. ∎

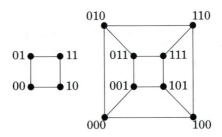

By the next corollary, the number of people in the world who have met an odd number of people is even. Applications of this and the Degree-sum Formula appear in Exercises 11.6-8 and Solutions 11.68-69.

11.16. Corollary. Every graph has an even number of vertices of odd degree.

Proof. By the Degree-sum Formula, the sum of the degrees is even. Hence the sum must have an even number of odd contributions. ∎

ISOMORPHISM OF GRAPHS

Consider the four cities {New York, Chicago, San Francisco, Champaign}. There are direct flights between Chicago and each of the other three cities, and direct flights between San Francisco and New York, but no direct flights between Champaign and either New York or San Francisco. We summarize this information by the graph below whose vertices are the four cities, and whose edges represent direct service.

Consider also the four integers {7, 10, 15, 42}. We define a graph with these integers as vertices, with two vertices forming an edge when they have a common factor larger than 1.

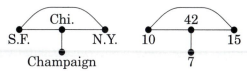

The picture shows that these two graphs have the same structure. They are not the same graph, since their vertices have different names. In order to treat them as the same object, we define a relation on the set of (finite) graphs, prove that it is an equivalence relation, and observe that these two graphs are in the same equivalence class. This echoes our use of equivalence classes of fractions to define rational numbers.

To avoid complications, we define this relation only for simple graphs. In a simple graph, we name each edge by its endpoints and treat the edge set as a set of vertex pairs.

11.17. Definition. An *isomorphism* from a simple graph G to a simple graph H is a bijection $f\colon V(G) \to V(H)$ such that $uv \in E(G)$ if and only if $f(u)f(v) \in E(H)$. We say "G *is isomorphic to* H", written $G \cong H$, if there is an isomorphism from G to H. The set of pairs G, H such that G is isomorphic to H is the *isomorphism relation*.

When G is isomorphic to H, also H is isomorphic to G, so we may say "G and H are isomorphic". The adjective "isomorphic" applies only to pairs of graphs; the phrase "G is isomorphic" has no meaning.

11.18. Example. The two 4-vertex graphs drawn below are isomorphic, by an isomorphism that maps $1, 2, 3, 4$ to a, d, b, c, respectively. Making this substitution changes the edges $12, 23, 34$ into ad, db, bc, respectively. These are indeed the edges of the second graph, so the vertex bijection is an isomorphism. Another isomorphism maps $1, 2, 3, 4$ to c, b, d, a, respectively. ∎

We can describe isomorphism for simple graphs more concisely using a natural relation on the vertex set.

11.19. Definition. Vertices u and v in a graph G are *adjacent* and are *neighbors* if they are the endpoints of an edge. The *adjacency relation* of G (defined on $V(G)$) is the set of ordered pairs (u, v) such that u and v are adjacent.

The adjacency relation is symmetric, and every symmetric relation is the adjacency relation of a graph. In the language of adjacency, simple graphs G and H are isomorphic if and only if there is a bijection $f: V(G) \rightarrow V(H)$ that preserves the adjacency relation.

11.20. Proposition. The isomorphism relation is an equivalence relation on the set of simple graphs.

Proof. The identity map on $V(G)$ is an isomorphism from G to itself. If $f: V(G) \rightarrow V(H)$ is a isomorphism from G to H, then f^{-1} is an isomorphism from H to G. If $f: V(F) \rightarrow V(G)$ and $g: V(G) \rightarrow V(H)$ are isomorphisms, then $g \circ f$ is a bijection from $V(F)$ to $V(H)$ that preserves the adjacency relation and hence is an isomorphism from F to H. Thus the isomorphism relation is reflexive, symmetric, and transitive. ■

11.21. Definition. An *isomorphism class* of graphs is an equivalence class of graphs under the isomorphism relation.

11.22. Remark. *Isomorphism classes.* Comments about the structure of a graph G also apply to every graph isomorphic to G. Some authors use the informal expression "unlabeled graph" instead of "isomorphism class of graphs". The vertices of a graph drawn on paper are named by their physical location; hence every drawing of a graph is a representative of its isomorphism class. Drawing a graph to illuminate its structure is choosing a convenient member of its isomorphism class.

Asking whether a given graph "is" G is asking whether it is isomorphic to G. Similarly, we use the phrase "H is a subgraph of G" to mean that H is isomorphic to a subgraph of G. In this sense, the 2-dimensional cube Q_2 is a subgraph of the 3-dimensional cube Q_3 (see Exercise 11.11), even though the 2-tuples used as vertices in Q_2 are shorter than the 3-tuples used as vertices in G_3. ■

We usually prove that two graphs are isomorphic by presenting a bijection f and showing that it preserves the adjacency relation. Since structural properties are determined by the adjacency relation, we can prove that G and H are not isomorphic by finding some structural property of one that fails for the other. They may have different numbers of edges, different vertex degrees, different subgraphs, etc. Exhibiting a difference in structure proves that no vertex bijection preserves the adjacency relation.

11.23. Example. *Testing isomorphism.* Vertex degrees are easy to compute, and an isomorphism from G to H must map every vertex $v \in V(G)$ to a vertex of H whose degree in H is $d_G(v)$. Hence the vertex degrees of isomorphic graphs (counted with repetition) must be the same. For

example, a graph whose vertices have degrees $1, 1, 1, 3$ cannot be isomorphic to a graph whose vertices have degrees $1, 1, 2, 2$, even though
each has four vertices and three edges.

Nevertheless, two graphs may have the same list of vertex degrees
and not be isomorphic. In each of the four graphs drawn below, each
vertex has degree 3. Only graph C has three vertices that are pairwise
adjacent, so it cannot be isomorphic to any of the others. The others are
pairwise isomorphic.

To show that $A \cong B$, we can verify that the bijection mapping
u, v, w, x, y, z to $1, 3, 5, 2, 4, 6$, respectively, is an isomorphism. Sending
u, v, w, x, y, z to $6, 4, 2, 1, 3, 5$ yields another isomorphism.

Graphs A and D have the same vertex set but different adjacency
relations; $xw \in E(A)$, but $xw \notin E(D)$. Thus they are different graphs.
They are isomorphic, though, by an isomorphism that maps
u, v, w, x, y, z in $V(A)$ to u, v, z, x, y, w in $V(D)$, respectively. ∎

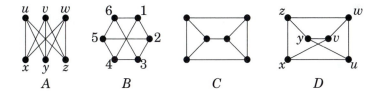

When two simple graphs have many edges and have corresponding
vertex degrees, looking at the nonadjacent pairs of vertices may make it
easier to tell whether the graphs are isomorphic.

11.24. Definition. The *complement* $\bar{G}$ of a simple graph G is the graph
with vertex set $V(G)$ and edge set $\{\{u, v\}: uv \notin E(G)\}$.

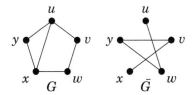

11.25. Example. The graphs below have large vertex degrees, so we
test isomorphism by considering the complements; graphs G and H are
isomorphic if and only if $\bar{G}$ and $\bar{H}$ are isomorphic (Exercise 11.14). The
vertices of these graphs have degree 5; in the complements the vertices
have degree 2. The complement of one of these graphs is Eulerian (it is
a single closed trail of length 8). The complement of the other graph is
not Eulerian (it consists of two disjoint closed trails of length 4). Therefore, the graphs are not isomorphic. ∎

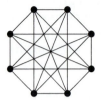 

11.26. Example. *Counting graphs.* The number of pairs of distinct vertices in a set of size n is $\binom{n}{2}$. Since each vertex pair may or may not form an edge, there are $2^{\binom{n}{2}}$ simple graphs having a fixed set of n vertices. We can also compute this by thinking of how many choices we must make to form an adjacency matrix. After ignoring the n zeros on the diagonal, we have one choice for each pair (i, j) and (j, i) of positions, so again the number of choices is $\binom{n}{2}$.

Hence there are 64 graphs having a fixed set of four vertices, but these fall into only 11 isomorphism classes. Representatives of these classes are drawn below; only one is isomorphic to its complement. ∎

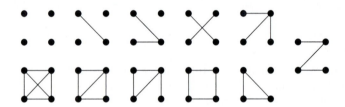

11.27. Remark. (optional) "Isomorphism" in mathematics generally describes a map between "equivalent" mathematical structures. An isomorphism between structures defined on sets S and T is a bijection between S and T that preserves the essential properties of the structure. For graphs, the sets S and T are the vertex sets, and the property to be preserved is the adjacency relation.

To define isomorphism for graphs that are not simple, we introduce an alternative description of a graph. Given a finite graph G, the *multiplicity* of an unordered pair $\{u, v\} \in V(G)$ is the number of edges in G with endpoints $\{u, v\}$. Two graphs G, H are *isomorphic* if there is a bijection $f \colon V(G) \to V(H)$ that preserves multiplicity.

This agrees with our earlier definition for simple graphs, since uv is an edge in a simple graph if and only if has multiplicity 1. Describing a graph by its vertex set and multiplicities ignores the names of edges but includes all information about the structure of the graph.

In the language of Definition 11.6, an isomorphism is two bijections $f \colon V(G) \to V(H)$ and $\tilde{f} \colon E(G) \to E(H)$ such that for all $v \in V(G)$ and $e \in E(G)$, e is incident to v if and only if $\tilde{f}(e)$ is incident to $f(v)$. ∎

CONNECTION AND TREES

For most concepts in this chapter, the distinction between simple graphs and general graphs is unimportant. We now confine our attention to simple graphs, viewing the edge set of a graph as a set of unordered pairs of vertices. In this setting we specify a trail by its ordered list of vertices, since a simple graph has (at most) one edge with specified endpoints v_{i-1} and v_i. We consider special types of trails.

11.28. Definition. A *path* is a trail with no repeated vertex. A u,v-*path* is a path with endpoints u and v. A *cycle* is a closed trail in which "first = last" is the only vertex repetition. We use P_n and C_n, respectively, to denote any representative of the isomorphism class that is a path or cycle with n vertices.

The definitions of P_n and C_n make sense because paths with n vertices are pairwise isomorphic, as are cycles with n vertices. Since a path or cycle is also a trail, its *length* is its number of edges. We often treat paths and cycles in a graph G as subgraphs rather than as vertex lists, especially when the direction along the path or the starting vertex of the cycle is unimportant. In the context of discussing cycles in simple graphs, we can name a particular cycle by listing its vertices in order without repeating the first vertex at the end. This is consistent with the specification of cycles in functional digraphs and permutations (see Definition 3.46 and Example 3.49).

11.29. Example. *Paths, trails, cycles.* In the 3-dimensional cube Q_3 (Example 11.15), we find subgraphs that are paths of lengths 0 through 7 and subgraphs that are cycles of lengths 4, 6, and 8. The graph drawn below contains three cycles (as subgraphs). For each pair $s, t \in V(G)$, this graph contains an s, t-path. ∎

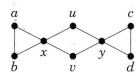

11.30. Definition. A graph G is *connected* if for every pair $u, v \in V(G)$, there is a u, v-path in G (otherwise, G is *disconnected*). A *component* of G is a connected subgraph of G that is not contained in any other connected subgraph. An *isolated vertex* has degree 0.

11.31. Example. A connected graph, like that of Example 11.29, has one component. The graph below has three components, and one of these is an isolated vertex. The vertex sets of the components are $\{r\}$,

$\{s, t, u, v, w\}$, and $\{x, y, z\}$. The subgraph consisting of the two components that are not isolated vertices is a disconnected graph with no isolated vertex. In an Eulerian graph, only one component has edges. ∎

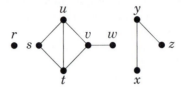

When studying paths in graphs, we say "u is connected to v" or "u and v are connected by a path" when G has a u, v-path. The *connection relation* on $V(G)$ is the set of ordered pairs (u, v) such that G has a u, v-path. For the stronger statement that u and v are adjacent, we say "u and v are joined by an edge", **not** "u and v are connected".

11.32. Example. A u, v-path and a v, w-path together need not form a u, w-path. The concatenation of the u, v-path u, x, y, v and the v, w-path v, z, y, w is the trail u, x, y, v, z, y, w, which is not a path. Nevertheless, this trail *contains* the u, w-path u, x, y, w. ∎

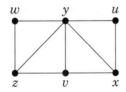

11.33. Proposition. If P is a u, v-path and P' is a v, w-path, then P and P' together contain a u, w-path.

Proof. We use extremality. At least one vertex of P appears in P', since both contain v. Let x be the first vertex of P that appears in P'. Following P from u to x and then P' from x to w yields a u, w path, since no vertex of P before x belongs to P'. ∎

11.34. Proposition. Let G be a graph. The connection relation on $V(G)$ is an equivalence relation, and its equivalence classes are the vertex sets of the components of G. If G has paths from one vertex to all others, then G is connected.

Proof. Reflexive property: v is connected to v by a path of length 0. Symmetric property: if P is a u, v-path, then reversing P yields a v, u-path. Transitive property: this is proved in Proposition 11.33.

Two vertices are in the same equivalence class if and only if they belong to a path; a path is a connected subgraph and hence appears in one component. If all vertices have paths to v, then Proposition 11.33 yields paths connecting all other pairs of vertices. ∎

We next introduce notation for subgraphs obtained by deleting a vertex or an edge of a graph.

11.35. Definition. The subgraph of G obtained by deleting an edge e is $G - e$. The subgraph obtained by deleting a vertex v and all edges containing v is $G - v$. The subgraph obtained by keeping all vertices but deleting the edges of a subgraph H is $G - E(H)$.

For example, if G is a cycle of length n with $e \in E(G)$ and $v \in V(G)$, then $G - e$ is a path of length $n - 1$, and $G - v$ is a path of length $n - 2$.

11.36. Lemma. If e is an edge of a connected graph G, then $G - e$ is connected if and only if e belongs to a cycle in G.

Proof. Suppose $e = xy \in E(G)$, and let $G' = G - e$. If $G - e$ is connected, then x and y belong to the same component in G', so G' contains an x, y-path, which completes a cycle with e in G.

Conversely, suppose e belongs to a cycle C. Choose $u, v \in V(G)$. Being connected, G has a u, v-path P. If P does not contain e, then P also exists in G'. If P contains e, suppose by symmetry that P reaches x before y when traveled from u to v. Since G' contains a u, x-path along P, an x, y-path along C, and a y, v-path along P, the transitivity of the connection relation implies that $G - e$ has a u, v-path. Since u, v were chosen arbitrarily from $V(G)$, we have proved that $G - e$ is connected. ∎

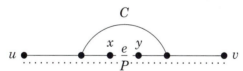

A *maximal* object of a particular type is one that is not contained in any other object of that type. Earlier we considered a maximal trail in a graph. A maximal path in a graph is one that cannot be extended by adding a vertex at either end. Every path of maximum length is a maximal path, but maximal paths need not have maximum length: in Example 11.29, a, x, b is a maximal path that does not have maximum length. Maximal paths often lead to short proofs by extremality.

11.37. Lemma. If G is a finite graph in which every vertex has degree at least two, then G contains a cycle.

Proof. Since $V(G)$ is finite, we can choose a maximal path P. Let v be an endpoint of P. Since $d(v) \geq 2$, v has a neighbor u that is not a neighbor of v on P. Since we cannot extend P to reach a new vertex from v, the vertex u already belongs to P, and the edge vu completes a cycle with the u, v-portion of P. ∎

If we allowed infinite vertex sets, this proposition would not hold. Suppose $V(G) = \mathbb{Z}$ and $E(G) = \{xy: y - x = 1\}$. This infinite graph contains no cycle (it is a single "path" that extends infinitely in both directions), but every vertex has degree 2.

How many edges must a graph with n vertices have in order to be connected? Because deleting an edge of a cycle cannot disconnect a graph (Lemma 11.36), the minimal connected graphs have no cycles.

11.38. Definition. A *tree* is a connected graph with no cycles. A *leaf* is a vertex of degree 1. A *spanning tree* of a graph G is a subgraph of G that is a tree containing all vertices of G.

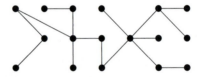

Gustav Kirchhoff (1824-1887) introduced spanning trees in connection with his work on electrical networks. Every connected graph has a spanning tree. This follows from Lemma 11.36: if G is connected, then deleting edges of cycles until no cycles remain produces a subgraph of G that is connected, has no cycles, and contains all vertices of G.

11.39. Lemma. Every tree with at least two vertices has a leaf, and deleting a leaf from a tree yields a tree with one less vertex.

Proof. Let G be a tree with n vertices, where $n \geq 2$. By the contrapositive of Lemma 11.37, a graph with no cycles has a vertex of degree less than two. Since G is connected and has more than one vertex, it has no vertex of degree 0, so it has a leaf x. Let $G' = G - x$.

We claim that G' is a tree with $n - 1$ vertices. We cannot create a cycle by deleting a vertex, so we need only show that G' is connected. Consider distinct vertices $u, v \in V(G')$. Because G is connected, there is a u, v-path P in G. Since internal vertices along a path have degree at least 2, P cannot contain x. Hence P is contained in G'. ∎

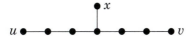

11.40. Theorem. Every tree with n vertices has $n-1$ edges.

Proof. We use induction on n. A tree with 1 vertex has no edges. For the induction step, we consider $n > 1$ and assume that trees with $n-1$ edges have $n-2$ vertices. If G is a tree with n vertices, then Lemma 11.39 yields a leaf x and a tree $G' = G - x$ with $n-1$ vertices. By the induction hypothesis, G' has $n-2$ edges. Since x appears in only one edge, we conclude that G has $n-1$ edges. ∎

Since deleting a leaf yields a smaller tree, each tree with $n+1$ vertices arises from some tree with n vertices by adding an edge to a new vertex. Hence an inductive proof about trees using "growing a new leaf" (from an arbitrary vertex) does not fall into the induction trap.

BIPARTITE GRAPHS

Our next class of graphs includes all trees and cubes.

11.41. Definition. A set $S \subseteq V(G)$ is an *independent set* in a graph G if $uv \notin E(G)$ for all $u, v \in S$ (S may be empty). A *bipartite graph* with *bipartition* X, Y is a graph G such that $V(G) = X \cup Y$ and X, Y are disjoint (possibly empty) independent sets. We call X and Y the *partite sets* or *parts* of the bipartition.

11.42. Example. The d-dimensional cube Q_d is bipartite. Let X be the set of vertices whose encoding as a binary d-tuple has an odd number of ones. Let Y consist of those with an even number of ones. In each edge of Q_d, the parity of the number of ones in the encoding is different at the two endpoints. Hence X and Y are independent sets. ∎

11.43. Proposition. Every tree is bipartite.

Proof. We use induction on the number of vertices. A tree with one vertex has a bipartition with one set empty. For the induction step, suppose that every tree with n vertices is bipartite, and let T be a tree with $n+1$ vertices. By Lemma 11.39, we can construct T by adding a leaf x adjacent to some vertex y of a tree T'. By the induction hypothesis, we can partition $V(T')$ into two independent sets X and Y, with $y \in Y$. We obtain the desired bipartition of $V(T)$ by placing x in X, since the only neighbor of x is in Y. ∎

A disconnected bipartite graph has more than one bipartition, but a connected bipartite graph has only one. The parts or partite sets of a bipartition are not themselves called "partitions", just as the teams in a sports league are not themselves called "leagues".

Bipartite graphs have a simple structural characterization using an obvious necessary condition that we prove is also sufficient.

11.44. Theorem. A graph is bipartite if and only if it contains no cycle of odd length.

Proof. We use "odd cycle" for "cycle of odd length". To prove that the condition is necessary, consider a bipartite graph G. Every trail in a bipartite graph alternates between the two partite sets of a bipartition. Hence returning to the original partite set (or original vertex) happens only after an even number of steps. In particular, G has no odd cycle.

For sufficiency, consider a graph G with no odd cycles. We prove that each component H is bipartite. Since H is connected, it has a spanning tree T. By Proposition 11.43, T is bipartite with some bipartition X, Y. If u and v belong to the same partite set in T, then T has an u, v-path, since T is connected. This path has even length, since it alternates between X and Y. An edge uv would complete an odd cycle, which does not exist in G; hence u and v are not adjacent. Since u, v were arbitrary vertices in the same partite set in T, we have proved that the bipartition of T is a bipartition of H. ∎

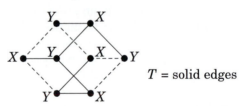

T = solid edges

11.45. Definition. A *complete graph* is a simple graph in which every pair of vertices forms an edge. We use K_n to denote any representative of the isomorphism class of complete graphs with n vertices.

11.46. Application. *The Airline Problem.* Suppose that an air traffic system has k airlines and n cities, and direct service between two cities includes flights in both directions. Suppose that each pair of cities has direct service from some airline, and suppose that no airline can offer a cycle through an odd number of cities. As a function of k, what is the maximum possible value of n?

The answer is 2^k. In light of Theorem 11.44, the direct flights for each airline must form the edge set of a bipartite graph, and the question asks for the largest n such that all edges of K_n can be obtained using k bipartite subgraphs. Suppose this can be done, with the ith subgraph G_i having bipartition X_i, Y_i. We may assume that $X_i \cup Y_i = V(G)$, since adding isolated vertices does not introduce odd cycles. For each vertex v, define a binary k-tuple a by setting $a_i = 0$ if $v \in X_i$ and $a_i = 1$ if $v \in Y_i$. There are only 2^k binary k-tuples. If there are more than 2^k vertices, then the pigeonhole principle implies that two vertices receive the same k-tuple. Hence these two vertices belong to the same partite set in each bipartite subgraph, and the edge between them belongs to none of the subgraphs. Assuming that n cities can be accommodated, we have proved that $n \leq 2^k$.

Conversely, when $n \leq 2^k$ we can assign distinct binary k-tuples to the n vertices. Let $E(G_i)$ consist of all edges between vertices whose ith coordinate is 0 and vertices whose ith coordinate is 1. This constructs k bipartite subgraphs. Since distinct k-tuples differ in some coordinate, every edge belongs to some G_i, and we have constructed $G_1, \ldots, G_k$ covering the edges of K_n (illustrated below for $k = 2$ and $n = 4$). Hence the upper bound of 2^k is achievable. ∎

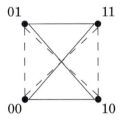

We cannot prove the upper bound in Application 11.46 by showing that a particular successful construction with 2^k vertices cannot accept another city. That argument does not consider all possible schedules for a system with $2^k + 1$ cities; it considers only those containing the special schedule with 2^k cities, and thus it falls into the induction trap. We must consider all possible schedules (see Exercise 11.35).

Next we consider the Marriage Problem (Problem 11.2). We first develop an obvious necessary condition. Let $X = \{x_1, \ldots, x_n\}$ be the set of girls, and let A_i be the set of boys liked by x_i. Giving x_i a partner requires $|A_i| \geq 1$, for each i. We also require $|A_i \cup A_j| \geq 2$ when $i \neq j$, because two girls can't have the same boy as a partner. In general, for each set of k girls, a solution selects k distinct partners from the union of their sets; this requires $|\cup_{i \in J} A_i| \geq |J|$ for every set of indices $J \subseteq [n]$.

This condition is called *Hall's Condition*; it is also sufficient. Let the set of boys be $Y = \{y_1, \ldots, y_n\}$. We can form a bipartite graph G with bipartition X, Y by putting $x_i y_j \in E(G)$ if and only if $y_j \in A_i$. A selection of n distinct boys as partners for the girls corresponds to a selection of n pairwise disjoint edges G. In the example below, the pairing that results is $\{x_1y_1, x_2y_3, x_3y_4, x_4y_2\}$.

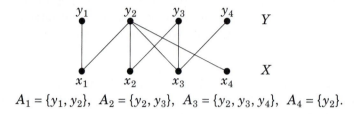

$A_1 = \{y_1, y_2\}, \quad A_2 = \{y_2, y_3\}, \quad A_3 = \{y_2, y_3, y_4\}, \quad A_4 = \{y_2\}.$

11.47. Definition. A *matching* in a graph is a set of pairwise disjoint edges; a *perfect matching* includes edges incident to each vertex.

11.48. Theorem. (Hall's Theorem) Given sets $A_1, \ldots, A_n$, there exist distinct elements $z_1, \ldots, z_n$ such that $z_i \in A_i$ for each i if and only if $|\cup_{i \in J} A_i| \geq |J|$ for every $J \subseteq [n]$.

Proof. We have observed that Hall's condition is necessary. For sufficiency, suppose that the condition holds. We prove by induction on n that the corresponding bipartite graph has a perfect matching. Let X, Y be the two partite sets, and put $x_i y_j \in E(G)$ if and only if $y_j \in A_i$. Next we express Hall's condition in the language of bipartite graphs. Given a set $S \subseteq X$, let $J(S) = \{i: x_i \in S\}$, and let $N(S) = \cup_{i \in J(S)} A_i$. Hence $N(S)$ is the set of vertices in Y that have neighbors in S. Hall's condition states that $|N(S)| \geq |S|$ for all $S \subseteq X$.

We prove by induction on n that the condition is sufficient; the statement is obvious for $n = 1$. For the induction step, suppose $n > 1$, and suppose that Hall's condition implies existence of perfect matchings in bipartite graphs with smaller partite sets. If $|N(S)| > |S|$ for every non-empty proper subset $S \subset X$, then we choose an arbitrary partner y for x_1 from A_1 and form $G' = G - x_1 - y$. Since this deletes at most one vertex of $N(S)$ for each $S \subseteq (X - \{x_1\})$, the graph G' satisfies Hall's Condition. By the induction hypothesis, G' has a perfect matching, which combines with $x_1 y$ to form a perfect matching in G.

Hence we may assume that $|N(S)| = |S|$ for some non-empty $S \subset X$. For all $S' \subseteq S$, we have $N(S') \subseteq N(S)$. Therefore, the subgraph consisting of S, $N(S)$, and the edges between them satisfies Hall's condition. By the induction hypothesis, it has a perfect matching. It suffices to show that the graph G' obtained by deleting S and $N(S)$ also has a perfect matching (see illustration below).

Let $X' = X - S$ and $Y' = Y - N(S)$, so X', Y' is the bipartition of G'. For $T \subseteq X'$, let $N'(T)$ be the set of vertices in Y' with neighbors in T; we have $N'(T) = N(T) - N(S)$. By the induction hypothesis, it suffices to show that $|N'(T)| \geq |T|$ for all $T \subset X'$. Note that T and S are disjoint, as are $N'(T)$ and $N(S)$. From this and the three relationships

$$N(T \cup S) = N'(T) \cup N(S) \qquad |N(S)| = |S| \qquad |N(T \cup S)| \geq |T \cup S|,$$

we compute

$$|N'(T)| = |N(T \cup S)| - |N(S)| \geq |T \cup S| - |S| = |T| \quad \blacksquare$$

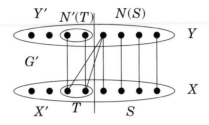

COLORING PROBLEMS

We use graph coloring to model questions about avoiding conflicts.

11.49. Problem. *Scheduling.* Suppose we want to schedule committee meetings in the Senate. Each committee needs one time period during the week, but we can't assign two committees to the same time period if some Senator belongs to both of them. How can we determine the minimum number of time periods needed? ∎

11.50. Definition. A *k-coloring* of a graph G is a function $f: V(G) \to S$, where S is a set of k elements called *colors* (the colors need not all be used). A *k*-coloring is *proper* if no pair of adjacent vertices receive the same color. The *chromatic number* of G, written $\chi(G)$, is the minimum k such that G has a proper k-coloring.

We use the Greek letter "χ" for chromatic number because it begins the Greek word for color.

11.51. Example. *Coloring of bipartite graphs, cycles, and complete graphs.* No two vertices receiving a given color can be adjacent, so $\chi(G)$ equals the minimum number of independent sets whose union is $V(G)$. Hence G is 2-colorable if and only if G is bipartite.

Thus odd cycles have chromatic number at least 3. We can also see this as follows. If C_{2k+1} were 2-colorable, the two colors would have to alternate as we follow the cycle. Since the number of vertices is odd, we wind up with two adjacent vertices of the same color. Changing one of these to a third color produces a 3-coloring.

A proper coloring of a complete graph must give the vertices distinct colors, and distinct colors suffice. Hence $\chi(K_n) = n$. Furthermore, if $K_n \subseteq G$, then $\chi(G) \geq n$. Always $\chi(G)$ is at least the size of its largest complete subgraph, but odd cycles show that equality need not hold. ∎

Finding time slots to schedule committee meetings is a graph coloring problem. Introducing a vertex for each committee, we let two vertices form an edge if their committees have a common member, since this means they cannot have the same time slot. The chromatic number of the resulting graph is the number of time slots needed.

This provides a mathematical model for the scheduling problem but not a solution; we have no general procedure to compute the chromatic number. We can compute it for graphs in some special classes. To prove that $\chi(G) = k$, we provide a proper k-coloring of G (this proves $\chi(G) \le k$), and we prove that G is not $k - 1$-colorable (this proves $\chi(G) \ge k$). We consider a class of graphs that includes all cycles.

11.52. Example. *Coloring of generalized cycles.* Place n points around a circle. For $k \le \lceil n/2 \rceil$, let $G_{n,k}$ be the graph obtained by making each point adjacent to the $k - 1$ nearest points in each direction. The graph $G_{n,1}$ is an independent set, which is 1-colorable. The graph $G_{n,2}$ is the ordinary cycle C_n, which has chromatic number 2 when n is even and 3 when n is odd. The graph $G_{8,3}$ below contains K_3, so $\chi(G_{8,3}) \ge 3$. In fact, $G_{8,3}$ is not 3-colorable. ∎

11.53. Theorem. If $n \ge k(k - 1)$, then the chromatic number of the generalized cycle $G_{n,k}$ is given by

$$\chi(G_{n,k}) = \begin{cases} k & \text{if } k \text{ divides } n \\ k + 1 & \text{if } k \text{ does not divide } n. \end{cases}$$

Proof. Because every k consecutive points on the circle form a complete subgraph, we know that $\chi(G_{n,k}) \ge k$. If $G_{n,k}$ has a proper k-coloring, then each set of k consecutive points must receive k distinct colors. By symmetry, we may assume that the first k labels are $1, \ldots, k$ in order. Since the next point is adjacent to the $k - 1$ most recent points as we proceed, it must have a color different from those. If we use only k colors, the $k + 1$st point must have color 1, and as we continue the colors must cycle through $[k]$ in order repeatedly. This will be a proper coloring if and only if the last k vertices have colors $1, \ldots, k$ before restarting with color 1. Hence $\chi(G_{n,k}) = k$ if and only if $k|n$.

To complete the computation of $G_{n,k}$ for $n \ge k(k - 1)$, we need only prove that $G_{n,k}$ is $k + 1$-colorable. If we can partition the points around the circle into consecutive sets of sizes k and $k+1$, then we can use stretches of colors $1, \ldots, k$ and $1, \ldots, k+1$ to complete a proper coloring. Thus it suffices to express n as $mk + l(k+1)$ for nonnegative integers m, l. The numbers $a = k$ and $b = k+1$ are relatively prime. Solution 6.17 (the Dart Board Problem) guarantees that a solution exists when $n \geq ab - a - b + 1$, which here means $n \geq (k-1)k$.

Alternatively, we can give an explicit formula for l and m. The Division Algorithm yields $n = qk + r$, with $0 \leq r < k$. With $n \geq k(k-1)$, we have $q \geq k-1$. We set $l = r \geq 0$ and $m = q - r \geq 0$. ∎

When $n < k(k-1)$, more colors are needed (see Exercise 11.48).

In general, computing $\chi(G)$ is difficult. The chromatic number is the smallest positive integer k such that the number of proper k-colorings is nonzero; we next consider the more general problem of counting the proper k-colorings. We will show that for each graph G this is a polynomial in k. Our discussion of this problem uses the Inclusion-Exclusion Principle and is optional.

11.54. Definition. Let $\chi(G; k)$ be the number of proper k-colorings of G. As a function of k, this is the *chromatic polynomial* of G.

11.55. Example. *Chromatic polynomials of complete graphs, their complements, and trees.* For some graphs, we can compute chromatic polynomials by using the rules of sum and product (Definitions 5.21 and 5.23) to count the ways of constructing proper colorings. When G is an independent set of size n, we can choose colors independently at the vertices; with k colors available, we have $\chi(G; k) = k^n$. For the clique K_n, the colors must be distinct. Choosing colors for vertices 1 through n in turn yields $\chi(K_n; k) = k(k-1) \cdots (k-n+1)$. We count k-colorings that differ by permuting the labels of the colors as distinct colorings.

Every tree arises from K_1 by iteratively adding a new vertex with one edge to an old vertex. If we color the vertices in the order added, then the color on the first vertex can be chosen in k ways. Subsequently there are $k-1$ ways to choose the color for the each new vertex, no matter how the choices have been made so far. By the rule of product, the chromatic polynomial of a tree with n vertices is $k(k-1)^{n-1}$. ∎

We next obtain a formula for the chromatic polynomial of an arbitrary graph. It does not provide a good algorithm to compute $\chi(G)$, because there are too many subsets of the edge set. It expresses $\chi(G; k)$ as an integer combination of terms of the form k^c with $c \leq n$, where $n = |V(G)|$. Hence $\chi(G; k)$ is a polynomial in k of degree $|V(G)|$.

11.56. Theorem. Let $c(G)$ denote the number of components of a graph G. Given a set $S \subseteq E(G)$ of edges in G, let G_S denote the subgraph of G consisting of all the vertices of G plus the edge set S. The number $\chi(G; k)$ of proper k-colorings of G is

$$\chi(G; k) = \Sigma_{S \subseteq E(G)}(-1)^{|S|}k^{c(G_S)}.$$

Proof. We want to count k-colorings that violate no edges, where an edge is "violated" if both endpoints have the same color. This suggests inclusion-exclusion. We define $|E(G)|$ subsets of the k-colorings; the set corresponding to edge e contains the colorings that violate edge e. Using the Inclusion-Exclusion formula (Theorem 10.12), it remains only to show that $k^{c(G_S)}$ is the number of k-colorings that violate the edges in S. To violate all the edges in S, every vertex we can reach from x by a path of edges in S must have the same color as x. Hence all the vertices within a component of G_S must have the same color, which we can pick in k ways. The choices for the various components are independent; by the rule of product, there are $k^{c(G_S)}$ ways to make all the choices. ∎

11.57. Example. *A chromatic polynomial.* When we apply Theorem 11.56 to a graph with n vertices and m edges, every subset with 0, 1, or 2 edges yields a subgraph with n, $n-1$, or $n-2$ components, respectively, so the contributions from the inclusion-exclusion sum always begin $k^n - mk^{n-1} + \binom{m}{2}k^{n-2}$.

When $|S| = 3$, the number of components is again $n-2$ if the three edges form a triangle; otherwise it is $n-3$. The graph drawn below has two triangles, and the remaining $\binom{5}{3} - 2 = 8$ sets of three edges yield one component. All subgraphs with four or five edges have only one component. Hence the inclusion-exclusion computation is

$$\chi(G; k) = k^4 - 5k^3 + 10k^2 - (2k^2 + 8k^1) + 5k - k = k^4 - 5k^3 + 8k^2 - 4k.$$

By ad hoc counting, one can also see that $\chi(G; k) = k(k-1)(k-2)(k-2)$. This is 0 when k is 1 or 2, but $\chi(G; 3) = 6$. ∎

PLANAR GRAPHS

Consider three hermits A, B, C living in the woods. We must cut paths from each house to three utilities (traditionally gas, water, and electricity). Can we do this without allowing the paths to cross? We will see that we cannot. Graphs that *can* be drawn in the plane have many applications.

11.58. Example. *The Four Color Problem.* Can the regions of every map drawn in the plane (or on the surface of a globe) be colored with four colors so that neighboring regions receive different colors?

This question about a map M becomes a question about a graph G when we create a vertex for each region and join two vertices by an edge if the corresponding regions share a boundary of nonzero length. The number of colors needed for M is $\chi(G)$. The famous Four Color Conjecture, posed in 1852, was that four colors would suffice for every map. Kenneth Appel and Wolfgang Haken proved this in 1976 at the University of Illinois, assisted by a computer.

When we view the vertex for each region as the "capital" and draw roads from the capital to the midpoints of borders with neighboring regions, the roads combine to draw G in the plane with no crossing edges. Below we draw a map with five regions separated by dashed boundaries. The resulting graph G with solid edges is 3-colorable. ■

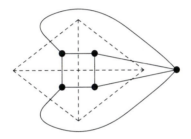

Until now, we have treated edges as abstract pairs of vertices. When we consider geometric properties of drawings of graphs, we treat the edges as curves in the plane. In order to do so, we assume some intuitive geometric properties about regions and curves in the plane. In particular, we do not give a rigorous definition of a continuous function from $[0, 1]$ to $\mathbb{R}^2$; we think of it as a function whose image can be traced by a pencil without lifting the pencil from the paper.

11.59. Definition. A *curve* from u to v in $\mathbb{R}^2$ is the image of a continuous function $f\colon [0, 1] \to \mathbb{R}^2$ such that $f(0) = u$ and $f(1) = v$; it is *simple* if f is injective, with the possible exception that $u = v$ is allowed in a simple curve. When $u = v$, the curve is *closed*.

11.60. Definition. A *drawing* of a graph G is a graph H isomorphic to G such that each vertex of H is a point in $\mathbb{R}^2$ and each edge e of H with endpoints u, v is a simple curve from u to v. In a drawing of a graph, distinct edges e_1 and e_2 *cross* if they intersect other than at a common endpoint. A *planar graph* is a graph that has a drawing without crossings. Such a drawing is called a *plane graph*.

A plane graph $\hat{G}$ that is a drawing of a planar graph G is a convenient representative of the isomorphism class of G.

11.61. Definition. A set $R \subseteq \mathbb{R}^2$ is *path-connected* if for all $u, v \in R$, there is a curve contained in R that has endpoints u, v. A *face* of a plane graph G is a maximal path-connected subset of $\mathbb{R}^2$ that intersects no edge or vertex of G.

11.62. Example. A plane graph with one vertex and no edges has one face. More generally, every plane graph G that is a tree has one face. If p, q are points in $\mathbb{R}^2$ that are not vertices and are not contained in edges of G, then there is a curve from p to q that intersects no edge or vertex of G (Exercise 11.51). ∎

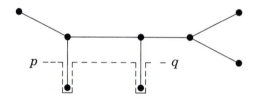

Our intuitive understanding of geometry in the plane suggests the next theorem, whose proof is surprisingly difficult and will be omitted. We use this theorem only in the proof of Theorem 11.64.

11.63. Theorem. (Jordan Curve Theorem). Every simple closed curve in $\mathbb{R}^2$ partitions its complement into two regions known as the *interior* and the *exterior*. Equivalently, every plane graph isomorphic to a cycle has two faces, one bounded and one unbounded.

11.64. Theorem. (Euler's Formula) If G is a connected plane graph with **v** vertices, **e** edges, and **f** faces, then $\mathbf{v} - \mathbf{e} + \mathbf{f} = 2$.

Proof. We use (strong) induction on the number of cycles in G. If G is connected and has no cycles, then G is a tree and $\mathbf{f} = 1$ (Example 11.62). Since $\mathbf{e} = \mathbf{v} - 1$ for a tree, we have $\mathbf{v} - \mathbf{e} + \mathbf{f} = 2$.

Now suppose G has a cycle C containing an edge e. Because C is a cycle, the Jordan Curve Theorem implies that the two sides of e in the drawing of G belong to different faces of G; one inside C and one outside C. When we delete e from G, the union of these two faces is path-connected and becomes a single face in the resulting plane graph G'. Since e belongs to a cycle, Lemma 11.36 implies that G' is connected. Furthermore, G' has fewer cycles than G, since every cycle in G' appears in G, but C appears in G and not G'. Hence we can apply the induction hypothesis to G'; if its numbers of vertices, edges, and faces are $\mathbf{v'}, \mathbf{e'}, \mathbf{f'}$, this yields $\mathbf{v'} - \mathbf{e'} + \mathbf{f'} = 2$. Since $\mathbf{v} = \mathbf{v'}$, $\mathbf{e} = \mathbf{e'} + 1$, and $\mathbf{f} = \mathbf{f'} + 1$, we conclude that $\mathbf{v} - \mathbf{e} + \mathbf{f} = 2$. ∎

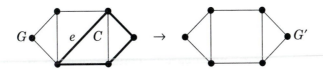

The statement and proof of Euler's Formula allows multiple edges. Our first application yields a necessary condition for planar graphs.

A *complete bipartite graph* is a simple bipartite graph in which two vertices are adjacent if and only if they belong to different partite sets. We use $K_{r,s}$ to denote such a graph with partite sets of sizes r and s. Example 11.23 shows three drawings of $K_{3,3}$. We will prove that K_5 and $K_{3,3}$ (the "gas-water-electricity" graph) are not planar. An edge e is in the *boundary* of a face F of a plane graph G if some segment with an endpoint in F crosses e and no other edge of G.

11.65. Theorem. Every simple planar graph with $n \geq 3$ vertices has at most $3n - 6$ edges. Every simple planar graph with $n \geq 3$ vertices and no copy of C_3 has at most $2n - 4$ edges.

Proof. For $n = 3$, both statements are true by inspection. Consider a maximal simple plane graph G with $n \geq 4$. We may assume that G is connected, since otherwise we can add at least one edge. We can now use Euler's formula to relate $\mathbf{v}$ and $\mathbf{e}$ if we can dispose of $\mathbf{f}$. Since G is simple, every face has at least three edges in its boundary. Every edge lies in the boundary of at most two faces. Summing the numbers of edges in the boundaries of the faces thus yields the inequality $2\mathbf{e} \geq 3\mathbf{f}$. Substituting this into $n - \mathbf{e} + \mathbf{f} = 2$ yields $\mathbf{e} \leq 3n - 6$.

For the second statement, we check the graphs with $n = 4$ individually (see Example 11.26). If also G has no C_3 and $n \geq 5$, then each face of G has at least four edges in its boundary. In this case the inequality becomes $2\mathbf{e} \geq 4\mathbf{f}$, and we obtain $\mathbf{e} \leq 2n - 4$. ∎

11.66. Example. K_5 *and* $K_{3,3}$. Theorem 11.65 implies that K_5 and $K_{3,3}$ are not planar graphs. For K_5, we have $\mathbf{e} = 10 > 9 = 3n - 6$. For the bipartite graph $K_{3,3}$, we have $\mathbf{e} = 9 > 8 = 2n - 4$. Each graph can be drawn using only one crossing. ∎

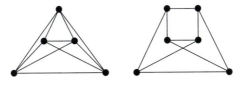

11.67. Remark. *Kuratowski's Theorem.* Example 11.66 yields the easy half of a characterization of planar graphs. Replacing an edge with a path having the same endpoints (and passing through new vertices) does not affect whether a graph is planar. Thus a graph containing a subgraph obtained from K_5 or $K_{3,3}$ by replacing edges with paths cannot be planar. Kuratowski's Theorem states that a graph is planar if and only if it contains no subgraph obtainable from K_5 or $K_{3,3}$ by replacing edges with paths. ∎

We next apply Euler's Formula to the Platonic Solid Problem. We describe solids and their relationship to planar graphs informally. A solid S is bounded by planes in space. The portion of the boundary of S belonging to one of these planes is a "face" of S. The edges of S are intersections of neighboring faces, and the vertices of S are intersections of edges. The vertices and edges of S are the vertices and edges of a graph, called the "skeleton" of S.

Suppose we translate and rotate S to rest on one of its faces in the plane. If we puncture another face and gradually spread the surface out into the plane, we obtain a plane graph G having the same intersection relationships as S among vertices, edges, and faces; it is a drawing (in the plane) of the skeleton of S. Each face bounded by l edges in S becomes a face of G whose boundary is a cycle of length l.

11.68. Solution. *The Platonic Solid Problem.* By definition, a Platonic solid has a graph in which every vertex has the same degree k and every face has the same length l. Also the physical properties require $k, l \geq 3$. We show that there are only five such solids by showing that there are only five planar graphs with these properties. Consider a planar drawing of such a graph. By the Degree-sum Formula, $2\mathbf{e} = \mathbf{v}k$. Since every edge belongs to exactly two faces, we also have $2\mathbf{e} = \mathbf{f}l$. By Euler's Formula, we have $\mathbf{v} - \mathbf{e} + \mathbf{f} = 2$. Substituting for $\mathbf{v}$ and $\mathbf{f}$ into Euler's Formula, we have $\mathbf{e}(2/k - 1 + 2/l) = 2$. Since $\mathbf{e}$ and 2 are positive, the other factor must also be positive, which yields $(2/k) + (2/l) > 1$, and hence $2l + 2k > kl$. This inequality is equivalent to $(k-2)(l-2) < 4$. Since $k, l \geq 3$, we find that there are only 5 integer solution pairs. Once we specify the vertex degrees and the cycle lengths, there is essentially only one way to form the planar graph (we omit the details of this). Hence there are no more than the five known Platonic solids. ∎

k	l	$(k-2)(l-2)$	$\mathbf{e}$	$\mathbf{v}$	$\mathbf{f}$	name
3	3	1	6	4	4	tetrahedron
3	4	2	12	8	6	cube
4	3	2	12	6	8	octahedron
3	5	3	30	20	12	dodecahedron
5	3	3	30	12	20	icosahedron

Euler's Formula applies to some geometric counting problems.

11.69. Solution. *Regions in a Circle.* We prove that the chords determined by n points on a circle cut the interior into $1 + \binom{n}{2} + \binom{n}{4}$ regions, if no three chords have a common intersection. We obtain a plane graph G by viewing the points on the circle and the intersections of two chords as vertices. Since each intersection of chords is determined by four points on the circle and each set of four points determine exactly one intersection, G has $\binom{n}{4} + n$ vertices. Since each interior intersection has degree 4 and each vertex on the circle has degree $n + 1$, the Degree-sum Formula implies that G has $\frac{1}{2} 4\binom{n}{4} + \frac{1}{2} n(n + 1)$ edges. By Euler's Formula, the number of faces is $2 + \mathbf{e} - \mathbf{v} = 2 + \binom{n}{2} + \binom{n}{4}$. Subtracting 1 for the unbounded face leaves $1 + \binom{n}{2} + \binom{n}{4}$ regions inside the circle. ■

We return to the issue of coloring planar graphs. The Four Color Theorem is difficult, but 6-colorability is not hard (Exercise 11.53). Also we can solve the Art Gallery Problem using an easier theorem about coloring a special class of planar graphs.

11.70. Definition. A graph is *outerplanar* if it has a drawing in the plane with every vertex on the boundary of the unbounded face.

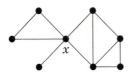
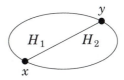

11.71. Theorem. Every outerplanar graph is 3-colorable.

Proof. We use induction on the number of vertices; every graph with at most three vertices is 3-colorable. For the induction step, suppose G is outerplanar with $n > 3$ vertices. Every subgraph of an outerplanar graph is outerplanar. We may assume that G is connected, else we apply the induction hypothesis to each component. Suppose G has a vertex x such that $G - x$ is disconnected. Let $G_1, \dots, G_k$ be the components of $G - x$, and let G_i' be the graph consisting of G_i together with x and the edges from x to $V(G_i)$. The graph G_i' is outerplanar and has fewer vertices than G. By the induction hypothesis, each G_i' is 3-colorable. We can permute the names of colors to make these colors agree at x and obtain a proper 3-coloring of G.

If G has no such vertex x, then the boundary of the unbounded face of G is a cycle C (drawn as a simple closed curve). If C is all of G, then G is a cycle and is 3-colorable. Otherwise, C has a chord xy. Let H_1 be the subgraph consisting of xy, an x, y-path on C, and all chords of the cycle so formed. Let H_2 be the subgraph formed using the other x, y-path on C. By the induction hypothesis, H_1 and H_2 are 3-colorable. Again we can permute the names of the colors in these 3-colorings to agree on $\{x, y\}$; together this yields a proper 3-coloring of G. ∎

11.72. Solution. *The Art Gallery Problem.* We prove that $\lfloor n/3 \rfloor$ guards suffice to watch art galleries with n walls (this is best possible - Exercise 11.57). Consider a simple polygon in the plane representing the art gallery. By adding chords between vertices, we obtain an outerplanar graph in which every bounded face is a triangle. By Theorem 11.71, this graph is 3-colorable. Below we show a triangulation and coloring of the art gallery in Problem 11.4.

Given a proper 3-coloring, the Pigeonhole Principle implies that one of the colors is used on at most $\lfloor n/3 \rfloor$ vertices (color 1 in the example below). We claim that guards placed at the vertices with that color can watch the entire gallery. Because every bounded face is a triangle, each bounded face receives all three colors on its vertices and hence receives a guard at one vertex. This guard can watch this entire triangle. ∎

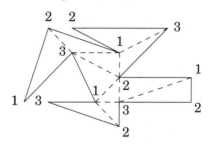

EXERCISES

11.1. Let G be the graph with vertex set $[12]$ in which vertices u, v are adjacent if and only if u and v are relatively prime. Count the edges of G.

11.2. Let G be the graph with vertex set $\mathbf{Z}_n$ in which vertices u, v satisfy the adjacency relation if and only if u and v differ by 6. For each $n \geq 1$, determine the number of components of G.

11.3. Prove or disprove: there is no Eulerian graph with an even number of vertices and an odd number of edges.

11.4. (!) Suppose G is a connected non-Eulerian graph. Prove that the minimum number of trails that together traverse each edge of G exactly once is half the number of vertices having odd degree. (Hint: transform G into a new graph G' by adding edges and/or vertices.)

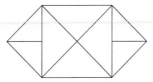

11.5. Can the vertices of a simple graph have distinct degrees?

11.6. In a league with two divisions of 11 teams each, is it possible to schedule a season with each team playing 7 games within its division and 4 games against teams in the other division?

11.7. (!) Suppose G is a connected graph in which every vertex has even degree. Prove that G has no edge whose deletion leaves a disconnected subgaph.

11.8. Suppose l, m, n are nonnegative integers with $l + m = n$. Find necessary and sufficient conditions on l, m, n such that there exists a connected n-vertex simple graph with l vertices of even degree and m vertices of odd degree.

11.9. Suppose G is a graph. Prove or disprove:
   a) Deleting a vertex of maximum degree cannot raise the average degree.
   b) Deleting a vertex of minimum degree cannot reduce the average degree.

11.10. (!) Suppose x and y are vertices of degree at least $(n + k - 2)/2$ in a simple graph G with n vertices. Prove that x and y have at least k common neighbors.

11.11. Describe an inductive construction of the d-cube Q_d. Use it to prove that Q_d has $d2^{d-1}$ edges and has a cycle containing all its vertices (if $d \geq 2$).

11.12. Count the cycles of lengths 4 and 6 in the d-dimensional cube. (Hint: there is more than one "type" of 6-cycle.)

11.13. (−) Among the graphs below, which pairs are isomorphic?

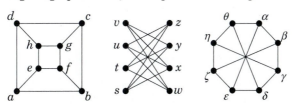

11.14. (−) For simple graphs G and H, prove that $G \cong H$ if and only if $\bar{G} \cong \bar{H}$.

11.15. What is the smallest value of n such that there are two n-vertex simple graphs with the same list of vertex degrees that are not isomorphic? (Hint: use the list in Example 11.26.)

11.16. Prove that there are exactly two isomorphism classes of 7-vertex simple graphs in which every vertex has degree 4. (Hint: consider the complements.)

11.17. Suppose G is a simple graph isomorphic to its complement $\bar{G}$. Prove that the number of vertices in G is congruent to 0 or 1 modulo 4.

11.18. (!) The *Petersen graph* is the graph drawn on the left below. Prove that the drawings below are all drawings of the Petersen graph; that is, show that these graphs are pairwise isomorphic.

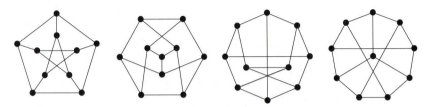

11.19. (!) Prove that there are exactly $2^{\binom{n-1}{2}}$ simple graphs with vertex set $v_1, \ldots, v_n$ in which every vertex has even degree. (Hint: establish a bijection between this set and the set of all simple graphs with vertex set $\{v_1, \ldots, v_{n-1}\}$.)

11.20. (!) Give combinatorial proofs of the statements below by using simple graphs. (Hint: interpret each quantity as counting something involving graphs.)
   a) If $n, k \in \mathbb{N}$ with $0 \le k \le n$, then $\binom{n}{2} = \binom{k}{2} + k(n-k) + \binom{n-k}{2}$.
   b) If $n_1, \ldots, n_k \in \mathbb{N}$ with $\Sigma_{i=1}^k n_i = n$, then $\Sigma_{i=1}^k \binom{n_i}{2} \le \binom{n}{2}$.

11.21. (!) Prove that a graph G is connected if and only if for every partition of $V(G)$ into non-empty sets S, T, there is an edge xy with $x \in S$ and $y \in T$.

11.22. Consider three buckets with integer capacities $l > m > n$ in gallons. Initially, the largest bucket is filled. We need to measure out k gallons, but there are no markers on the buckets, so the only operation we can perform is to pour as much water from one bucket into another as will fit; in this way, we can keep track of how much water is in each bucket. Use graph theory to describe a method for determining whether it is possible to measure k gallons.

11.23. Suppose G is a graph in which every vertex has degree at least k, where k is an integer at least 2. Prove that G has a path of length at least k and a cycle of length at least $k + 1$. (Hint: consider a maximal path.)

11.24. Suppose k is the maximum length of a path in a connected graph G. If P, Q are paths of length k in G, prove that P and Q have a common vertex.

11.25. (!) Suppose G is a simple graph having n vertices and no 3-vertex cycle. Prove that G has at most $n^2/4$ edges. (Hint: consider the subgraph consisting of the neighbors of a vertex of maximum degree and the edges among them.)

11.26. Suppose G is a simple n-vertex graph in which every vertex has at least $\lfloor n/2 \rfloor$ neighbors. Prove that G is connected. Furthermore, show that this bound is best possible by constructing for each $n \ge 2$ a disconnected n-vertex graph in which every vertex has at least $\lfloor n/2 \rfloor - 1$ neighbors.

11.27. (!) Prove that every graph with n vertices and $n - k$ edges has at least k components.

11.28. (!) Suppose G is a graph with n vertices and $n - 1$ edges. Prove that G is connected if and only if G has no cycles. (Comment: compare with Theorem 11.40.)

11.29. (!) Prove that a graph G is a tree if and only if for all $x, y \in V(G)$, there is exactly one x, y-path in G.

11.30. (−) Prove that every tree with maximum degree k has at least k leaves.

11.31. Prove that a connected graph with n vertices has exactly one cycle if and only if it has exactly n edges.

11.32. Let $d_1, \ldots, d_n$ be n natural numbers. Prove that there exists a tree with n vertices that has these as its vertex degrees if and only if $\Sigma d_i = 2n - 2$. (Hint: two implications are needed; use induction for one of them. Comment: It is not true that every n-vertex graph whose degrees sum to $2n - 2$ is a tree.)

11.33. (!) Suppose that T is a tree with m edges, and that G is a simple graph in which every vertex has degree at least m. Prove that G contains T as a subgraph. (Hint: use induction on m.)

11.34. (!) Prove that every graph has a bipartite subgraph with at least half of its edges. (Hint: prove that the largest bipartite subgraph has this property.)

11.35. Use induction on k (for both implications) to prove that $E(K_n)$ can be covered by k bipartite graphs if and only if $n \leq 2^k$. (Comment: this repeats the result of Application 11.46 (the Airline Problem).)

11.36. Suppose G is a graph having no cycle of even length. Prove that every edge of G appears in at most one cycle.

11.37. Prove that every tree has at most one perfect matching.

11.38. Determine the minimum size of a maximal matching in the cycle C_n.

11.39. Suppose that $k \in \mathbb{N}$ and that G is a bipartite graph with bipartition X, Y in which every vertex of G has degree k. Prove that $|X| = |Y|$.

11.40. (!) Suppose G is a bipartite graph in which all vertices have degree k, where $k \geq 1$. Prove that G has a perfect matching.

11.41. (!) Suppose G is a bipartite graph. Prove that G has a perfect matching if and only if the maximum size of an independent set of vertices in G is exactly half the total number of vertices.

11.42. (−) How many perfect matchings does $K_{n,n}$ contain? How many cycles of length $2n$ does $K_{n,n}$ contain?

11.43. (−) The *wheel* with n vertices consists of a cycle with $n - 1$ vertices and one additional vertex adjacent to all the vertices on that cycle. Determine the chromatic number of the wheel with n vertices.

11.44. Is it always true that $\chi(G) \le 1 + d$, where d is the average degree of the vertices? Give a proof or a counterexample.

11.45. Prove that if G does not have two disjoint odd cycles in G, then $\chi(G) \le 5$.

11.46. (!) Suppose every vertex of a graph G has degree at most k. Prove that $\chi(G) \le k + 1$. For each k, construct a graph where the maximum vertex degee is k and the chromatic number equals $k + 1$.

11.47. Given a collection of lines in the plane with no three intersecting at a point, form a graph G whose vertices are the points of intersection of the lines and whose edges are the segments on the lines joining two points of intersection. Prove that $\chi(G) \le 3$.

11.48. Let $G_{n,k}$ be the generalized cycle defined in Example 11.52. Use the pigeonhole principle to prove that $\chi(G_{n,k}) > k + 1$ when $n = k(k-1) - 1$.

11.49. Prove that the coefficients of $\chi(G; k)$ sum to 0 unless G has no edges.

11.50. (+) Suppose G is a simple graph with n vertices and m edges. Prove that G has at most $\frac{1}{3} \binom{m}{2}$ triangles. Conclude that the coefficient of k^{n-2} in $\chi(G; k)$ is positive, unless G has at most one edge.

11.51. Without using Euler's formula, prove that a plane graph that is a tree has one face. (Hint: use induction on the number of vertices.)

11.52. (–) Suppose G is a simple planar graph with at least 11 vertices. Prove that $\bar{G}$ is not planar.

11.53. (!) Prove that every planar graph has a vertex of degree at most 5, and use this to prove that every planar graph has chromatic number at most 6.

11.54. Suppose that G is an n-vertex simple planar graph with no cycle of length less than k. Prove that G has at most $(n-2)k/(k-2)$ edges, and use this to prove that the Petersen graph of Exercise 11.18 is not planar.

11.55. (!) Use Euler's Formula to prove that an outerplanar graph with n vertices has at most $2n - 3$ edges.

11.56. Prove by induction that every simple outerplanar graph with at least four vertices has two nonadjacent vertices with degree at most 2.

11.57. (+) For each n, construct an art gallery with n walls to prove that the bound of $\lfloor n/3 \rfloor$ in Solution 11.72 is best possible. (Hint: use groups of three vertices to build "rooms" such that no guard can see into more than one room.)

Chapter 12

Recurrence Relations

Let a_n be the number of binary lists of length n. We first express a_n in terms of a_{n-1}. The number of lists doubles when we increase the length by one. To each list of length $n-1$, we may append a 0 or a 1. The resulting $2a_{n-1}$ lists are distinct, and every list of length n arises from some list of length $n-1$ in this way. Therefore, $a_n = 2a_{n-1}$. Since $a_1 = 2$, we see inductively that $a_n = 2^n$.

We have viewed the counting of binary lists in two ways. The formula gives a_n explicitly as a function of n; the recursive definition also determines a_n. Recursive definitions can provide more insight.

In this chapter, we consider combinatorial problems whose analysis leads to recursive definitions of sequences. We also develop techniques for obtaining explicit formulas from these recursive definitions. Recurrence relations enable us to analyze problems inductively, particularly when it is difficult to see a general pattern.

12.1. Problem. *The Tower of Hanoi.* The French mathematician Edouard Lucas (1842-1891) constructed a puzzle with three pegs and seven rings of different sizes that could slide onto the pegs. Legend has it that an order of monks had a similar puzzle with 64 large golden disks. Starting with all rings on one peg in order by size, the problem is to transfer the pile to another peg subject to two conditions: rings are moved one by one, and no ring is ever placed on top of a smaller ring. The monks supposedly believed that the world would crumble when the job was finished. How many moves are required? ∎

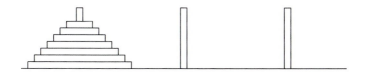

12.2. Problem. *The Fibonacci numbers.* Suppose n spaces are available for parking along a curb. We can fill the spaces using Rabbits, which take one space, and/or Cadillacs, which take two spaces. In how many ways can we fill the spaces? In other words, how many lists of 1's and 2's sum to n? The answer arises in many natural phenomena. ∎

12.3. Problem. *The number of triangles in a triangular grid.* How many triangles are there in the equilateral triangular grid T_n with side length n? Below we illustrate the grid for $n = 3$. This has nine triangles of size 1, three of size 2, and one of size 3, for 13 altogether. ∎

12.4. Problem. *The Polygon Problem.* A *triangulation* of a convex n-sided polygon is a partition of the interior into triangles formed by adding $n - 3$ noncrossing diagonals between corners. A triangle has 1 triangulation, a quadrilateral has 2, a pentagon has 5, and a hexagon has 14. How many triangulations does a convex n-gon have? ∎

FIRST-ORDER RECURRENCES

Recall that a sequence $\langle a \rangle$ of real numbers is a function from $\mathbb{N}$ or $\mathbb{N} \cup \{0\}$ to $\mathbb{R}$. We write a_n for the value at the integer n. In this chapter we start sequences at $n = 0$. We can specify a sequence $\langle a \rangle$ by a formula for a_n or by a recursive expression for a_n. For example, the sequence defined by the formula $a_n = 3(-1)^n$ can also be defined recursively by $a_0 = 3$ and $a_n = -a_{n-1}$ for $n \geq 1$. It is easy to prove by induction that these two definitions produce the same sequence. To begin our investigation of recursive definitions, we study an easy case.

12.5. Definition. A *first-order recurrence relation* for a sequence $\langle a \rangle$ is an equation $a_n = g(a_{n-1}, n)$ valid for $n \geq 1$, with $g: \mathbb{R} \times \mathbb{N} \to \mathbb{R}$.

12.6. Proposition. A first-order recurrence relation for $\langle a \rangle$, together with an initial value $a_0 = s$, uniquely determines a_n for all $n \geq 0$.

Proof. We use induction on n. Basis step: the hypothesis specifies that $a_0 = s$. Induction step (where $n \geq 1$): the induction hypothesis states that a_{n-1} has been uniquely determined. The recurrence $a_n = g(a_{n-1}, n)$ then determines a_n uniquely, since g is a function. ∎

12.7. Example. Consider the first-order recurrence $a_n = a_{n-1} + n$ for $n \geq 1$, with $a_0 = s$. We can verify by induction that $a_n = s + (n+1)n/2$ for $n \geq 0$. ∎

12.8. Definition. A first-order recurrence relation for $\langle a \rangle$ is *linear* if there are functions f, h such that $a_n = h(n)a_{n-1} + f(n)$ for $n \geq 1$.

12.9. Example. The recurrence relations $a_n = 2a_{n-1}$ and $b_n = 2b_{n-1} + 1$ are linear. With initial conditions $a_0 = s$ and $b_0 = t$, their solutions are $a_n = s2^n$ and $b_n = (1+t)2^n - 1$, again proved by induction. ∎

12.10. Example. *Recurrences that become summations.* Consider the first-order linear recurrence $a_n = a_{n-1} + f(n)$ for $n \geq 1$. We can "iterate" the recurrence to express a_n as a summation. Iteration yields

$$a_n = a_{n-1} + f(n) = a_{n-2} + f(n) + f(n-1) = \cdots = a_0 + \Sigma_{i=1}^{n} f(i).$$

When we can evaluate the sum, we obtain a formula for a_n. We did this for $f(n) = n$ in Example 12.7. ∎

We consider two combinatorial problems that lead to first-order linear recurrence relations. There are two steps in solving a problem using recurrence relations. First we derive the recurrence relation, then we solve the recurrence to obtain a formula.

12.11. Example. *Regions in the plane.* A *configuration* of lines is a finite collection of lines in the plane such each pair of lines has one common point and no three lines have a common point. Let a_n be the number of regions created by a configuration of n lines. It is not obvious that every configuration of n lines creates the same number of regions; this follows inductively when we establish a recurrence for a_n.

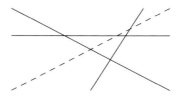

We begin with no lines and one region, so $a_0 = 1$. We prove that $a_n = a_{n-1} + n$ if $n \geq 1$. Consider a configuration of n lines, with $n \geq 1$, and let L be one of these lines. The other lines form a configuration of

$n-1$ lines. We argue that adding L increases the number of regions by n. The intersections of L with the other $n-1$ lines partition L into n portions. Each of these portions cuts a region into two. Thus adding L increases the number of regions by n. Since this holds for all configurations we have $a_n = a_{n-1} + n$ for $n \geq 1$. By Proposition 12.6, this determines a unique sequence starting with $a_0 = 1$, and hence every configuration of n lines creates the same number of regions.

This is the recurrence of Example 12.7 with the initial condition $a_0 = 1$. The solution is $a_n = 1 + (n+1)n/2$. ∎

12.12. Solution. *The Tower of Hanoi.* Consider the Tower of Hanoi problem for n rings; let a_n be the number of moves required to move the pile to another peg. To move the pile, we must move the bottom ring. In order to move the bottom ring, we must first move the top $n-1$ rings to another peg. By the definition of a_{n-1}, this requires a_{n-1} moves. Then we move the bottom ring to the empty peg and must solve the smaller problem again to put the other rings back on top of it. This yields the recurrence $a_n = 2a_{n-1} + 1$ for $n \geq 1$. We have argued that this many moves are required, and also this many moves are sufficient. For the initial condition we have $a_0 = 0$, since it takes no moves to move no rings. In Example 12.9, we found the solution $a_n = 2^n - 1$. Don't worry about the world crumbling; $2^{64} - 1$ seconds is more than 10^{11} years. ∎

Up to now, we have presented solutions to recurrence relations and proved them by induction. Computing several terms of a sequence sometimes suggests a general formula, but we might guess wrong.

12.13. Example. *A cautionary remark.* Consider n points on a circle so that no three of the segments connecting them intersect at a single point. Let a_n be the number of regions into which such segments cut the region enclosed by the circle. For $n \geq 1$, the sequence begins 1, 2, 4, 8, 16, but $a_6 = 31$. Thus the answer is not the powers of 2 (see also Solution 11.69, Exercise 5.31, and Exercise 12.13. ∎

We cannot rely on guessing the answer, so we need systematic methods for solving recurrences. We have seen that $a_n = a_{n-1} + f(n)$ can be solved by evaluating $\Sigma_{i=1}^{n} f(i)$. We can also solve $a_n = ca_{n-1} + f(n)$ when c is a constant and f is a polynomial. The recurrence we derived for the Tower of Hanoi problem has this form.

12.14. Theorem. Suppose f is a polynomial of degree d. The solution of the recurrence $a_n = ca_{n-1} + f(n)$ for $n \geq 1$ has the form $Ac^n + p(n)$, where p is a polynomial. If $c \neq 1$, then p has degree d. If $c = 1$, then p has degree $d + 1$. The polynomial p is independent of the initial condition; the initial condition determines A.

Proof. When $c = 1$, we have $a_n = a_0 + \sum_{i=1}^{n} f(i)$, as in Example 12.10. When f is a polynomial of degree d, the value of this sum is a polynomial in n of degree $d + 1$, by Theorem 9.12.

Suppose $c \neq 1$. We prove the existence of p by induction on the degree d of f. For the basis step $d = 0$, suppose $f(n) = b$ for all n. Let $p(n) = b/(1 - c)$ for all n. We verify for all $A \in \mathbb{R}$ that $a_n = Ac^n + p(n)$ satisfies the recurrence:

$$ca_{n-1} + f(n) = c[Ac^{n-1} + \frac{b}{1-c}] + b = Ac^n + \frac{b}{1-c} = a_n$$

When we set $A = a_0 - \frac{b}{1-c}$, our formula agrees with the initial condition. By Proposition 12.6, this yields the unique solution.

For the induction step, we assume that the claim holds for all polynomials of degree less than d. Suppose $f(n) = Dn^d + g(n)$, where g has degree less than d. Define a new sequence $\langle \alpha \rangle$ by $\alpha_n = a_n - \frac{D}{1-c} n^d$. We prove that $\langle \alpha \rangle$ satisfies a first-order linear recurrence $\alpha_n = c\alpha_{n-1} + h(n)$ with h a polynomial of degree less than d. Using the formula for α_n in terms of a_n, and using $-D/(1-c) = -D - cD/(1-c)$, we obtain

$$\alpha_n - c\alpha_{n-1} = a_n - \frac{D}{1-c} n^d - c[a_{n-1} - \frac{D}{1-c}(n-1)^d]$$

$$= [a_n - ca_{n-1} - Dn^d] + \frac{cD}{1-c}[(n-1)^d - n^d]$$

The first bracketed expression is $g(n)$, a polynomial of degree less than d. The second is a polynomial of degree $d - 1$. We let $h(n)$ be their sum.

By the induction hypothesis, $\alpha_n = Bc^n + q(n)$, where q is a polynomial of degree $d - 1$ and B is determined by the initial condition $\alpha_0 = a_0$. Using the definition of α, we obtain a formula for a of the desired form.

$$a_n = \alpha_n + \frac{D}{1-c} n^d = Bc^n + \frac{D}{1-c} n^d + q(n). \qquad \blacksquare$$

This theorem justifies using the method of undetermined coefficients to find the polynomial p.

12.15. Example. Consider the recurrence $a_n = 2a_{n-1} + n^2 - 1$ for $n \geq 1$, with $a_0 = 1$. We have proved that the solution has the form $a_n = A2^n + p(n)$, where p has degree 2. We write $p(n) = b_0 + b_1 n + b_2 n^2$ and substitute the resulting formula for a_n into the recurrence to determine the coefficients.

$$A2^n + b_0 + b_1 n + b_2 n^2 = 2A2^{n-1} + 2b_0 + 2b_1(n-1) + 2b_2(n-1)^2 + n^2 - 1$$

We equate coefficients of corresponding powers of n in this equation (see Corollary 4.19).

exponent on n	left side	right side
0	b_0	$2b_0 - 2b_1 + 2b_2 - 1$
1	b_1	$2b_1 - 4b_2$
2	b_2	$2b_2 + 1$

The solution is $b_2 = -1$, $b_1 = -4$, $b_0 = -5$. We determine A to satisfy the initial condition, $1 = A \cdot 2^0 + (-5)$. Thus $a_n = 6 \cdot 2^n - 5 - 4n - n^2$. ∎

In Theorem 12.14, the reader may wonder why we obtain a polynomial of higher degree when $c = 1$. Notice that the recurrence specifies $\langle a_n \rangle$ using $d + 3$ constants: the $d + 1$ coefficients from f, the constant c, and the initial condition. Thus the solution should also involve $d + 3$ constants. When $c \neq 1$, these are c, A, and the $d + 1$ coefficients in p. When $c = 1$, the information in the constant term in p is absorbed by the coefficient A. To have $d + 1$ independent coefficients in p, we use the coefficients of $n, \ldots, n^{d+1}$.

The techniques we have developed in this section apply more generally. Often a substitution reduces a recurrence to a simpler form. The introduction of α_n in the proof of Theorem 12.14 illustrates this technique; we close this section with another example.

12.16. Corollary. Suppose f is a polynomial of degree d. The solution of the recurrence $a_n = ca_{n-1} + f(n)\beta^n$ for $n \geq 1$ has the form $Ac^n + p(n)$, where p is a polynomial. If $c \neq \beta$, then p has degree d. If $c = \beta$, then p has degree $d + 1$. The polynomial p is independent of the initial condition; the initial condition determines A.

Proof. We define b_n by setting $a_n = \beta^n b_n$. By substituting into the recurrence for $\langle a \rangle$ and canceling β^n, we obtain $b_n = (c/\beta)b_{n-1} + f(n)$. This recurrence has the form specified in Theorem 12.14. We take the solution given by Theorem 12.14 and multiply by β^n to obtain the formula for a_n. We leave the details to Exercise 12.14. ∎

SECOND-ORDER RECURRENCES

In this section, we present the most famous example of a second-order recurrence, and we develop techniques for solving some second-order recurrences.

12.17. Definition. A *second-order recurrence relation* for $\langle a \rangle$ is an equation $a_n = g(a_{n-1}, a_{n-2}, n)$, valid for $n \geq 2$, with $g \colon \mathbb{R} \times \mathbb{R} \times \mathbb{N} \to \mathbb{R}$. The relation is *linear* if there are functions h_1, h_2, f such that

$g(a_{n-1}, a_{n-2}, n) = h_1(n)a_{n-1} + h_2(n)a_{n-2} + f(n)$. The linear recurrence relation *has constant coefficients* if h_1 and h_2 are constant.

12.18. Proposition. A second-order recurrence relation for $\langle a \rangle$, together with initial values $a_0 = s$ and $a_1 = t$, uniquely determines a_n for all $n \geq 0$.

Proof. We use induction on n to show that, for all $n \geq 1$, a_{n-1} and a_n are uniquely determined. Basis step: for $n = 1$, the hypothesis specifies the values of a_0 and a_1. Induction step: for $n \geq 2$, the induction hypothesis states that a_{n-2} and a_{n-1} are known. The recurrence $a_n = g(a_{n-1}, a_{n-2}, n)$ then determines a_n uniquely, since g is a function. ■

The proof of Proposition 12.18 illustrates a general principle: when using induction to prove a statement about the solution to a second-order recurrence, we must check two values in the basis step. Otherwise, our argument for the induction step is not valid where we first try to apply it. This occurred also in the inductive solution to the L-Tiling Problem (Solution 4.18), because the induction step used the induction hypothesis for the two preceding values.

12.19. Example. Consider the second-order recurrence relation $a_n = 4a_{n-1} - 4a_{n-2}$ for $n \geq 2$, with $a_0 = 1$ and $a_1 = 4$. The solution is $a_n = (n + 1)2^n$. Checking the formula for $n = 0$ and $n = 1$ means verifying that $1 \cdot 2^0 = 1 = a_0$ and $2 \cdot 2^1 = 4 = a_1$. In the induction step we verify the formula when n is at least 2 given that it holds for $n - 1$ and $n - 2$. Using the recurrence for $\langle a \rangle$, we compute

$$a_n = 4a_{n-1} - 4a_{n-2} = 4n2^{n-1}4(n-1)2^{n-2} = 2n2^n - (n-1)2^n = (n+1)2^n$$

The validity of this computation when $n = 2$ depends on having checked the formula for both $n = 0$ and $n = 1$. ■

Now we consider how a second-order recurrence relation can arise.

12.20. Example. *Fibonacci numbers.* The recurrence $a_n = a_{n-1} + a_{n-2}$ was studied by Leonardo of Pisa (1170?-1250), known as Fibonacci. The *Fibonacci numbers* F_n are the numbers determined by the *Fibonacci recurrence* $F_n = F_{n-1} + F_{n-2}$ with the initial conditions $F_0 = F_1 = 1$.[†]

The Fibonacci sequence occurs in many applications; the mathematical journal *The Fibonacci Quarterly* is devoted to its study. Fibonacci studied a model of a rabbit farm where rabbits mature and reproduce rapidly. Suppose that every pair of rabbits produces a new pair of

---

[†] The initial conditions $F_0 = 0$ and $F_1 = 1$ are also commonly used. This merely shifts the index of the sequence by one. We choose the indexing $F_0 = F_1 = 1$ for its direct relationships to combinatorial models and to the generating function.

rabbits every month as soon as it is two months old. If the farm starts with one pair of rabbits born at time 0, then there is also one pair at time 1, and beginning at time 2 the rabbits alive two months earlier all give birth. Hence the number of pairs alive at time n are all those alive at time $n-1$ plus those newly born, which is the number alive two months earlier. Hence the total number satisfies the recurrence $F_n = F_{n-1} + F_{n-2}$, with the initial conditions $F_0 = F_1 = 1$. ∎

12.21. Solution. *Rabbits and Cadillacs.* Suppose n spaces in a row are available for parking. We want to count the ways to fill the spaces with Rabbits, which take one space, or Cadillacs, which take two spaces; let a_n be the number of ways to do this. There is one way if $n = 0$ or $n = 1$. If $n \geq 2$, then we can partition the configurations into those having a Rabbit at the end and those having a Cadillac at the end. By the definition of the sequence, there are a_{n-1} of the former and a_{n-2} of the latter. Since this accounts for all the configurations, there are $a_{n-1} + a_{n-2}$ ways to fill the n spaces. Thus a_n satisfies the same recurrence as the Fibonacci sequence, with the same initial conditions; hence $a_n = F_n$. In Solution 12.25, we obtain an explicit formula for F_n. ∎

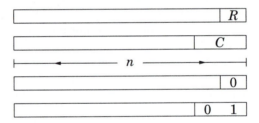

12.22. Example. *Fibonacci models.* In Solution 12.21 we have a canonical combinatorial model for the Fibonacci numbers; the Fibonacci number F_n counts the (ordered) lists of 1's and 2's that sum to n.

Another model for the Fibonacci numbers arises from binary n-tuples. Let a_n be the number of binary n-tuples with no consecutive 1's. There is one such list of length 0 and two of length 1. For $n > 1$, consider the last item in the list. If it is 0, then the first $n-1$ digits can be any binary $n-1$-tuple with no consecutive 1's. If the last item is 1, then position $n-1$ must be 0, but the first $n-2$ digits can be any list of length $n-2$ with no consecutive 1's. Hence the number a_n of ways to build such a list of length n satisfies the recurrence $a_n = a_{n-1} + a_{n-2}$ for $n \geq 2$. With the initial conditions $a_0 = 1$ and $a_1 = 2$, we conclude by induction that $a_n = F_{n+1}$. ∎

The model of 1, 2-lists summing to n yields combinatorial arguments for identities involving the Fibonacci numbers, just as the subset-selection or block-walking models yield combinatorial arguments concerning binomial coefficients. It can also solve counting problems directly. For example, we can prove that there are F_{n+1} binary n-tuples with no consecutive 1's by establishing a bijection between the set of these n-tuples and the set of 1, 2-lists summing to n (Exercise 12.20).

Having obtained a recurrence for the Fibonacci numbers, we want to solve it to obtain an explicit formula for F_n. More generally, we consider second-order linear recurrence relations of the form $a_n = c_1 a_{n-1} + c_2 a_{n-2}$.

12.23. Lemma. (Linearity) If both $\langle x \rangle$ and $\langle y \rangle$ satisfy the recurrence $a_n = c_1 a_{n-1} + c_2 a_{n-2}$, and A and B are constants, then $\langle Ax + By \rangle$ also satisfies the recurrence.

Proof. We have $x_n = c_1 x_{n-1} + c_2 x_{n-2}$ and $y_n = c_1 y_{n-1} + c_2 y_{n-2}$ for $n \geq 1$. We multiply the first by A and the second by B and then add them. ∎

12.24. Theorem. Suppose $\langle a \rangle$ satisfies the recurrence relation $a_n = c_1 a_{n-1} + c_2 a_{n-2}$, with $a_0 = s$ and $a_1 = t$. If the equation $x^2 - c_1 x - c_2 = 0$ has distinct solutions α, β, then there exist constants A, B such that $a_n = A\alpha^n + B\beta^n$. If $x^2 - c_1 x - c_2 = (x - \alpha)^2$, then there exist constants A, B such that $a_n = A\alpha^n + Bn\alpha^n$. In either case, the constants are determined by the initial conditions.

Proof. Suppose first that $x^2 - c_1 x - c_2 = 0$ has distinct solutions α, β. For $\gamma \in \{\alpha, \beta\}$, we have $\gamma^2 = c_1 \gamma + c_2$. For $n \geq 2$, we can multiply this by γ^{n-2} to obtain $\gamma^n = c_1 \gamma^{n-1} + c_2 \gamma^{n-2}$. Hence the sequences defined by $a_n = \alpha^n$ and $a_n = \beta^n$ satisfy the recurrence. By linearity (Lemma 12.23), $a_n = A\alpha^n + B\beta^n$ also satisfies the recurrence.

By Proposition 12.18, it suffices to show that we can choose A, B to make this formula satisfy the initial conditions. From $n = 0$ and $n = 1$, we obtain the two requirements $s = A + B$ and $t = \alpha A + \beta B$. Since $\alpha \neq \beta$, we can solve this system of two linear equations in two unknowns to obtain $A = \dfrac{t - \beta s}{\alpha - \beta}$ and $B = \dfrac{t - \alpha s}{\beta - \alpha}$.

Now suppose that $x^2 - c_1 x - c_2 = (x - \alpha)^2$. The argument that α^n is a solution is as above. For $n\alpha^n$, we observe that the factorization requires $c_1 = 2\alpha$ and $c_2 = -\alpha^2$, and hence $c_1 \alpha + 2c_2 = 0$. Thus,

$$c_1(n-1)\alpha^{n-1} + c_2(n-2)\alpha^{n-2} = n(c_1\alpha + c_2)\alpha^{n-2} - (c_1\alpha + 2c_2)\alpha^{n-2}.$$

Since $c_1\alpha + c_2 = \alpha^2$, the first term becomes $n\alpha^n$; since $c_1\alpha + 2c_2 = 0$, the second term is 0. Hence $c_1(n-1)\alpha^{n-1} + c_2(n-2)\alpha^{n-2} = n\alpha^n$, and $n\alpha^n$ satisfies the recurrence. By linearity, $a_n = A\alpha^n + Bn\alpha^n$ also satisfies it.

Again we can choose A, B to make this formula satisfy the initial conditions. From $n = 0$ and $n = 1$, we obtain the two requirements $s = A$ and $t = \alpha A + \alpha B$. We satisfy these by setting $A = s$ and $B = (t/\alpha) - s$. ∎

The polynomial defined by $x^2 - c_1 x - c_2$ is the *characteristic polynomial* for the recurrence $a_n = c_1 a_{n-1} + c_2 a_{n-2}$. The resulting formula $A\alpha^n + B\beta^n$ or $A\alpha^n + Bn\beta^n$ is the *general solution*. The constants A, B are determined by the initial conditions.

12.25. Solution. *Formula for the Fibonacci numbers.* Factoring the characteristic polynomial $x^2 - c_1 x - c_2$ of the recurrence introduces irrational numbers; the roots are $\alpha = \frac{1}{2}(1 + \sqrt{5})$ and $\beta = \frac{1}{2}(1 - \sqrt{5})$. We use $F_0 = 1$ and $F_1 = 1$ to determine A, B in the general solution $F_n = A\alpha^n + B\beta^n$, solving the linear equations as described in the proof of the theorem. The resulting formula (see Exercise 12.25) is

$$F_n = \frac{1}{\sqrt{5}} \left(\frac{1+\sqrt{5}}{2}\right)^{n+1} - \frac{1}{\sqrt{5}} \left(\frac{1-\sqrt{5}}{2}\right)^{n+1}$$

Since the recurrence produces a sequence of integers, the value of this strange formula involving irrational numbers is an integer for every nonnegative integer n. ∎

GENERAL LINEAR RECURRENCES

Combining the ideas in Theorems 12.14 and 12.24 leads to solution techniques for higher-order linear recurrence relations.

12.26. Definition. A *kth-order linear recurrence relation* is an equation of the form $a_n - h_1(n)a_{n-1} - h_2(n)a_{n-2} - \cdots - h_k(n)a_{n-k} = f(n)$, valid for $n \geq k$. The recurrence has *constant coefficients* when each h_i is constant. The recurrence is *homogeneous* when $f(n) = 0$ for all n. The expression $f(n)$ is the *inhomogeneous term*.

To determine the sequence $\langle a \rangle$, a recurrence of order k must be supplied with k *initial values* $a_0, \ldots, a_{k-1}$. As in Proposition 12.6 and Proposition 12.18, the recurrence for $n \geq k$ and the k initial values together specify a unique sequence. Inductive proofs of statements about a_n usually check k initial values in the basis step and use the recurrence relation in the induction step for $n \geq k$.

One general approach to solving linear recurrence relations with constant coefficients is the *Characteristic Equation Method*. The general technique extends what we have done for $k = 1$ and $k = 2$.

12.27. Definition. Let $a_n - c_1 a_{n-1} - c_2 a_{n-2} - \cdots - c_k a_{n-k} = f(n)$ be a *k*th-order linear recurrence relation with constant coefficients. The polynomial p defined by $p(x) = x^n - \sum_{i=1}^{k} c_i a_{n-i}$ is its *characteristic polynomial*, and its *characteristic equation* is $p(x) = 0$.

12.28. Theorem. Suppose $\langle a \rangle$ satisfies a kth-order homogeneous linear recurrence relation with constant coefficients. If the characteristic polynomial factors as $p(x) = \Pi_{i=1}^{r}(x - \alpha_i)^{d_i}$ for distinct $\alpha_1, \ldots, \alpha_r$, then the solution of the recurrence is $a_n = \Sigma_{i=1}^{r} q_i(n)\alpha_i^n$, where each q_i is a polynomial of degree $d_i - 1$. The k coefficients of these polynomials are determined by the initial conditions.

Proof. (Exercise 12.29). ∎

Before satisfying the initial conditions, we write the solution in Theorem 12.28 using unknown variables for the coefficients in each q_i. This expression is the *general solution* to the homogeneous recurrence relation. The k coefficients are determined by solving k linear equations obtained from the initial conditions; the value of the expression must agree with the initial condition a_i for $0 \le i \le k - 1$.

Solving the characteristic equation is also the first step in solving an inhomogeneous relation. We combine this with a particular solution of the inhomogenous relation (ignoring initial conditions) to obtain the general solution of the inhomogeneous relation.

12.29. Theorem. If $\langle y \rangle$ satisfies a kth-order linear recurrence relation with coefficients $c_1, \ldots, c_k$ and inhomogeneous term $f(n)$, then every sequence $\langle a \rangle$ that satisfies this recurrence relation has the form $a_n = x_n + y_n$, where $\langle x \rangle$ satisfies the homogeneous linear recurrence relation with coefficients $c_1, \ldots, c_k$.

Proof. (sketch) Summing the recursive expressions for x_n and y_n shows that every sequence of this form satisfies the recurrence. Conversely, a sequence $\langle a \rangle$ satisfying the recurrence is determined by k initial conditions. It suffices to show that the coefficients in the general solution to the homogeneous relation can be chosen so that the resulting $\langle x \rangle + \langle y \rangle$ agrees with the initial conditions $a_0, \ldots, a_{k-1}$. ∎

The task that remains is finding a particular solution when the inhomogeneous term is $f(n)$. When $f(n) = q(n)\beta^n$, where q is a polynomial and β is a constant, the method of Theorem 12.14 and Corollary 12.16 yields a particular solution of the form $p(n)\beta^n$, where p is a polynomial. Furthermore, the degree of p exceeds the degree of q by the multiplicity of β as a root of the characteristic polynomial. When $f(n) = f_1(n) + f_2(n)$, we can find particular solutions for the inhomogeneous terms $f_1(n)$ and $f_2(n)$ separately and then sum them (Exercise 12.30); this is called the *superposition principle*.

In Chapter 4, we counted the squares determined by a square grid. Now we solve the more subtle problem of counting the triangles determined by a triangular grid. We use combinatorial arguments to derive a recurrence and use the Characteristic Equation Method to solve it.

12.30. Solution. *The number of triangles in a triangular grid.* Let a_n be the number of triangles (of all sizes) in the equilateral triangular grid T_n with side length n, as illustrated below for $n = 3$. Note that $a_1 = 1$, $a_2 = 5$, and $a_3 = 13$ (we are requiring sides of positive integer length). We obtain a recurrence for $\langle a \rangle$.

The grid T_n contains three copies of T_{n-1}. Each triangle T in T_n occurs in at least one of these copies of T_{n-1}, unless it touches all three sides. Let $f(n)$ be the number of triangles touching all three sides. Always there is one upright triangle touching all three sides (the full region), and when n is even there is one inverted triangle with sides of length $n/2$ that touches all three sides. Thus $f(n) = 2$ when n is even and $f(n) = 1$ when n is odd, which we can write as $f(n) = \frac{3}{2} + \frac{1}{2}(-1)^n$.

The term $3a_{n-1}$ correctly counts all the triangles touching exactly two sides, but it doubly counts the triangles touching one side and triply counts the triangles touching no sides. Using the inclusion-exclusion principle, the number of triangles contained in at least one of the copies of T_{n-1} is $3a_{n-1} - 3a_{n-2} + a_{n-3}$. Since there are $f(n)$ triangles not counted by this, we have

$$a_n - 3a_{n-1} + 3a_{n-2} - a_{n-3} = \frac{3}{2} + \frac{1}{2}(-1)^n \quad \text{for } n \ge 3,$$

with initial conditions $a_0 = 0$, $a_1 = 1$, $a_2 = 5$.

The characteristic polynomial $x^3 - 3x^2 + 3x - 1 = (x-1)^3$ has 1 as a triple root. Thus the general solution to the homogeneous recurrence is a quadratic polynomial in n. To find the particular solution, we sum solutions for the term $\frac{3}{2}$ and the term $\frac{1}{2} \cdot (-1)^n$.

Since -1 is not a characteristic root, the particular solution corresponding to the term $\frac{1}{2}(-1)^n$ has the form $A(-1)^n$. Satisfying the recurrence requires $A(-1)^n - 3A(-1)^{n-1} + 3A(-1)^{n-2} - 1A(-1)^{n-3} = \frac{1}{2}(-1)^n$. This simplifies to $A + 3A + 3A + A = 1/2$, and hence $A = 1/16$.

Since 1 is a characteristic root of multiplicity three and the inhomogenous term $\frac{3}{2}$ is 1^n times a polynomial of degree 0, the solution corresponding to this term is a polynomial of degree three. We need only determine the leading coefficient, because the lower-order terms belong to the homogeneous solution. We thus let $y_n = Bn^3$ and have

$$Bn^3 - 3B(n-1)^3 + 3B(n-2)^3 - 1B(n-3)^3 = 3/2.$$

The coefficients of the nonzero powers of n on the left (must!) cancel. The constant term yields $3B - 24B + 27B = 3/2$, with solution $B = 1/4$.

Now we determine coefficients in the general solution to satisfy the initial conditions. We evaluate

$$a_n = C_0 + C_1 n + C_2 n^2 + (1/16)(-1)^n + (1/4)n^3,$$

at $n = 0, 1, 2$ to obtain

$$0 = C_0 + (1/16)$$
$$1 = C_0 + C_1 + C_2 - (1/16) + (1/4)$$
$$5 = C_0 + 2C_1 + 4C_2 + (1/16) + 2$$

The solution of this system is $C_0 = -1/16$, $C_1 = 1/4$, $C_2 = 5/8$. The resulting formula for a_n is

$$a_n = \frac{1}{16}[4n^3 + 10n^2 + 4n - 1 + (-1)^n] \quad \blacksquare$$

Straightforward inductive arguments verify the solution of fixed-order linear constant-coefficient recurrences when the characteristic roots are known and the inhomogeneous term has the specified form. Essentially, we have guessed a formula and verified that it works. The method of generating functions provides a deeper explanation and applies in more general situations (see Application 12.34).

OTHER CLASSICAL RECURRENCES (optional)

We present several interesting recurrences and methods of solution. We begin with an easy example of the method of substitution.

12.31. Example. Consider the first-order recurrence $a_n = (1 - \frac{1}{n+1})a_{n-1}$ for $n \geq 1$, with $a_0 = 1$. Multiplying the recurrence by $n + 1$ yields $(n + 1)a_n = na_{n-1}$. This suggests making the substitution $b_n = (n + 1)a_n$ to simplify the recurrence. Indeed, the new sequence $\langle b \rangle$ then satisfies $b_n = b_{n-1}$ for $n \geq 1$, with $b_0 = (0 + 1)a_0 = 1$. The solution to this recurrence is $b_n = 1$ for all $n \geq 0$. Undoing the substitution yields $a_n = b_n/(n + 1) = 1/(n + 1)$ for $n \geq 0$. $\blacksquare$

We illustrate the techniques of substitution and reduction of order by giving another solution of the Derangements Problem (Problem 10.9). We obtain a second-order linear recurrence relation with nonconstant coefficients. (Exercise 12.35 obtains another recurrence.)

12.32. Solution. *Derangements.* Recall that D_n counts the permutations of $[n]$ without fixed points. With $D_1 = 0$ and $D_2 = 1$, we derive a second-order recurrence valid for $n \geq 3$. We partition the set of derangements of $[n]$ according to the position $k \in [n - 1]$ where n is placed. For each such k, we further partition the set according to whether k appears in position n. If k appears in position n, then we can complete the derangement in D_{n-2} ways by deranging the other $n-2$ objects. Hence there are D_{n-2} derangements of this type for each choice of k.

n	j
k	n

If k does not appear in position n, then some element other than k appears there, and we can interchange the elements in positions n and k to obtain a derangement of $[n-1]$, with n appended at the end. Conversely, given any derangement of $[n-1]$ with n appended at the end, making the same interchange restores the original configuration. Hence the number of derangements of this type (element n in position k, and element k not in position n) is D_{n-1}, for each choice of k.

Since there are $n-1$ choice for the position k where n is placed, we obtain the recurrence $D_n = (n-1)(D_{n-1} + D_{n-2})$ for $n \geq 3$. We can make the recurrence valid also for $n = 2$ by defining $D_0 = 1$. This is natural; there is one permutation with no elements, and it has no fixed points.

To solve this recurrence, we make successive substitutions to simplify its form until we obtain a recurrence that we can solve. The first substitution $f_n = D_n/n!$ eliminates the nonlinear factor $n-1$; also f_n is the probability that a random permutation is a derangement. The second substitution reduces the order of the recurrence.

To dispose of the factor $n-1$, we introduce the auxiliary sequence f_n defined by $f_n = D_n/n!$; note that $f_0 = 1$. Substitution yields

$$n!f_n = (n-1)(n-1)!f_{n-1} + (n-1)!f_{n-2},$$

and then we divide by $n!$ to obtain $f_n = (1-1/n)f_{n-1} + (1/n)f_{n-2}$.

Next we rewrite the recurrence for f_n as

$$f_n - f_{n-1} = (-1/n)(f_{n-1} - f_{n-2}).$$

Having done this, it becomes natural to define $g_n = f_n - f_{n-1}$ for $n \geq 1$ and obtain the recurrence $g_n = (-1/n)g_{n-1}$. This helps because $\Sigma_{k=1}^n g_k = \Sigma_{k=1}^n (f_k - f_{k-1}) = f_n - f_0$; this is called a *telescoping sum*, because all the intermediate terms cancel out or "collapse" as a telescope collapses. Since we defined g_n for $n \geq 1$, and $g_1 = f_1 - f_0 = -1$, we can iterate the recurrence for g_n to write $g_n = (-1)^{n-1}g_1/n! = (-1)^n/n!$. Using the telescoping sum and the value of f_0, we have $f_n = 1 + \Sigma_{k=1}^n (-1)^k/k! = \Sigma_{k=0}^n (-1)^k/k!$. Thus $D_n = n!\Sigma_{k=0}^n (-1)^k/k!$. ∎

We next discuss the use of generating functions to solve recurrence relations. The recurrence leads to an equation for the generating function of the sequence. We try to solve the equation to find the generating function explicitly and extract the coefficients. The technique applies to both linear and nonlinear recurrences. We illustrate it by sketching an alternative derivation of the Fibonacci numbers.

12.33. Example. *Generating function for Fibonacci numbers.* Let $F(x) = \sum_{n=0}^{\infty} F_n x^n$. We multiply the recurrence by x^n and sum over the values of n where the recurrence is valid. This yields

$$\sum_{n=2}^{\infty} F_n x^n = \sum_{n=2}^{\infty} F_{n-1} x^n + \sum_{n=2}^{\infty} F_{n-2} x^n.$$

With $F_0 = F_1 = 1$, this becomes $F(x) - 1 - x = x(F(x) - 1) + x^2 F(x)$. Hence $F(x) = 1/(1 - x - x^2)$. The denominator factors as $(1 - \alpha x)(1 - \beta x)$, where $\{\alpha, \beta\} = (1 \pm \sqrt{5})/2$. The method of partial fractions then yields constants A, B such that

$$F(x) = \frac{A}{1 - \alpha x} + \frac{B}{1 - \beta x}.$$

Using the geometric series (see Theorem 9.42) now yields $F_n = A\alpha^n + B\beta^n$, the same formula obtained previously for the Fibonacci numbers (Exercise 12.25). ∎

12.34. Application. We sketch the application of the method of generating functions to kth order linear recurrence relations with constant coefficients. After multiplying the recurrence by x^n and summing over the region of validity $n \geq k$, we obtain the generating function $A(x)$ for the sequence $\langle a \rangle$ as the ratio of two polynomials. The coefficients of the denominator are the reverse of the coefficients of the characteristic polynomial. We have $(x - \alpha)$ as a factor of the characteristic polynomial if and only if $(1 - \alpha x)$ is a factor in the denominator polynomial of $A(x)$, with the same multiplicity.

When the characteristic polynomial factors as $p(x) = \prod_{i=1}^{r}(x - \alpha_i)^{d_i}$ for distinct $\alpha_1, \ldots, \alpha_r$, the denominator of $A(x)$ thus factors as $\prod_{i=1}^{r}(1 - \alpha_i x)^{d_i}$. Using partial fractions, we write $A(x) = \sum_{i=1}^{r} \frac{q_i(x)}{(1 - \alpha_i x)^{d_i}}$, where $q_i(x)$ is a polynomial of degree less than d_i.

The coefficients of these polynomials are determined by the initial conditions. Because $(1 - \alpha x)^{-d} = \sum_{n=0}^{\infty} \binom{n+d-1}{d-1} \alpha^n x^n$ (Theorem 9.42), we obtain a formula for a_n of the form claimed in Theorem 12.29. This formula emerges from the generating function expansions without guesses.

Furthermore, suppose the recurrence has an inhomogeneous term that is a polynomial of degree d in n times β^n. Because $\sum_{n=0}^{\infty} \beta^n x^n = 1/(1 - \beta x)$, this adds d factors of $(1 - \beta x)$ to the denominator of $A(x)$. Thus the particular solution also emerges from the generating function method. ∎

We close this chapter by discussing the recursive aspects of the Catalan numbers obtained in Solution 9.17. The numbers $C_n = \frac{1}{n+1}\binom{2n}{n}$ are named for Eugène Charles Catalan (1814-1894). In 1838 he discovered that C_n counts the ways to multiply together $n + 1$ factors by a non-associative binary product (Exercise 12.39). Euler had shown in 1758 that C_n also counts the triangulations of a convex polygon with $n + 2$

sides (Exercise 12.40). We have seen that C_n also counts the good elections in the Ballot Problem. We use the ballot model to obtain a nonlinear recurrence for $\{C_n\}$ that does not have a fixed order; the value of C_n depends on all values $C_0, \ldots, C_{n-1}$.

12.35. Example. *The Ballot Problem.* The *ballot paths* to (n, n) are the lattice paths of length $2n$ that do not step above the diagonal. Suppose there are a_n of these; we derive a recurrence for a_n. Every ballot path to (n, n) has some first return to the diagonal; suppose it occurs at position (k, k). The portion of the path to this point starts by stepping right, then does not go above $y = x - 1$ until it reaches $(k, k - 1)$, then steps up. Hence there is a bijection between the possible initial portions of the path and the ballot paths of length $2(k - 1)$. To complete the path from (k, k) to (n, n), we add a translation of a ballot path of length $2(n - k)$. Hence the number of ballot paths of length $2n$ that first return to the diagonal at (k, k) is $a_{k-1}a_{n-k}$. Summing over the possible locations for the first return, we have $a_n = \Sigma_{k=1}^n a_{k-1}a_{n-k}$ for $n \geq 1$, with the initial condition $a_0 = 1$. ∎

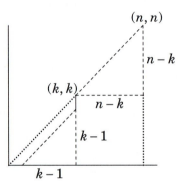

From Solution 9.17, we know that a_n in Example 12.35 equals the Catalan number $C_n = \frac{1}{n+1}\binom{2n}{n}$. We next prove that the solution to Problem 12.2 satisfies the same recurrence relation and has the same initial condition. Hence it must also be the Catalan sequence. In Exercises 12.39-42, we consider additional counting problems that lead to the same recurrence and hence to the Catalan numbers.

12.36. Solution. *The Polygon Problem.* We prove that the counting sequence for triangulations of a convex polygon satisfies the same recurrence as the Catalan numbers. Let a_n be the number of triangulations of a convex polygon with $n + 2$ sides, and define $a_0 = 1$. Starting with $n = 0$, the sequence begins $1, 1, 2, 5, 14, \cdots$.

Consider a convex polygon with $n + 2$ sides, and call the vertices $v_0, \ldots, v_{n+1}$. In every triangulation, the edge $v_{n+1}v_0$ belongs to some triangle; suppose v_k is the third corner of this triangle. To complete the triangulation, we must triangulate the polygon formed by $v_0, \ldots, v_k$ and the polygon formed by $v_k, \ldots, v_{n+1}$, which have $k+1$ sides and $n-k+2$ sides, respectively. These are smaller polygons, which can be triangulated in a_{k-1} and a_{n-k} ways, respectively. Summing over the possible values of k, we obtain $a_n = \Sigma_{k=1}^n a_{k-1} a_{n-k}$ for $n \geq 1$, with $a_0 = 1$. Hence $\langle a \rangle$ must be the Catalan sequence, since it satisfies the same recurrence and initial conditions as the sequence C_n. We conclude that there are $\frac{1}{n+1} \binom{2n}{n}$ ways to triangulate a convex polygon with $n+2$ sides. ■

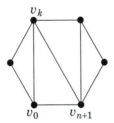

We also use the method of generating functions to solve the Catalan recurrence, obtain the formula for C_n yet again.

12.37. Solution. *Solution of the Catalan recurrence.* Our sequence $\langle C \rangle$ satisfies $C_n = \Sigma_{k=1}^n C_{k-1} C_{n-k}$ for $n \geq 1$, with $C_0 = 1$. Let $A(x) = \Sigma_{n=0}^{\infty} C_n x^n$ be the generating function. Multiplying the recurrence by x^n and summing over the region $n \geq 1$ where the recurrence is valid yields

$$A(x) - C_0 = x\Sigma_{n=1}^{\infty}\Sigma_{l=0}^{n-1}C_l C_{n-1-l}x^{n-1} = x\Sigma_{m=0}^{\infty}\Sigma_{l=0}^{m}C_l C_{m-l}x^m = x[A(x)]^2.$$

The resulting equation for A is $xA^2 - A + 1 = 0$. Using the quadratic formula, we have $A(x) = (1 \pm (1-4x)^{1/2})/2x$. The formula for the coefficients of $A(x)$ can now be extracted by using an extended form of the binomial theorem: $(1-4x)^{1/2} = \Sigma_{n=0}^{\infty}\binom{1/2}{n}(-4x)^n$. The extended binomial coefficient $\binom{1/2}{k}$ is defined to be the value of the polynomial expression $\binom{u}{n} = u(u-1)\cdots(u-n+1)/n!$ at $u = 1/2$ (see Remark 9.11). In the formula for $A(x)$, we choose the negative square root because by definition the coefficient of x^{-1} in $A(x)$ is 0. Thus the Catalan number for $n \geq 1$ is the coefficient of x^n in $-(1-4x)^{1/2}/2x$. We obtain $C_n = -\binom{1/2}{n+1}(-4)^{n+1}/2 = \frac{1}{n+1}\binom{2n}{n}$ (Exercise 12.43 requests the details). ■

EXERCISES

"Obtain" requests proof, and obtaining a recurrence relation includes finding appropriate initial conditions. Solve these recurrences only when asked.

12.1. (–) Suppose $a_n = 3a_{n-1} - 2$ for $n \geq 1$, with $a_0 = 1$. Determine a_n.

12.2. (–) Suppose $a_n = a_{n-1} + 2a_{n-2}$ for $n \geq 2$, with $a_0 = 1$ and $a_1 = 8$. Prove that $a_n = 3 \cdot 2^n - 2(-1)^n$. Describe how to find this solution if it were not given.

12.3. Suppose $a_n = 2a_{n-1} + 3a_{n-2}$ for $n \geq 2$.
    a) Prove that a_n is odd for all $n \geq 0$ if a_0, a_1 are odd.
    b) Prove that $a_n = \frac{1}{2}(3^n + (-1)^n)$ for all $n \geq 0$ if $a_0 = a_1 = 1$.

12.4. Suppose $a_n = 5a_{n-1} - 6a_{n-2}$ for $n \geq 2$, with $a_0 = 1$ and $a_1 = 3$. Determine a formula for a_n.

12.5. Suppose $a_n = 3a_{n-1} - 1$ for $n \geq 1$, with $a_0 = 1$. Determine a formula for a_n.

12.6. Suppose $a_n = \frac{1}{2}(a_{n-1} + 2/a_{n-2})$ for $n \geq 2$, with $a_0 = a_1 = 1$. Prove that $1 \leq a_n \leq 2$ for each $n \in \mathbb{N}$.

12.7. Obtain a recurrence relation to count the distributions of n distinct objects into ten distinct boxes.

12.8. Consider a collection of n circles in the plane, such that each circle intersects every other circle and no three circles meet at a point. Obtain a recurrence for the number of regions into which the circles cut the plane. Solve the recurrence to obtain a formula for the number of regions.

12.9. Use Euler's formula to count the regions determined by a configuration of n lines in the plane, where no three lines have a common point. (Hint: add a circle that encloses all the points of intersection, and count vertices and edges.)

12.10. At the start of each year, \$100 is added to a savings account. At the end of each year, interest equal to 5 percent of the amount in the account is added by the bank. Let a_n be the amount of money in the account after the interest payment in the nth year. Obtain a recurrence for a_n.

12.11. On a particular \$50,000 mortgage, interest is calculated each year as 5 percent of the unpaid amount, and afterwards a payment of \$5,000 ends the year. Obtain a recurrence for the amount outstanding at the end of the nth year. Using a calculator, determine the number of years needed to pay off the mortgage. What happens if the interest rate is 10 percent instead of 5 percent?

12.12. Obtain a recurrence relation to count the pairings of $2n$ people.

12.13. Consider the sequence $\langle a \rangle$ defined in Example 12.13, where a_n is the number of regions inside the circle when all $\binom{n}{2}$ chords are drawn among n points on a circle and no three chords have a common intersection.
    a) Obtain the recurrence relation $a_n = a_{n-1} + f(n)$ for $n \geq 1$, where $f(n) = n - 1 + \sum_{i=1}^{n-1}(i-1)(n-1-i)$, with initial condition $a_0 = 1$.
    b) Using the methods of Chapter 9, solve the recurrence of part (a) to obtain an explicit formula for a_n.

12.14. Complete the proof of Corollary 12.16, solving the recurrence $a_n = ca_{n-1} + f(n)\beta^n$, where f is a polynomial and β is a constant.

12.15. Obtain a recurrence relation to count the ways to move a marker exactly n spaces, given that the number of spaces moved at one time can be 1 or 2 or 3.

12.16. Obtain a recurrence relation to count the ways to fill up the parking along a curb with n spaces, given that there are three types of cars, of which one type takes one space and two types take two spaces.

12.17. Let a_n be the number of ways to tile a two-by-n checkerboard using n identical dominoes (see Example 10.7 for definitions). Obtain a recurrence relation for a_n.

12.18. A shopkeeper makes change for n cents by placing one coin at a time on the counter, keeping a running total; pennies, nickels, and dimes are available. Let a_n be the number of ways the shopkeeper can make change for n cents; for example, $a_6 = 3$, counting the lists 111111, 51, and 15. Obtain a recurrence relation and initial conditions for a_n.

12.19. Prove by induction that the Fibonacci numbers satisfy the following relations. (For all problems about Fibonacci numbers, we use the sequence $\{F_n\}$ defined by $F_0 = F_1 = 1$ and $F_n = F_{n-1} + F_{n-2}$ for $n \geq 2$.)

a) $\sum_{i=0}^{n} F_i^2 = F_n F_{n+1}$.
b) $\sum_{i=0}^{n} F_{2i} = F_{2n+1}$.
c) $\sum_{i=0}^{2n-1} (-1)^i F_{2n-i} = F_{2n-1}$.

12.20. Establish a bijection between the set of 1,2-lists that sum to n and the set of 0,1-lists of length $n - 1$ that have no consecutive 1's.

12.21. Give two proofs (by induction and by combinatorial argument) that $1 + \sum_{i=0}^{n} F_i = F_{n+2}$.

12.22. Give two proofs (by induction and by combinatorial argument) that $F_n = \sum_{i=0}^{n} \binom{n-i}{i}$.

12.23. Prove bijectively (using lists of 1's and 2's) that $F_{n+m} = F_n F_m + F_{n-1} F_{m-1}$. Conclude for each $k \in \mathbb{N}$ that F_{n-1} divides F_{kn-1}.

12.24. Prove that every natural number can be written as a sum of distinct numbers that are Fibonacci numbers.

12.25. (−) Complete the details of computing the formula for the Fibonacci numbers in Solution 12.25. Also complete the details of computing the formula for the Fibonacci numbers in Example 12.33.

12.26. Build the generating function for the Fibonacci sequence directly, using the model that F_n is the number of 1,2-lists that sum to n.

12.27. *Efficiency of the Euclidean algorithm.* When the Euclidean algorithm takes many steps, the sum of the inputs must be large. Suppose the Euclidean algorithm takes k steps to operate on the pair (a_0, a_1) with $a_0 > a_1$, producing a list $(a_0, a_1), (a_1, a_2), \ldots, (a_k, 0)$ with $a_0 > a_1 > \cdots > a_k > 0$ by applying the Division Algorithm k times. For example, $(3, 2)$ takes two steps, and $(5, 3)$ takes three steps. Prove that $a_0 + a_1 \geq F_{k+2}$ when $k \geq 2$. Prove also that this is best possible: for each $k \geq 2$ there is a pair with sum F_{k+2} that takes k steps.

12.28. (+) Consider a pile of cards labeled 1 through n. If the top card is m, we reverse the order of the first m cards. The process stops only when card 1 is at the top. Prove that for any initial order of the cards, the process does stop. Let a_n be the maximum (over all initial orderings of the n cards) of the number of steps in the process. Prove that $a_n \le F_n - 1$ (Fibonacci number). (Hint: prove inductively that if k distinct cards appear at the top during the process, then there are at most $F_k - 1$ steps.)

12.29. Prove Theorem 12.28, describing the general solution to a kth-order homogeneous linear recurrence relation with constant coefficients.

12.30. Suppose that $\langle b \rangle$ and $\langle d \rangle$ are solutions to the inhomogeneous linear kth-order recurrences $x_n = (\Sigma_{i=1}^{k} h_i(n) x_{n-i}) + f(n)$ and $x_n = (\Sigma_{i=1}^{k} h_i(n) x_{n-i}) + g(n)$, respectively. Prove that $\langle b \rangle + \langle d \rangle$ is a solution to the recurrence $x_n = (\Sigma_{i=1}^{k} h_i(n) x_{n-i}) + f(n) + g(n)$.

12.31. Suppose that $\langle a \rangle$ is a solution of $x_n = c_1 x_{n-1} + c_2 x_{n-2} + c\alpha^n$, where $c_1, c_2, c, \alpha \in \mathbb{R}$. Prove that $\langle a \rangle$ and $C\alpha^n$ are solutions of the homogeneous third-order recurrence $x_n = (c_1 + \alpha) x_{n-1} + (c_2 - \alpha c_1) x_{n-2} - \alpha c_2 x_{n-3}$.

12.32. Suppose $a_n = a_{n-1} + a_{n-2} + a_{n-3}$ for $n > 3$. Prove that $a_n \le 2^{n-2}$ if $a_i = 1$ for $i \in \{1, 2, 3\}$, and $a_n < 2^n$ if $a_i = i$ for $i \in \{1, 2, 3\}$.

12.33. Solve the recurrence $a_n = \frac{2}{3}(1 + \frac{2}{3^n+1}) a_{n-1}$ for $n \ge 1$, with $a_0 = 1$. (Hint: substitute $b_n = (3^n + 1) a_n$.)

12.34. Consider the following algorithm for finding the largest and smallest in a set of n numbers. If $n = 2$, compare the two numbers. If $n > 2$, (1) split the numbers into sets of sizes $\lfloor n/2 \rfloor$ and $\lceil n/2 \rceil$, (2) apply the algorithm inductively to determine the largest and smallest numbers (x and y) in the first set and the largest and smallest numbers (u and v) in the second set, (3) use u, v, x, y to compute the largest and smallest in the original set. Let a_n be the number of comparisons used on a set of size n. Obtain a recurrence relation for a_n. Use the substitution method to obtain a formula for a_n when n is a power of 2.

12.35. Consider the recurrence $D_n = (n-1)(D_{n-1} + D_{n-2})$ for derangements (with $D_0 = 1$). Substitute $f_n = D_n - nD_{n-1}$ to obtain the first-order recurrence $D_n = nD_{n-1} + (-1)^n$. Use this and induction to prove that $D_n = n! \Sigma_{k=0}^{n} (-1)^k / k!$.

12.36. Let B_n be the number of equivalence relations on n elements; this equals the number of partitions of the set $[n]$. Prove that $B_n = \Sigma_{k=1}^{n} \binom{n-1}{k-1} B_{n-k}$ for $n \ge 1$, with the initial condition $B_0 = 1$. (Comment: These are the *Bell numbers*.)

12.37. Let a_n be the number of sets $\{x, y, z\} \subseteq \mathbb{N}$ such that x, y, z are the lengths of the sides of a triangle with perimeter n.

    a) (!) Obtain a recurrence relation for a_n. (Hint: the recurrence is different for odd n and for even n.)

    b) (+) Using any method, find a nonrecursive formula for a_n.

12.38. Suppose $\langle a \rangle$ satisfies the recurrence $a_n = c_1 a_{n-1} + c_2 a_{n-2}$, with initial values a_0, a_1. Express the generating function for a as a ratio of two polynomials (see Example 12.33 for a special case).

12.39. When a list of numbers is combined using a non-associative binary operator, the order of operations matters. There is one way to combine two numbers $(a \circ b)$, but three numbers can be combined using $a \circ (b \circ c)$ or $(a \circ b) \circ c$. With four numbers, there are five ways: $a(b(cd))$, $a((bc)d)$, $(a(bc))d$, $((ab)c)d$, $(ab)(cd)$. Each such grouping is a *parenthesization* of the list. Let a_n be the number of parenthesizations of an ordered list of $n + 1$ elements. Prove recursively that a_n equals the Catalan number C_n.

12.40. Establish a bijection between the parenthesizations of $n + 1$ distinct elements and the triangulations of a convex $n + 2$-gon. (Hint: The bijection is suggested by the figure below.)

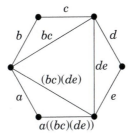

12.41. Suppose $2n$ people are seated in order around a circular table. A *non-crossing pairing* is a matching of these people to shake hands such that no two pairs cross when drawn as chords of the circle. Let a_n be the number of non-crossing pairings of the $2n$ people. As shown below, $a_3 = 5$. Prove recursively that a_n equals the Catalan number C_n.

12.42. Consider arrangements of pennies built on a row of n pennies. Each penny not in the base row rests on two pennies immediately below it, as illustrated below. Prove that the number of arrangements that can be built on a row of n pennies is the Catalan number C_n.

12.43. Add the details in Solution 12.37 to complete the derivation of the formula for the Catalan numbers using the method of generating functions.

12.44. Let a_n denote the number of lattice paths of length n that never step above the diagonal (these end at some point $(n - k, k)$ with $k \leq n/2$). Prove that $a_n = \binom{n}{\lfloor n/2 \rfloor}$. (Hint: use the reflection argument.)

12.45. Suppose f is a polynomial of degree n. The *first difference* of f is the function $g = \Delta f$ defined by $g(x) = f(x + 1) - f(x)$. The *kth difference* of f is the function $g^{(k)}$ defined inductively by $g^{(0)} = f$ and $g^{(k)} = \Delta g^{(k-1)}$ for $k \geq 1$. Obtain a formula for the nth difference of f.

12.46. Let $s(n, k)$ be the number of surjective functions from an n-element set to a k-element set. Derive a recurrence relation for $s(n, k)$ in terms of $s(n-1, k)$ and $s(n-1, k-1)$. Don't forget to specify initial conditions.

12.47. The partitions of an integer n (the configurations in the Penny Problem) are the nonincreasing lists of positive integers that sum to n. Let $p_{n,k}$ be the number of partitions of n having k parts (for example, there are two partitions of 5 with three parts, 311 and 221, so $p_{5,3} = 2$). Prove that $p_{n,k} = p_{n-1,k-1} + p_{n-k,k}$ for appropriate values of n and k, and specify initial conditions that enable the recurrence to determine $p_{n,k}$ for all $n \geq 0$ and $k \geq 1$.

12.48. When counting distinguishable ways to distribute identical balls into distinct boxes, it matters only how many balls go in each box. Obtain a recurrence for the number of ways to distribute n identical balls into k distinct boxes with at least 2 balls and at most 4 balls in each box.

12.49. Let G_n be the graph pictured below, consisting of a path with n vertices and one additional vertex adjacent to every vertex of the path. Let a_n be the number of spanning trees in G_n.
  a) Prove that $a_n = a_{n-1} + \sum_{i=0}^{n-1} a_i$ for $n \geq 2$, where $a_0 = a_1 = 1$.
  b) Use part (a) to prove that $a_n = 3a_{n-1} - a_{n-2}$ for $n \geq 3$.

12.50. Let G_n be the graph on $2n$ vertices and $3n - 2$ edges pictured below, for $n \geq 1$. Prove that the chromatic polynomial of G_n is $(k^2 - 3k + 3)^{n-1}k(k-1)$.

12.51. *Gambler's Ruin.* Two people gamble by flipping a fair coin until one player goes broke. If the coin comes up heads, then A pays B \$1, otherwise B pays A \$1. Suppose A starts with r dollars and B starts with s dollars. Let $a_n(r, s)$ be the probability that A goes broke on the nth flip. Obtain a recurrence relation for $a_n(r, s)$. (There are three parameters in this recurrence; be careful about the initial conditions.)

PART IV

CONTINUOUS MATHEMATICS

Chapter 13

The Real Numbers

We now begin to study the consequences of the Completeness Axiom. After recalling its statement, we derive elementary consequences and then define convergence of sequences. We also develop the decimal representation of real numbers and use it to prove that the set of real numbers is uncountable.

THE COMPLETENESS AXIOM

The Completeness Axiom is implicit in our understanding of the decimal expansion of a real number. Recall that a sequence of real numbers is a function from $\mathbb{N}$ to $\mathbb{R}$, and that we name such a function using angled brackets and write its successive values using subscripts: $\langle a \rangle = \{a_1, a_2, \cdots\}$. Consider the sequence $\langle a \rangle = \frac{1}{1}, \frac{14}{10}, \frac{141}{100}, \frac{1414}{1000}, \frac{14142}{10000}, \cdots$. We think of the decimal expansion for $\sqrt{2} = 1.414235\ldots$ as an abbreviation for the sequence $\langle a \rangle$. Each term in $\langle a \rangle$ is rational. The further we go, the more accurately the terms approximate $\sqrt{2}$. The sequence is nondecreasing, and each term is smaller than 3/2. Thus, 3/2 is an upper bound on the set of terms of $\langle a \rangle$, as are 142/100 and 1415/1000. Our intuition suggests that $\sqrt{2}$ is the *smallest* upper bound.

The Completeness Axiom declares the existence of a real number that is the least upper bound of the set of rational numbers in the sequence. We call that real number $\sqrt{2}$. We recall from Chapter 1 the precise statement of the axiom and terminology about bounds.

13.1. Definition. Let S be a set of real numbers. A number $\alpha \in \mathbb{R}$ is an *upper bound* for S if $x \leq \alpha$ for all $x \in S$. Also α is the *least upper bound* or *supremum* for S if α is an upper bound for S and S has no upper bound less than α. Similarly, α is a *lower bound* for S if $x \geq \alpha$ for all $x \in S$, and α is the *greatest lower bound* or *infimum* for

S if S has no lower bound greater than α. We use $\sup(S)$ and $\inf(S)$ to denote the supremum and infimum of S, when they exist.

13.2. Axiom. (The Completeness Axiom for $\mathbb{R}$). Every non-empty subset of $\mathbb{R}$ that has an upper bound has a least upper bound.

Similarly, every non-empty subset of $\mathbb{R}$ having a lower bound has a greatest lower bound; this is the "Greatest Lower Bound Property". It is equivalent to the Completeness Axiom (Exercise 13.2), which we also call the "Least Upper Bound Property". The Greatest Lower Bound Property is the analogue for $\mathbb{R}$ of the Well-Ordering Property of $\mathbb{N}$.

13.3. Remark. Using "the" for least upper bound (or greatest lower bound) is justified by order properties. Since $\alpha < \beta$ or $\alpha > \beta$ whenever α, β are distinct real numbers, a set cannot have more than one supremum or infimum. ∎

The least upper bound of a set need not be an element of the set. The word "has" in Axiom 13.2 does not mean "contains".

13.4. Example. *Sups and infs.* If $S = \{x \in \mathbb{R} : 0 < x < 1\}$, then the infimum and supremum of S are 0 and 1, which belong to $\mathbb{R}$ but do not belong to S. This set S has neither a minimum nor a maximum. A set of real numbers has a *minimum* if it has an infimum and the infimum belongs to the set, in which case the infimum is the minimum (similarly for maximum and supremum). ∎

13.5. Example. *The rationals are "incomplete".* If S is the set of rational numbers whose square is less than 2, then $S \subseteq \mathbb{Q}$ and S has an upper bound. By the definition of S, $\sqrt{2}$ is an upper bound and S is non-empty. Because a rational number can be found between any two real numbers (Exercise 13.8), S has no smaller upper bound. Hence $\sqrt{2} = \sup(S)$. The set S has upper bounds in $\mathbb{Q}$, but its least upper bound is not in $\mathbb{Q}$. The supremum or infimum of a set of rational numbers need not be a rational number; the Completeness Axiom does not hold for $\mathbb{Q}$. ∎

In Appendix A, we construct the real numbers from the rational numbers. We define addition, multiplication, and order for the real numbers, and we prove that the result is a complete ordered field. We constructed $\mathbb{Q}$ from $\mathbb{Z}$ and can construct $\mathbb{R}$ from $\mathbb{Q}$; this prompted the mathematician Leopold Kronecker (1823-1891) to say, "God created the integers; all the rest is the work of man."

In Part IV, we assume this construction. Our results about $\mathbb{R}$ all follow from the axioms of a complete ordered field. In fact, $\mathbb{R}$ is (up to isomorphism) the only complete ordered field. This means that, given a complete ordered field **F**, we can define a bijection from $\mathbb{R}$ to **F** that preserves the arithmetic and order operations. Thus **F** behaves just like $\mathbb{R}$; only the names of the elements have been changed. We sketch the proof in Appendix A.

Archimedes (287?-212 BC) asked whether placing segments of unit length end to end would produce arbitrarily long segments. The content of the statement is "The natural numbers form an unbounded set of real numbers." This may seem obvious, like the observation in Example 2.28 that there is no largest real number. That argument does show that there is no largest natural number (each is exceeded by the next), but this proves only that $\mathbb{N}$ does not have a bound in $\mathbb{N}$. There may be a bound on $\mathbb{N}$ that is not a natural number, and indeed there are ordered fields in which $\mathbb{N}$ is a bounded set (Exercise 13.17). With the Completeness Axiom, this cannot happen.

13.6. Theorem. (The Archimedean Property) Given positive real numbers a, b, there exists a natural number n such that $na > b$. Equivalently, no real number is an upper bound for the set $\mathbb{N}$.

Proof. We prove first that $\mathbb{N}$ has no upper bound. If $\mathbb{N}$ has an upper bound, then by the Completeness Axiom $\mathbb{N}$ has a least upper bound α. Since α is the least upper bound, $\alpha - 1$ is not an upper bound, and hence there is a natural number n such that $n > \alpha - 1$. But now the properties of arithmetic yield $n + 1 > \alpha$, contradicting the choice of α as an upper bound. Thus $\mathbb{N}$ has no upper bound. In particular, b/a is not an upper bound, so there exists $n \in \mathbb{N}$ such that $n > b/a$, and hence $na > b$. ∎

We need to define the distance between two real numbers before defining "limits". Recall from Chapter 3 that the distance between x and 0 is the "absolute value" $|x|$, defined by

$$|x| = \begin{cases} x & \text{if } x \geq 0 \\ -x & \text{if } x < 0 \end{cases}$$

In general, the distance between x and y is $|x - y|$. We will often use the *triangle inequality*: $|x + y| \leq |x| + |y|$ (Exercise 3.3). For example, it gives us a short proof that if two numbers are close together, then their absolute values are also close together.

13.7. Lemma. Suppose $a, b, c \in \mathbb{R}$ with $c > 0$. If $|a - b| < c$, then $|a| < c + |b|$, and also $\big| |a| - |b| \big| < c$.

Proof. Applying the triangle inequality after adding and subtracting b, we have $|a| = |a - b + b| \leq |a - b| + |b| < c + |b|$. Since $|b - a| = |a - b|$, the same argument implies $|b| < c + |a|$. We have proved that $|a| - |b| < c$ and $|b| - |a| < c$, so the result follows. ∎

13.8. Definition. A sequence $\langle a \rangle$ of real numbers has *limit* $L \in \mathbb{R}$ if, for every $\varepsilon > 0$, there exists $N \in \mathbb{N}$ (depending on ε) such that $n > N$ implies $|a_n - L| < \varepsilon$. A sequence *converges* if it has a limit. We write $a_n \to L$ to mean "a_n converges to (the limit) L".

We also write $a_n \to L$ as $\lim a_n = L$ or $\lim_{n \to \infty} a_n = L$. We read the latter as "the limit of a_n as n goes to infinity is L". The phrase "as n goes to infinity" (written "$n \to \infty$") means that we study the behavior as n grows arbitrarily large; there is no element of $\mathbb{R}$ called "infinity".

The Greek letter ε denotes an arbitrary positive number, usually thought of as small. The order of quantification of N and ε in the definition is crucial: the choice of N generally depends on the value of ε.

13.9. Example. *Convergent sequences.* When using the definition of convergence to prove that a sequence converges, we choose N appropriately in terms of ε. Consider the sequence $\langle a \rangle$ defined by $a_n = 3 + 2/n$; we claim $\langle a \rangle$ converges to 3. Given $\varepsilon > 0$, we want $N \in \mathbb{N}$ such that $n > N$ implies $2/n < \varepsilon$. The Archimedean property allows us to choose N so that $N > 2/\varepsilon$. Then $n > N$ implies $2/n < 2/N < \varepsilon$, and hence $|a_n - 3| < \varepsilon$. This proves $a_n \to 3$. No one value of N works for all ε.

As another example, let a_n be the rational number closest to $\sqrt{2}$ among the rational numbers expressible by fractions with denominator n. Then $a_n \to \sqrt{2}$. Given $\varepsilon > 0$, choose $N \in \mathbb{N}$ such that $N \geq 1/(2\varepsilon)$. Then for all $n > N$, we have $|a_n - \sqrt{2}| < 1/(2n) < \varepsilon$. Below we illustrate this with $\varepsilon = .1$. Note that a_n may be closer than a_{n+1} to $\sqrt{2}$. ∎

n	1	2	3	4	5	6	7	8
a_n	1	$\frac{3}{2}$	$\frac{4}{3}$	$\frac{6}{4}$	$\frac{7}{5}$	$\frac{9}{6}$	$\frac{10}{7}$	$\frac{11}{8}$

We will henceforth abbreviate "sequence of real numbers" to "sequence". Many statements about sequences can be understood geometrically by viewing the sequence as a function on $\mathbb{N}$ and drawing its graph, as above. Geometrically, the statement of convergence to L is that for every fixed ε, the graph of the sequence "eventually" remains within a band of width 2ε centered on the horizontal line given by $y = L$. The precise meaning of "eventually" involves the "tail" of a sequence:

13.10. Definition. Given a sequence $\langle a \rangle$, the Nth *tail* of $\langle a \rangle$ is the sequence $a_{N+1}, a_{N+2}, \cdots$.

A statement about a_n valid for all $n > N$ is a statement about the Nth tail of $\langle a \rangle$. Hence we have $a_n \to L$ if and only if for every $\varepsilon > 0$, there is some tail of $\langle a \rangle$ whose elements all are within ε of L.

We speak of *the* limit of a sequence; the next lemma shows that this makes sense. Intuitively, if a sequence converges to both L and M, then eventually its elements must be arbitrarily close to both L and M. This cannot happen if L and M differ by a positive amount.

13.11. Lemma. Every convergent sequence of real numbers has a unique limit.

Proof. Suppose $a_n \to L$ and $a_n \to M$, but $L \neq M$; by symmetry, we may assume $M > L$. Using $\varepsilon = (M - L)/2$ in the definition of convergence, we are given the existence of $N_1, N_2 \in \mathbb{N}$ such that $n > N_1$ implies $|a_n - L| < \varepsilon$ and $n > N_2$ implies $|a_n - M| < \varepsilon$. Let $N = \max\{N_1, N_2\}$. If $n > N$, then $a_n < L + \varepsilon = M - \varepsilon < a_n$, which is a contradiction. ∎

$$
\begin{array}{ccc}
 & M - \varepsilon \quad M \quad M + \varepsilon & \\
\hline
L - \varepsilon \quad L \quad L + \varepsilon & &
\end{array}
$$

13.12. Remark. We can characterize the supremum (or infimum) of a set using convergent sequences. The number α is the supremum of the set S if and only if α is an upper bound for S and S contains a sequence converging to α (Exercise 13.3). For example, if $S = (0, 1)$, then setting $a_n = 1 - 1/n$ defines a sequence in $(0, 1)$ converging to the supremum. A set has a maximum when it contains its supremum α, and then the constant sequence $\alpha, \alpha, \alpha, \cdots$ is a sequence in S converging to $\sup(S)$. ∎

13.13. Theorem. (Monotone Convergence Theorem) Every bounded monotone sequence of real numbers has a limit: a bounded nondecreasing sequence converges to its supremum, and a bounded nonincreasing sequence converges to its infimum.

Proof. By symmetry, we need only consider the nondecreasing case. Since $\langle a \rangle$ is bounded, it has a supremum; call it L. Given a fixed $\varepsilon > 0$, it suffices to obtain a tail of $\langle a \rangle$ in which every element is within ε of L. Since L is the least upper bound for $\langle a \rangle$, we know that $L - \varepsilon$ is not an upper bound for $\langle a \rangle$. Hence there is some $N \in \mathbb{N}$ such that $a_N > L - \varepsilon$. Since $\langle a \rangle$ is increasing and L is an upper bound, $n > N$ implies $L - \varepsilon < a_n \leq L$. Hence $n > N$ implies $|L - a_n| < \varepsilon$, and this choice of N in terms of ε shows that $\langle a \rangle$ satisfies the definition of convergence. ∎

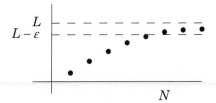

DECIMAL EXPANSION AND UNCOUNTABILITY

We can give a canonical representation of a rational number by writing it as a fraction in lowest terms; how do we represent real numbers? The most familiar description is the decimal expansion. As we did for the natural numbers, we can construct a canonical representation of a real number with respect to any natural-number base $k > 1$. For each $n \in \mathbb{N}$, we obtained a unique list of integers $c_0, \ldots, c_N$ with $0 \leq c_j \leq k - 1$ for $0 \leq j \leq N$ such that $n = \Sigma_{j=0}^{N} c_j k^j$. For each positive real number α, we define also the k-ary expansion of the "fractional part" $\alpha - \lfloor \alpha \rfloor$. As a bonus, k-ary expansions will allow us to prove that $\mathbb{R}$ is uncountable.

We want the canonical k-ary expansion to be a sequence $\langle a \rangle$ such that $a_n \in \{0, \ldots, k - 1\}$ for each $n \in \mathbb{N}$. Suppose we partition $[0, 1]$ into k equal subintervals. If α belongs to the ith of these intervals, then we set $a_1 = i - 1$. When we further subdivide the interval containing α into k subintervals, the location of α tells us a_2. Continuing this gives us as much of the expansion as we want. Below we illustrate the first two steps of this procedure for $\alpha = \sqrt{2} - 1$ and $k = 10$.

13.14. Definition. Suppose α is a real number with $0 \leq \alpha < 1$. The *canonical k-ary expansion* of α is the sequence $\langle a \rangle$ defined by: 1) a_1 is the integer j such that $j/k \leq \alpha < (j + 1)/k$, and 2) for $n \geq 1$, a_{n+1} is the integer j such that $b_n + j/k^{n+1} \leq \alpha < b_n + (j + 1)/k^{n+1}$, where $b_n = \Sigma_{i=1}^{n} a_i / k^i$. For $k = 2, 3, 10$, the sequence $\langle a \rangle$ is called the *binary, ternary,* or *decimal* expansion, respectively.

The number b_n described here is the rational number obtained by truncating the expansion to the first n digits. The definition makes sense only if $b_n \leq \alpha < b_n + 1/k^n$. These inequalities imply that $\langle b \rangle$ converges to α (see Exercise 13.5).

13.15. Theorem. Let k be an integer greater than 1. For every α with $0 \le \alpha < 1$, the canonical k-ary expansion $\langle a \rangle$ of α exists, and the sequence $\langle b \rangle$ defined by $b_n = \Sigma_{i=1}^{n} a_i/k^i$ converges to α.

Proof. To verify that each successive a_{n+1} is an integer between 0 and $k-1$, we must prove that $b_n \le \alpha < b_n + 1/k^n$. We prove these together by induction on n. Let $b_0 = 0$. Since $0 \le \alpha < 1$, the statement holds when $n = 0$. If $b_n \le \alpha < b_n + 1/k^n$, then the unique integer j such that $b_n + j/k^{n+1} \le \alpha < b_n + (j+1)/k^n$ is in the set $\{0, \ldots, k-1\}$, and we call this a_{n+1}. Now $b_{n+1} = b_n + a_{n+1}/k^{n+1}$, and our choice of a_{n+1} implies that $b_{n+1} \le \alpha < b_{n+1} + 1/k^{n+1}$. This completes the induction step.

Now we have shown that $\langle b \rangle$ is nondecreasing and is bounded above by α. Furthermore $|\alpha - b_n| < 1/k^n$. To prove $b_n \to \alpha$, consider an arbitrary $\varepsilon > 0$. We can choose $N \in \mathbb{N}$ such that $n > N$ implies $k^n > 1/\varepsilon$ (Exercise 13.5). Then for all $n > N$, we have $|\alpha - b_n| < 1/k^n < \varepsilon$. ∎

Theorem 13.15 allows us to "represent" real numbers by their decimal expansions, but a philosophical question remains. What does it mean to "know" a number? We all feel comfortable with integers and perhaps with fractions, but real numbers are another matter. To what extent do we understand $\sqrt{2}$ or π? We can compute their decimal expansions with arbitrarily high accuracy, but no one will ever know their full expansions. Nevertheless, we accept that they have a precise meaning.

Rational numbers representable as fractions with a power of k in the denominator have two k-ary expansions. For example, 1/2 equals both 0.1000000... and 0.0111111... in binary, just as 0.999999... equals 1 in decimal. These are the only ways to obtain multiple k-ary representations for a single number (Exercise 13.9). The algorithm used in Theorem 13.15 always chooses the expansion with infinitely many 0's when there is an alternative expansion ending in repeating $k-1$'s.

We have proved that $\mathbb{Q}$ is countable. We will show that there is no bijection from $\mathbb{N}$ to $\mathbb{R}$, and hence $\mathbb{R}$ is uncountable (Definition 5.16). The proof, due to Georg Cantor (1845-1918), is one instance of "Cantor's Diagonalization Argument".

13.16. Theorem. (Cantor) The set of real numbers is uncountable.

Proof. It suffices to show that $[0,1]$ is uncountable. If not, then we have a bijection from $\mathbb{N}$ to $[0,1]$. This is a sequence $\langle x \rangle$ that lists all numbers in $[0,1]$, in some order. By considering the canonical decimal expansions of these numbers, we will construct a number not on the list.

$$x_1 = .a_{1,1}a_{1,2}a_{1,3}\cdots$$
$$x_2 = .a_{2,1}a_{2,2}a_{2,3}\cdots$$
$$x_3 = .a_{3,1}a_{3,2}a_{3,3}\cdots$$
$$\vdots$$

Suppose that the expansions appear in order as indicated above. We build a number whose canonical expansion disagrees with every expansion in our list in some position. Let $c_n = 1$ if $a_{n,n} = 0$, and $c_n = 0$ if $a_{n,n} > 0$. Now $\langle c \rangle$ disagrees in position n with the expansion of x_n. Furthermore, since $\langle c \rangle$ has no 9, $\langle c \rangle$ cannot be the alternative expansion of any number in our list. Therefore, the number with decimal expansion $\langle c \rangle$ does not appear in our list, contradicting the hypothesis that we have them all. ∎

One could try to use this argument to prove that $\mathbb{Q}$ is uncountable. Proceeding as above, we list the expansions of numbers in $\mathbb{Q}$ and create an expansion for a number y not on our list. This does not give a contradiction, because y is not in $\mathbb{Q}$.

The set of real numbers whose k-ary expansions have only finitely many nonzero terms is countable; in fact, this is a subset of $\mathbb{Q}$. The number of nonzero terms can be "arbitrarily large" but not infinite. The possibility that more than one digit in the expansion appears infinitely often makes the set of real numbers bigger than $\mathbb{N}$.

EXERCISES

13.1. (–) Compute the first six places of the canonical 3-ary expansion of 1/10.

13.2. (–) Prove that the Least Upper Bound Property holds for an ordered field **F** if and only if the Greatest Lower Bound Property holds for **F**.

13.3. (!) Suppose $S \subseteq \mathbb{R}$. Prove that $\alpha = \sup(S)$ if and only if α is an upper bound for S and S contains a sequence converging to α.

13.4. Let A and B be non-empty sets of real numbers. Define $C = \{x + y : x \in A, y \in B\}$. Prove that if A and B have upper bounds, then C has a least upper bound, and $\sup C = \sup A + \sup B$.

13.5. Suppose k is a natural number greater than 1. Given $\varepsilon > 0$, prove that there exists $N \in \mathbb{N}$ such that $n > N$ implies $k^n > 1/\varepsilon$. (Hint: use the Archimedean property.)

13.6. (–) Let $x_n = (1 + n)/(1 + 2n)$. Prove that $\lim_{n \to \infty} x_n$ exists by using the Monotone Convergence Theorem. Prove that $\lim_{n \to \infty} x_n = 1/2$ by using the definition of limit.

13.7. Let $x_n = \dfrac{1}{n+1} + \dfrac{1}{n+2} + \cdots + \dfrac{1}{2n}$. Prove that $\lim_{n \to \infty} x_n$ exists. (Comment: in fact, the limit equals $\ln 2$, but this information is not needed for this exercise.)

13.8. Prove that there is a rational number between any two irrational real numbers, and an irrational number between any two rational numbers. (This does not prove that the cardinality of the two sets is the same; in fact, the set of rationals between any two irrationals is countable, but the set of irrationals between any two rationals is uncountable.)

13.9. (!) Prove that a real number has more than one k-ary expansion if and only if it is expressible as a fraction using a denominator that is a power of k.

13.10. (!) *k-ary expansions of rational numbers.* A k-ary expansion is *eventually periodic* if after some initial portion, the remainder is a repeating list of some finite length (this includes terminating expansions, where the repeating list is "0"). Prove that the k-ary expansion of every rational is eventually periodic. (Hint: First prove this for rational numbers of the form j/k with $0 \le j < k$. Then use this and k-ary expansions of integers to prove the claim in the general case. Comment: the converse is Exercise 14.19.)

13.11. Given $a, b \in \mathbb{N}$, use the pigeonhole principle to prove that the decimal expansion of a/b has a period of length less than b.

13.12. Find the flaw in the following argument that claims to prove that the set of real numbers between 0 and 1 is countable: "Using decimal expansions, we can list the numbers in the interval $(0, 1)$ as follows: .1, .2, .3, ..., .9, .01, ..., .09, .11, ..., .19, ..., .99, .001, ..., .009, .011, ..., .019, .021, ..., etc." In other words, we first list the numbers whose last nonzero digit is in the tenths place, then those whose last nonzero digit is in the hundredths place, and so on.

13.13. (–) Prove that every infinite subset of a countable set is countable. Prove that every set that contains an uncountable set is uncountable. Conclude that $\mathbb{R}$ is uncountable if $[0, 1]$ is uncountable.

13.14. Let S be the set of subsets of $\mathbb{N}$. Let $T = \{x \in \mathbb{R}: 0 \le x < 1\}$. Prove that S and T have the same cardinality.

13.15. (!) Prove that $\mathbb{R} \times \mathbb{R}$ has the same cardinality as $\mathbb{R}$.

13.16. Prove that the union of countably many countable sets is countable.

13.17. (+) *An ordered field in which $\mathbb{N}$ is a bounded set.* Let F be the set of all formal expressions of the form $a = \sum_{i \in \mathbb{Z}} a_i x^i$ such that each a_i belongs to $\mathbb{R}$ and $\{i < 0: a_i \ne 0\}$ is finite. (Here x is a formal symbol, not a number.) An element $a \in F$ is *positive* if the least-indexed nonzero coefficient a_k in the expression for a is a positive real number. The *sum* of $a \in F$ and $b \in F$ is the element $c \in F$ defined by $c_i = a_i + b_i$ for all $i \in \mathbb{Z}$. The *product* of $a \in F$ and $b \in F$ is the element $c \in F$ defined by $c_j = \sum_{i \in \mathbb{Z}} a_i b_{j-i}$ for all $j \in \mathbb{Z}$.

a) Prove that the sum and product of two elements of F is an element of F.

b) We have defined addition, multiplication, and order on F. Prove that with these operations, F is an ordered field.

c) We interpret each real number α as the element $a \in F$ with $a_i = 0$ for all i, except $a_0 = \alpha$; hence we can think of $\mathbb{R}$ as a subset of F. Prove that $\mathbb{N}$ is a bounded set in F. Conclude that F does not satisfy the Archimedean property. (Hint: Compare n with x^{-1}.)

Comment: The elements of F that have nonzero coefficients only for non-negative powers in x are formal power series (Definition 9.39). Formal power series do not form a field; we need to allow (finitely many) negative powers of x so that elements will have multiplicative inverses.

Chapter 14

Sequences and Series

In Chapter 13 we defined convergent sequences and used them to explain decimal expansions for real numbers. We regard a decimal expansion as an infinite sum or "series". The usefulness of this approach suggests developing the theory of sequences and series.

A sequence converges when its values cluster around one limiting value; hence the values of a convergent sequence must eventually be close together. The central result of this chapter is the converse: when the terms of a sequence are eventually close together, the sequence must have a limit. Thus the real number system has no gaps. This intuition leads to a deep understanding of the Completeness Axiom.

In this chapter we also prove related results that we apply to the theory of calculus in later chapters. We obtain several criteria for the convergence of infinite series, and we solve several problems where understanding an infinite series is the main issue.

14.1. Problem. *"Rationalization" of repeating decimals.* Find a simple expression as a rational number for the number with decimal expansion $.abcabcabc\ldots$. Why is every repeating decimal a rational number? ∎

14.2. Problem. *The Tennis Problem.* Suppose that the points in a tennis game are independent and that the server wins each point with probability p. The first player to have at least four points and at least two more points than the opponent wins the game. What is the probability that the server wins the game? ∎

CAUCHY SEQUENCES

Many proofs about limits use a type of argument suggested by a discussion of errors in measurements. Suppose we measure quantities in a laboratory and have bounds on the experimental error. For example, we may measure two quantities L, M with accuracies $L \pm 2$ and $M \pm 3$. Because the errors might have the same sign, when we sum these quantities we write $(L + M) \pm 5$. For this reason, we determine each of L, M within accuracy $\varepsilon/2$ in order to ensure that the error in the total is at most ε. Bounding an error by bounding two constituent errors is a standard technique of proof called an $\varepsilon/2$ *argument*.

Sometimes we want to know whether a sequence converges, without knowing the limit. Our criteria so far are the definition of convergence (which requires that we know the limit), and the Monotone Convergence Theorem (which applies only to monotone sequences). We will prove in this chapter that it suffices to find a tail of the sequence in which the elements are close together. We define this more precisely:

14.3. Definition. A sequence $\langle a \rangle$ is a *Cauchy sequence* if for every $\varepsilon > 0$, there exists an $N \in \mathbb{N}$ (depending on ε) such that $n, m > N$ implies $|a_n - a_m| < \varepsilon$.

This property is named for Augustin Cauchy (1789-1857). The triangle inequality tells us that if each of two numbers is within $\varepsilon/2$ of a fixed number L, then the distance between them is at most ε. This is the basis of the next proof.

14.4. Proposition. Every convergent sequence is a Cauchy sequence.

Proof. Suppose $\langle a \rangle$ is a convergent sequence. To prove that $\langle a \rangle$ is a Cauchy sequence, for each $\varepsilon > 0$ we must choose an $N \in \mathbb{N}$ so that every two terms in the Nth tail differ by at most ε. Let $L = \lim a_n$. Given $\varepsilon > 0$, we can apply the definition of convergence of $\langle a \rangle$ for the number $\varepsilon/2$. This yields an $N \in \mathbb{N}$ such that $n > N$ implies $|a_n - L| < \varepsilon/2$. If we choose $n, m > N$, then we have the desired bound

$$|a_m - a_n| = |a_m - L + L - a_n| \le |a_m - L| + |L - a_n| < \varepsilon/2 + \varepsilon/2 = \varepsilon. \qquad \blacksquare$$

The converse of Proposition 14.4 is the fundamental result about convergence, called the *Cauchy Convergence Criterion*. It is equivalent to the Completeness Axiom in the following sense. If we had taken the Cauchy Convergence Criterion as an axiom, then we could have derived the Completeness Axiom from it as a theorem. Each of these is a precise mathematical formulation of the intuitive notion that the real numbers have no gaps. We will prove the Cauchy Convergence Criterion as Theorem 14.19 after developing a number of needed results.

14.5. Lemma. Every Cauchy sequence is bounded.

Proof. Let $\langle a \rangle$ be a Cauchy sequence. Using $\varepsilon = 1$ in the definition, we obtain an $N \in \mathbb{N}$ such that $m, n > N$ implies $|a_n - a_m| < 1$. In particular,

$n > N$ implies $|a_n - a_{N+1}| < 1$, which in turn implies $|a_n| < |a_{N+1}| + 1$ (see Lemma 13.7). Now let $M = \max\{|a_{N+1}| + 1, |a_1|, |a_2|, \ldots, |a_N|\}$. We conclude that $|a_n| < M$ for all $n \in \mathbb{N}$. ∎

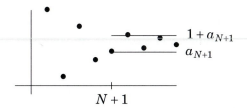

We defined the sum or product of two sequences termwise; this is a special case of the pointwise sum or product of two functions.

14.6. Lemma. If $\langle a \rangle$ and $\langle b \rangle$ converge to 0, then $\langle a + b \rangle$ converges to 0.

Proof. Suppose $a_n \to 0$ and $b_n \to 0$; we want to choose N in terms of ε so that $n > N$ implies $|a_n + b_n| < \varepsilon$. The hypotheses allow us to find tails in which $|a_n|$ and $|b_n|$ (respectively) are bounded by $\varepsilon/2$. Precisely, there exist N_1, N_2 such that $n > N_1$ implies $|a_n| < \varepsilon/2$ and $n > N_2$ implies $|b_n| < \varepsilon/2$. If we choose $N = \max\{N_1, N_2\}$, then for every $n > N$ the triangle inequality yields

$$|a_n + b_n| \le |a_n| + |b_n| < \varepsilon/2 + \varepsilon/2 = \varepsilon. \qquad ∎$$

14.7. Remark. *The form of the $\varepsilon/2$-argument.* In these arguments we prove statements about the convergence of some sequence, often using a hypothesis about convergence of other sequences. Our treatment of the "ε" in the definition of convergence depends on whether we are proving convergence or using convergence. When we want to prove that $c_n \to L$, we are given an arbitrary positive number, say ε, and we must construct $N \in \mathbb{N}$ such that $n > N$ implies $|c_n - L| < \varepsilon$. To construct this tail, we may use known convergence of another sequence $\langle a \rangle$. When we already know that $a_n \to M$, we are guaranteed for every positive number ε' the existence of some $N' \in \mathbb{N}$ such that $|a_n - M| < \varepsilon'$. We have the flexibility to choose ε' as we wish. Since we want to draw a conclusion about the number ε we were given, we will choose ε' as a function of ε, obtaining a tail of $\langle a \rangle$ in which some contribution to the difference between c_n and L will be small. Often, as in the proof above, we choose $\varepsilon' = \varepsilon/2$.

We then construct N in terms of the resulting N'. Obtaining the desired bound of ε on the total contribution to the error involves guaranteeing several conditions that each occurs when the index is sufficiently large. When N is the maximum of the numbers that guarantee these conditions, all the conditions will occur. In some arguments the contributions to the error may be more complicated. ∎

14.8. Lemma. The product of a bounded sequence and a sequence converging to 0 converges to 0.

Proof. Suppose $a_n \to 0$, and suppose $|b_n| < M$ for every $n \in \mathbb{N}$ (this implies $M > 0$). Define $c_n = a_n b_n$. Given $\varepsilon > 0$, we obtain a tail of $\langle c \rangle$ in which $|c_n| < \varepsilon$. The definition of convergence to 0 allows us to choose N such that $n > N$ implies $|a_n| < \varepsilon/M$. If we choose N this way (in terms of ε and M), then $n > N$ implies $|c_n| = |b_n| |a_n| \leq M\varepsilon/M = \varepsilon$. We conclude that $\langle c \rangle$ converges to 0. ∎

Often a statement follows either from an $\varepsilon/2$ argument *or* from previously-proved results about convergence; we will illustrate both.

14.9. Lemma. If $a_n \to L$ and $b_n \to M$, then
    1) $a_n + b_n \to L + M$.
    2) $a_n b_n \to LM$ (special case: $ca_n \to cL$).
    3) $a_n/b_n \to L/M$ (if $M \neq 0$).

Proof. Suppose $a_n \to L$ and $b_n \to M$. (1) Given $\varepsilon > 0$, we want to define N so that $n > N$ implies $|a_n + b_n - (L + M)| < \varepsilon$. Since $a_n \to L$ and $b_n \to M$, we can make $|a_n - L|$ and $|b_n - M|$ small by making n large; we choose $N_1, N_2 \in \mathbb{N}$ so that $n > N_1$ implies $|a_n - L| < \varepsilon/2$ and $n > N_2$ implies $|b_n - M| < \varepsilon/2$. If $N = \max\{N_1, N_2\}$, then $n > N$ implies

$$|a_n + b_n - (L + M)| = |a_n - L + b_n - M| \leq |a_n - L| + |b_n - M| < \varepsilon.$$

By the definition of convergence, we have proved that $a_n + b_n \to L + M$.

Alternatively, we could conclude that $(\langle a \rangle + \langle b \rangle) - (L + M) \to 0$ by expressing this sequence as $(\langle a \rangle - L) + (\langle b \rangle - M)$. This is the sum of two sequences converging to 0, which by Lemma 14.6 converges to 0.

(2) Continuing the alternative approach, we prove $a_n b_n \to LM$ by expressing $a_n b_n - LM$ in terms of quantities we can make small by making n large. We write $a_n b_n - LM = a_n b_n - a_n M + a_n M - LM = a_n(b_n - M) + (a_n - L)M$. Since $\langle a \rangle$ is a Cauchy sequence, it is bounded (Lemma 14.5); we also view the constant M as a bounded sequence. Hence we have expressed $\langle a \rangle \langle b \rangle - LM$ as the sum of two sequences, each of which is the product of a bounded sequence and a sequence converging to 0. We have seen that such a sequence converges to 0 (Lemma 14.8), and the sum of two sequences converging to 0 converges to 0 (Lemma 14.6), so $a_n b_n - LM \to 0$.

We also can prove this from the definition of convergence. We seek N in terms of ε so that $n > N$ implies $|a_n b_n - LM| < \varepsilon$. Again we use $a_n b_n - LM = a_n(b_n - M) + (a_n - L)M$. Since $a_n \to L$ and $b_n \to M$, we can choose N_1 and N_2 so that $n > N_1$ implies $|a_n - L| < \dfrac{\varepsilon}{2(1 + |M|)}$ and $n > N_2$ implies $|b_n - M| < \dfrac{\varepsilon}{2(\varepsilon + |L|)}$. If $N = \max\{N_1, N_2\}$, then $n > N$ implies

$$|a_n b_n - LM| < |a_n| |b_n - M| + |a_n - L| |M| < (\varepsilon + |L|) \frac{\varepsilon}{2(\varepsilon + |L|)} + \frac{\varepsilon}{2(1 + |M|)} |M| \leq \varepsilon.$$

Letting b_n be a constant sequence in (2) yields the special case $ca_n \to cL$. We leave the proof of (3) as Exercise 14.3. ∎

We next introduce a technique for finding limits. Given an equation involving terms of a convergent sequence, letting the index tend to infinity on both sides of the equation yields an equation for the limit.

14.10. Example. *Convergence to $\sqrt{2}$.* We construct a sequence of rational numbers that converges to $\sqrt{2}$. We know that $x^2 = 2$ if and only if $x = 2/x$. If we start with a positive number x_1, then one of $\{x_1, 2/x_1\}$ is larger than $\sqrt{2}$ and the other is smaller than $\sqrt{2}$. We hope that the average of x_1 and $2/x_1$ will be a better approximation to $\sqrt{2}$ than x_1 is. Given $x_1 > 0$, this suggests that we define the sequence $\langle x \rangle$ by setting $x_{n+1} = \frac{1}{2}(x_n + 2/x_n)$ for $n \geq 1$.

The two sides of the recurrence are different names for the same sequence. Hence if $\langle x \rangle$ converges, we have $\lim x_{n+1} = \lim(x_n/2 + 1/x_n)$. The elementary properties of limits then tell us that the limit L must satisfy $L = L/2 + 1/L$, which requires $L^2 = 2$. If $x_n > 0$, then x_{n+1} is the average of two positive numbers and is also positive, so when $x_1 > 0$ the only possible limit is the positive square root of 2. We still must show that the sequence does have a limit, for every positive initial guess.

If $x_1 = 1$, then $x_2 = 3/2$, $x_3 = 17/12$, etc., so with this initial condition the sequence is not monotone. Nevertheless, we claim that after the first step the sequence is decreasing and bounded below by $\sqrt{2}$, so the Monotone Convergence Theorem does imply that the limit exists.

To apply the Monotone Convergence Theorem, we prove for all $n > 1$ that $x_{n+1} > \sqrt{2}$ and that $x_{n+1} < x_n$. Since $x_2 = 3/2 > \sqrt{2}$, it suffices to show that both inequalities hold when $x_n > \sqrt{2}$. The first inequality follows from the AGM Inequality (Proposition 1.1), which states that the average of two positive real numbers is at least the square root of their product. Because x_{n+1} is the average of x_n and $2/x_n$, we conclude that $x_{n+1} \geq \sqrt{x_n(2/x_n)} = \sqrt{2}$.

To prove monotonicity, we compute $x_n - x_{n+1}$. Using the recurrence, we have $x_n - x_{n+1} = x_n - \frac{1}{2}(x_n + 2/x_n) = \frac{1}{2}(x_n - 2/x_n)$. Since $x_n > \sqrt{2}$, we have $x_n > 2/x_n$, and thus $x_n - x_{n+1} > 0$.

We have proved that if $x_1 > 0$, then the tail beginning at x_2 is monotone decreasing and bounded below by $\sqrt{2}$, so it converges. ∎

We can use this recurrence to compute an expansion for $\sqrt{2}$ to any desired accuracy, but it does not yield an exact "value" for $\sqrt{2}$. Unless a real number is rational, its decimal expansion is infinite and never repeats (Exercise 14.19). We must therefore be content with other descriptions of real numbers. This difficulty is in the nature of the real number system and can be overcome only by understanding limits.

Our next proof employs a technique similar to that in Example 14.10. First we establish that $\langle b \rangle$ converges. Once we know that $\langle b \rangle$ converges to some value L, we find an equation for L.

14.11. Proposition. If $\langle b \rangle$ is a sequence such that $|b_{n+1}|/|b_n| \to x$ with $0 \leq x < 1$, then $b_n \to 0$. Thus $\lim_{n \to \infty} x^n = 0$ when $|x| < 1$.

Proof. It suffices to show that $|b_n| \to 0$, so we may assume that $b_n > 0$ for all n. Since $b_{n+1}/b_n \to x$, we can use $\varepsilon = 1 - x$ in the definition of convergence to obtain N such that $n > N$ implies $b_{n+1}/b_n < 1$. Hence the Nth tail is a decreasing positive sequence, and the Monotone Convergence Theorem guarantees that it has a limit L. We find an equation that L must satisfy. Since 0 is a lower bound for $\langle b \rangle$, we have $L \geq 0$. If $L \neq 0$, then we have $x = \lim \dfrac{b_{n+1}}{b_n} = \dfrac{\lim b_{n+1}}{\lim b_n} = \dfrac{L}{L} = 1$. This contradicts our hypothesis that $x < 1$, so $L \neq 0$ is impossible. ∎

Our next property of limits is geometrically intuitive and easy to prove. It rests on the simple observation that if two numbers are each within ε of L, then every number between them is also within ε of L.

14.12. Theorem. (The Squeeze Theorem). Suppose $a_n \leq b_n \leq c_n$ for all n. If $a_n \to L$ and $c_n \to L$, then also $b_n \to L$.

Proof. Given $\varepsilon > 0$, we need to define N so that $n > N$ implies $|b_n - L| < \varepsilon$. Since b_n is between a_n and c_n, the number b_n is between L and one of $\{a_n, c_n\}$. Hence b_n is as close to L as one of $\{a_n, c_n\}$, and we conclude that $|b_n - L| \leq \max\{|a_n - L|, |c_n - L|\}$. Since $a_n \to L$ and $c_n \to L$, we can choose N_1, N_2 so that $n > N_1$ implies $|a_n - L| < \varepsilon$ and $n > N_2$ implies $|c_n - L| < \varepsilon$. If we set $N = \max\{N_1, N_2\}$, then $n > N$ implies $|b_n - L| < \varepsilon$. With this choice of N, we have proved that $n > N$ implies $|b_n - L| < \varepsilon$, so $b_n \to L$. ∎

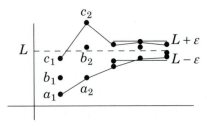

If $b \leq L + \varepsilon$ for all $\varepsilon > 0$, then $b \leq L$; we next generalize this.

14.13. Lemma. 1) If $b \leq c_n$ for all n, and $c_n \to L$, then $b \leq L$.
2) If $c_n \to L$ and $b_n - c_n \to 0$, then $b_n \to L$.

Proof. (Exercise 14.4). ∎

Suppose $\langle a \rangle, \langle c \rangle$ are sequences whose difference converges to 0. If one of these converges to L, then the other must also converge to L. We prove next that if we know one sequence is increasing and the other is decreasing, then we can drop the hypothesis of convergence and obtain the same conclusion. (Of course, $c_n - a_n \rightarrow 0$ does not alone guarantee convergence, as shown by the pair $\langle a \rangle, \langle c \rangle$ with $a_n = (-1)^n + 1/n$ and $c_n = (-1)^n - 1/n$.)

14.14. Lemma. If $\langle a \rangle$ is a nondecreasing sequence and $\langle c \rangle$ is a nonincreasing sequence with $c_n - a_n \rightarrow 0$, then $\langle a \rangle$ and $\langle c \rangle$ converge and have the same limit.

Proof. We first show that $a_m \le c_{m'}$ for all m, m'. If this fails, then $a_m > c_{m'}$ for some m, m'. If $n > \max\{m, m'\}$, then $a_n - c_n > a_m - c_{m'}$, since $\langle a \rangle$ is nondecreasing and $\langle c \rangle$ is nonincreasing. Letting ε be the positive number $a_m - c_{m'}$, we have shown that there is no $N \in \mathbb{N}$ such that $n > N$ implies $|a_n - c_n| < \varepsilon$, which contradicts $c_n - a_n \rightarrow 0$.

Hence $\langle a \rangle$ is bounded above by c_1 and $\langle c \rangle$ is bounded below by a_1, and the Monotone Convergence Theorem implies that each converges. If $c_n \rightarrow L$ and $a_n \rightarrow M$, then $L - M = 0$, because the difference of two convergent sequences converges to the difference of the limits. ∎

We know that a bounded monotone sequence is convergent, but boundedness alone does not guarantee convergence (consider $a_n = (-1)^n + 1/n$). In this example, the terms with even index form a convergent subsequence. We prove next that every bounded sequence has a convergent subsequence.

14.15. Definition. A *subsequence* of the sequence $\langle a \rangle$ is a sequence $\langle b \rangle$ defined by $b_k = a_{n_k}$, where $n_1 < n_2 < \cdots$ is an increasing sequence of indices.

14.16. Example. *Subsequences.* If $a_n = 2n - 1$ and $n_k = k^2$, then $b_k = 2k^2 - 1$. If $a_n = (-1)^n$ and $n_k = 2k$, then $b_k = 1$ for all k. ∎

n	1	2	3	4	5	6	7	8	9
a_n	1	3	5	7	9	11	13	15	17
k	1			2					3
b_k	1			7					17

We can interpret subsequences as composite functions. When $\langle b \rangle$ is a subsequence of $\langle a \rangle$, the function $b: \mathbb{N} \rightarrow \mathbb{R}$ is the composition of the function $a: \mathbb{N} \rightarrow \mathbb{R}$ with the increasing function $n: \mathbb{N} \rightarrow \mathbb{N}$; using the notation of functions instead of subscripts, we obtain $b(k) = a(n(k))$.

The next theorem reveals the importance of subsequences. It uses a version of the pigeonhole principle for infinite sets. When the union of two sets is infinite, at least one of them must be infinite. We apply this to sets of real numbers obtained by repeatedly bisecting intervals.

14.17. Theorem. (Bolzano-Weierstrass Theorem) Every bounded sequence of real numbers has a convergent subsequence.

Proof. Suppose $\langle x \rangle$ is a sequence with $L < x_n < M$ for all $n \in \mathbb{N}$. We construct a convergent subsequence $\langle b \rangle$, where $b_k = x_{n_k}$. We will choose b_k so that $a_k \leq b_k \leq c_k$, where $\langle a \rangle$ and $\langle c \rangle$ are sequences converging to the same limit K. The squeeze theorem then implies $b_k \to K$.

Set $a_1 = L$ and $c_1 = M$. We construct $\langle a \rangle$ and $\langle c \rangle$ iteratively. Having specified a_k and c_k, let $z_k = (a_k + c_k)/2$ be the midpoint of the interval between them. If there are infinitely many terms of $\langle x \rangle$ in the lower half $[a_k, z_k]$, then we set $a_{k+1} = a_k$ and $c_{k+1} = z_k$. Otherwise, we set $a_{k+1} = z_k$ and $c_{k+1} = c_k$.

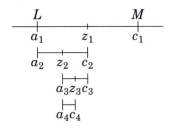

For each $k \in \mathbb{N}$, we claim that $[a_k, c_k]$ contains infinitely many terms of $\langle x \rangle$ and has length $(M - L)/2^{k-1}$. We prove this by induction on k. For $k = 1$, the interval $[a_1, c_1]$ contains all of $\langle x \rangle$ and has length $M - L$. For the induction step, suppose the claim holds for some $k \geq 1$. Each of $[a_k, z_k]$ and $[z_k, c_k]$ is half the length of $[a_k, c_k]$, so we have $c_{k+1} - a_{k+1} = (c_k - a_k)/2 = (M - L)/2^k$. Also, since $[a_k, c_k]$ contains infinitely many terms of $\langle x \rangle$ (by the induction hypothesis), there must be infinitely many elements of $\langle x \rangle$ in $[a_k, z_k]$ or in $[z_k, c_k]$.

By construction, $\langle a \rangle$ is increasing and $\langle c \rangle$ is decreasing, and we have proved $c_k - a_k \to 0$. Lemma 14.14 implies that these sequences converge and have the same limit, which we call K. It remains only to select a subsequence $\langle b \rangle$ of $\langle x \rangle$ such that $b_k \in [a_k, c_k]$; we do this iteratively. We must have $b_k = x_{n_k}$ with $\{n_k\}$ being an increasing sequence.

We choose $b_1 = x_1$, i.e. $n_1 = 1$. Suppose we have chosen $n_1, \ldots, n_{k-1}$ as the indices in $\langle x \rangle$ for the first $k - 1$ terms of $\langle b \rangle$. Because $[a_k, c_k]$ contains infinitely many terms of $\langle x \rangle$, we can choose such a term whose index is larger than all of those previously chosen. Let this index be n_k. With $b_k = x_{n_k}$ for all k, we have constructed a subsequence $\langle b \rangle$ with $a_k \leq b_k \leq c_k$ for all k, and hence $b_k \to K$. ∎

The method of proof used here, in which we successively select the upper or lower half of the current interval to continue the search, is called the *method of bisection*. Applied to a convergent sequence, this method develops the binary expansion of the limit.

14.18. Example. *Binary expansion by the method of bisection.* Suppose we want the binary expansion of a real number α in the interval $[0, 1]$. We can produce the sequence of zeroes and ones iteratively by the method of bisecting intervals. The original (first) interval is $[0, 1]$. If α is less than $1/2$, the first digit is 0, otherwise it is 1. In general, we let the nth digit be 0 when α is in the first half of the nth interval and 1 when α is in the second half. The next interval is obtained by bisecting the current interval and selecting the half containing α. More generally, we can find the k-ary expansion of the number by dividing the intervals into k equal pieces at each stage. Indeed, we did precisely this in Definition 13.14 to produce the canonical k-ary expansion. ∎

We can prove that a sequence converges by guessing the limit and applying the definition of convergence. Even when we cannot guess the limit, we sometimes can prove that a limit exists by showing that the sequence is monotone and bounded; this criterion is sufficient but not necessary. The next theorem captures the essence of the completeness axiom and is used as the definition of completeness in more general settings. Informally, the theorem says that if the terms of a sequence are eventually close to each other, then there is a real number to which the sequence converges.

14.19. Theorem. (Cauchy Convergence Criterion) A sequence of real numbers converges if and only if it is a Cauchy sequence.

Proof. We have already proved that a convergent sequence is a Cauchy sequence, using an $\varepsilon/2$-argument. For the converse, suppose that $\langle a \rangle$ is a Cauchy sequence. By Lemma 14.5, $\langle a \rangle$ is bounded. By the Bolzano-Weierstrass Theorem, $\langle a \rangle$ therefore has a convergent subsequence $\langle b \rangle$ with $b_k = a_{n_k}$. Let $L = \lim b_k$; we prove that also $a_n \to L$. Consider an arbitrary $\varepsilon > 0$. Since $\langle a \rangle$ is a Cauchy sequence, we can choose N_1 such that $n, m > N_1$ implies $|a_n - a_m| < \varepsilon/2$. Since $b_k \to L$, we can choose N_2 such that $k > N_2$ implies $|b_k - L| < \varepsilon/2$. Set $N = \max\{N_1, N_2\}$. Since $\{n_k\}$ is an increasing sequence of natural numbers, we have $n_k \geq k$. Hence $k > N$ implies $n_k > N$, and we compute

$$|a_k - L| = |a_k - b_k + b_k - L| \leq |a_k - a_{n_k}| + |b_k - L| < \varepsilon. \qquad ∎$$

We next prove that a sequence formed by averaging initial terms of a convergent sequence converges to the same limit.

14.20. Proposition. If $\langle a \rangle$ converges to L, and $\langle b \rangle$ is defined by $b_n = (1/n)\Sigma_{i=1}^{n} a_i$, then $\langle b \rangle$ also converges to L.

Proof. We define $c_n = a_n - b_n$ and prove that $c_n \to 0$; since $\langle a \rangle$ converges, this implies that $\langle b \rangle$ converges and has the same limit as $\langle a \rangle$. We consider c_n in the following form:

$$c_n = a_n - b_n = \frac{1}{n} \Sigma_{i=1}^n (a_n - a_i).$$

Given $\varepsilon > 0$, we will obtain an $N \in \mathbb{N}$ such that $n > N$ implies $|c_n| < \varepsilon$. Since $\langle a \rangle$ is bounded, there exists M such that $n \in \mathbb{N}$ implies $|a_n| < M$. Since $\langle a \rangle$ is a Cauchy sequence, there exists N' such that $n, m > N'$ implies $|a_n - a_m| < \varepsilon/2$. When we consider the first N' terms in the summation for c_n, we can bound them using $|a_n - a_i| \le |a_n| + |a_i| < 2M$. Since the triangle inequality implies $|\Sigma x_i| \le \Sigma |x_i|$, we have $|\frac{1}{n} \Sigma_{i=1}^{N'} (a_n - a_i)| < 2MN'/n$. In order to bound this contribution to $|c_n|$ by $\varepsilon/2$, we choose $N \ge \max\{N', 4MN'/\varepsilon\}$. Then $n > N$ implies

$$|c_n| \le \frac{1}{n} \Sigma_{i=1}^{N'} |a_n - a_i| + \frac{1}{n} \Sigma_{i=N'+1}^n |a_n - a_i|$$

$$< \frac{N'}{n} 2M + \frac{n - N'}{n} \frac{\varepsilon}{2} < \varepsilon. \quad \blacksquare$$

INFINITE SERIES

We have discussed how to evaluate various finite sums. Now we consider summing the terms in an infinite sequence. This is not generally possible. Even when it is possible, the value may depend on the order in which the terms are summed (Exercise 14.35). Therefore we need a precise definition for the sum of an infinite series.

14.21. Definition. Suppose $\langle a \rangle$ is a sequence of real numbers. The number $s_n = \Sigma_{k=1}^n a_k$ is the *nth partial sum* of the sequence $\langle a \rangle$. The formal expression $\Sigma_{k=1}^\infty a_k$ is an *infinite series*. The infinite series $\Sigma_{k=1}^\infty a_k$ *converges* if $\lim_{n \to \infty} s_n$ exists; otherwise the series *diverges*. When $\Sigma_{k=1}^\infty a_k$ converges, we write $L = \lim s_n = \Sigma_{k=1}^\infty a_k$ and say that the sum of the series equals L.

14.22. Example. *Sequence versus partial sums.* Let $a_n = 1/2^n$, and let $s_n = \Sigma_{k=1}^n a_k = 1 - 1/2^n$. Then $a_n \to 0$ and $s_n \to 1$. Thus $\Sigma_{k=1}^\infty 1/2^n = 1$.

As another example, we present a table of the terms and the partial sums when $a_n = (-1)^{n+1}/n$. This is an alternating series with terms of decreasing absolute value tending to zero. Every such series converges, as shown in Exercise 14.33. The sum of this series is .693147 to six decimal places; the table below suggests that it converges slowly. ∎

n	1	2	3	4	5	6	7	$\cdots$	99
a_n	1	$-.5$	$.333$	$-.25$	$.2$	$-.166$	$.143$	$\cdots$	$.010101$
$\Sigma_{k=1}^{n} a_k$	1	$.5$	$.833$	$.583$	$.783$	$.617$	$.760$	$\cdots$	$.698172$

14.23. Proposition. (The Distributive Law) If $\Sigma_{k=1}^{\infty} a_k$ converges and $c \in \mathbb{R}$, then $\Sigma_{k=1}^{\infty} ca_k$ also converges and equals $c\Sigma_{k=1}^{\infty} a_k$.

Proof. The nth partial sum for $\Sigma_{k=1}^{\infty} ca_k$ equals cs_n, where s_n is the nth partial sum for $\Sigma_{k=1}^{\infty} a_k$. By Lemma 14.9, $\lim cs_n = c \lim s_n$, and the result follows. ∎

14.24. Solution. *"Rationalization" of repeating decimals.* Let x equal the repeating decimal $.abcabcabcabc\ldots$. By the definition of decimal expansion, $x = (100a + 10b + c)/1000 + (100a + 10b + c)/1000000 + \cdots$. Therefore, $1000x = 100a + 10b + c + x$; here we have used Proposition 14.23. Solving for x, we obtain $x = (100a + 10b + c)/999$. ∎

This procedure generalizes to rationalize any decimal expansion that eventually repeats (Exercise 14.19). Alternatively, we could use Proposition 14.23 to write $x = (100a + 10b + c)\Sigma_{k=1}^{\infty} (1/1000)^k$ and obtain the same result by summing the geometric series.

14.25. Theorem. (The Geometric Series) Given $x \in \mathbb{R}$, the *geometric series* $\Sigma_{k=0}^{\infty} x^k$ converges to $\dfrac{1}{1-x}$ if $|x| < 1$ and diverges otherwise.

Proof. When $x \neq 1$, the partial sum $s_n = \Sigma_{k=0}^{n} x^k$ equals $(1 - x^{n+1})/(1 - x)$ (Corollary 4.10). Now we apply the properties of limits. Because $x^{n+1} \to 0$ if $|x| < 1$ and x^{n+1} does not converge if $x > 1$, the sequence of partial sums (and hence the series $\Sigma_{k=0}^{\infty} x^k$) converges to $1/(1-x)$ if $|x| < 1$ and diverges if $|x| > 1$. When $x = 1$, the nth partial sums is $n + 1$; when $x = -1$, the partial sums alternate between 1 and 0. Hence the series also diverges when $|x| = 1$. ∎

We have indexed the geometric series beginning at 0 instead of 1. When $|x| < 1$, the series $\Sigma_{k=1}^{\infty} x^k$ differs from $\Sigma_{k=0}^{\infty} x^k$ by 1; we have $\Sigma_{k=1}^{\infty} x^k = x/(1 - x)$.

14.26. Example. *The Multiplier Effect.* Suppose the typical individual in a society spends a fraction t of all new or extra income, where $0 < t < 1$. Economists call t the *marginal propensity to consume*. The sum of the geometric series describes the "multiplier effect" on the economy. When a typical individual receives an extra dollar in wages, he or she spends $\$t$, which is received by someone else. That person then spends $\$t^2$. The net overall increase in economic activity in the society is $\sum_{k=0}^{\infty} t^k = 1/(1-t)$ dollars. The higher the marginal propensity to consume, the greater the multiplier effect. ∎

14.27. Solution. *The Tennis Problem.* The server wins each particular point with probability p; let $q = 1 - p$. First we consider the ways the server can win with exactly four points. The other player may score zero, one, or two points before the server scores four, and the total probability for these mutually exclusive possibilities is $p^4 + \binom{4}{1}p^4 q + \binom{5}{2}p^4 q^2$. The server may also win after the game reaches a 3-3 tie; a 3-3 tie happens with probability $\binom{6}{3}p^3 q^3$. In this situation, the probability that the server wins after a total of exactly $2k + 2$ more points is $(2pq)^k p^2$. Summing this over all $k \geq 0$ yields the geometric series $p^2 \sum_{k=0}^{\infty} (2pq)^k = p^2/(1-2pq)$. Hence the total probability is $p^4(1 + 4q + 10q^2) + 20p^5 q^3/(1-2pq)$. This equals $.736$ when $p = .6$ and $.901$ when $p = .7$; this suggests the difficulty of "breaking serve".

The probability x of winning after the score is tied with at least three points each can be computed in another way. The server can win the next two points or can win after splitting the next two points, which repeats the tied situation. Hence $x = p^2 + 2pqx$, or $x = p^2/(1-2pq)$. This is essentially the same computation as summing a geometric series by using the distributive law. The sum converges because it represents a probability, and hence the series computation is not needed. ∎

Many students confuse the words "sequence" and "series" in mathematics, perhaps because the use of the word "series" in English is similar to the use of the word "sequence" in mathematics. In English, we generally use "series" to mean a *finite sequence* ("list") of events, such as a "series" of baseball games.

A series converges if its sequence of partial sums has a limit; otherwise it diverges. We say that a series "diverges to positive infinity" if, for every $M \in \mathbb{R}$, there is an $N \in \mathbb{N}$ such that $n > N$ implies $s_n > M$. By the Monotone Convergence Theorem, a divergent series of positive numbers must diverge to positive infinity. The definition of divergence to negative infinity is similar, with $s_n < M$ in place of $s_n > M$. Series may diverge in other ways. The series $1 - 2 + 3 - 4 = \sum_{k=1}^{\infty} k(-1)^{k+1}$ has arbitrarily large partial sums of both signs. The series $\sum_{k=1}^{\infty} (-1)^k$ has bounded partial sums, but the sequence of partial sums does not converge. We begin with a necessary condition for convergence.

14.28. Lemma. If the series $\sum_{k \geq 0} a_k$ converges, then $a_n \to 0$.

Proof. Let $s_n = \sum_{k=1}^{n} a_k$; thus $a_n = s_n - s_{n-1}$. Because the series converges, the sequence $\langle s \rangle$ has a limit L. Therefore, $a_n \to L - L = 0$.

Alternatively, if $\langle s \rangle$ converges, then $\langle s \rangle$ is a Cauchy sequence, so for every $\varepsilon > 0$ we have $|s_n - s_{n-1}| < \varepsilon$ for s_n in some tail. We again have $a_n = s_n - s_{n-1}$, and hence $a_n \to 0$ by the definition of convergence. ∎

The converse of Lemma 14.28 is false. In the next example, the terms go to 0, but more "slowly" than in the geometric series--slowly enough that the sum diverges. Convergence of a series of positive numbers requires that the terms go to 0 sufficiently rapidly.

14.29. Example. *The harmonic series.* Let $a_k = 1/k$; below we graph this sequence and its sequence of partial sums. To see that $\sum_{k=1}^{\infty} a_k$ diverges even though $a_k \to 0$, we compare $\langle a \rangle$ with an even simpler example of a divergent sequence with terms approaching 0. Let $\langle c \rangle = \frac{1}{2}, \frac{1}{4}, \frac{1}{4}, \frac{1}{8}, \frac{1}{8}, \frac{1}{8}, \frac{1}{8}, \frac{1}{16}, \cdots$; here there are 2^{j-1} copies of $1/2^j$ for each $j \geq 1$. Since the copies of $1/2^j$ for each fixed j sum to $1/2$, for each $M \in \mathbb{N}$ the partial sum $\sum_{k=1}^{n} c_k$ exceeds M for large enough n, and $\sum_{k=1}^{\infty} c_k$ diverges. When we compare $\langle a \rangle$ and $\langle c \rangle$, we see that $a_k > c_k$ for every k (the last copy of $1/2^j$ in $\langle c \rangle$ is compared with $1/(2^j - 1)$ in $\langle a \rangle$). For each n, summing n of these inequalities yields $\sum_{k=1}^{n} a_k > \sum_{k=1}^{n} c_k$. Hence $\sum_{k=1}^{\infty} a_k$ also diverges. ∎

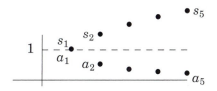

At first, we can only prove convergence of series by applying the definition. Applying the definition requires finding a formula for the partial sum $s_k = \sum_{n=1}^{k} a_n$ in terms of k, and then determining whether the sequence defined by this function converges. Seldom can we compute the limit of the partial sums directly, so we need other methods to test for convergence of a series. Example 14.29 illustrates the "comparison test". When the test applies, it settles the question of convergence.

14.30. Proposition. (Comparison test) Suppose $c_n \geq 0$ for all n. If $\sum_{n=1}^{\infty} c_n$ converges and $|a_n| \leq c_n$ for all $n > N$, then $\sum_{n=1}^{\infty} a_n$ converges. If $\sum_{n=1}^{\infty} c_n$ diverges to ∞ and $a_n \geq c_n$ for all n, then also $\sum_{n=1}^{\infty} a_n$ diverges to ∞.

Proof. Let $s_k = \sum_{n=1}^{k} a_n$, and let $S_k = \sum_{n=1}^{k} c_n$. Since $\sum_{n=1}^{\infty} c_n$ converges, $\langle S \rangle$ is a Cauchy sequence; we show that $\langle s \rangle$ is also a Cauchy sequence. Given $\varepsilon > 0$, choose N such that $m, n > N$ implies that $|S_m - S_n| < \varepsilon$. Given $m > n > N$, we have

$$|s_m - s_n| = |\Sigma_{i=n+1}^m a_i| \le \Sigma_{i=n+1}^m |a_i| \le \Sigma_{i=n+1}^m c_i = |S_m - S_n| < \varepsilon.$$

Hence $\langle s \rangle$ is a Cauchy sequence and converges. We leave the second statement to Exercise 14.31. ∎

14.31. Corollary. If $\Sigma |a_n|$ converges, then Σa_n converges.

Proof. Apply the comparison test with $c_n = |a_n|$. ∎

Applying the comparison test to prove convergence requires known convergent series for comparisons, such as the geometric series. The behavior of the geometric series suggests another general test for convergence. Consider a series $\Sigma_{k=1}^\infty a_k$ with positive terms, and let $c_k = a_{k+1}/a_k$. If $\langle a \rangle$ is a geometric series, then $\langle c \rangle$ is a constant. If $\langle c \rangle$ is not constant but does have a limit ρ, then the convergence criterion is the same: it converges if $\rho < 1$ and diverges if $\rho > 1$. Here we use "ρ" to suggest "ratio". The test is inconclusive when $\rho = 1$, but Exercise 14.42 develops a refinement (Raabe's test) that sometimes guarantees convergence even when $\rho = 1$.

14.32. Theorem. (Ratio test) Suppose $\langle a \rangle$ is a sequence such that $|a_{k+1}/a_k| \to \rho$. If $\rho < 1$, then $\Sigma_{k=1}^\infty a_k$ converges. If $\rho > 1$, then $\Sigma_{k=1}^\infty a_k$ diverges.

Proof. We leave the proof for $\rho > 1$ to Exercise 14.37; assume $\rho < 1$. By Corollary 14.31, it suffices to show that $\Sigma_{k=1}^\infty |a_k|$ converges. Since the hypothesis involves only the absolute values of the terms, we only need to consider the case where each a_k is positive. Let $\langle s \rangle$ be the sequence of partial sums: $s_n = \Sigma_{k=1}^n a_k$. We prove that $\langle s \rangle$ is a Cauchy sequence and hence converges.

Let ε be an arbitrary positive number, and choose β between ρ and 1. Since $a_{k+1}/a_k \to \rho$, we can choose N_1 such that $k > N_1$ implies $a_{k+1}/a_k < \beta$. In particular, this implies $a_{k+j} < a_k \beta^j$ for $k > N_1$ and $j \ge 1$. We have already proved that $a_{k+1}/a_k \to \rho < 1$ implies $a_k \to 0$ (Proposition 14.11). Hence there also exists N_2 such that $k > N_2$ implies $a_k < (1 - \beta)\varepsilon$. Choose $N = \max\{N_1, N_2\}$. Given any $k, l > N$ with $l \ge k$ we have $0 < s_l - s_k = \Sigma_{n=k+1}^l a_n < \Sigma_{j=1}^{l-k}(1 - \beta)\varepsilon \beta^j < (1 - \beta)\varepsilon\Sigma_{j=0}^\infty \beta^j = \varepsilon$. Hence the sequence of partial sums is a Cauchy sequence; by Theorem 14.9, it converges. ∎

The ratio test enables us to define the exponential function via series. We discuss its crucial properties in Chapter 17.

14.33. Example. *Exponential series.* Given $x \in \mathbb{R}$, the *exponential function* is defined by $\exp(x) = \Sigma_{n=0}^\infty x^n/n!$. For each $x \in \mathbb{R}$, this series converges by the ratio test, since $|\dfrac{x^{n+1}}{(n+1)!} / \dfrac{x^n}{n!}| = \dfrac{|x|}{n+1} \to 0$. ∎

Our last convergence test will be applied in Chapter 17.

14.34. Theorem. (Root test) Suppose $\langle a \rangle$ is a sequence such that $|a_n|^{1/n} \to \rho$. If $\rho < 1$, then $\Sigma_{k=1}^{\infty} a_k$ converges. If $\rho > 1$, then $\Sigma_{k=1}^{\infty} a_k$ diverges.

Proof. (Exercise 14.45.) ∎

EXERCISES

14.1. Compute the binary expansions of $2/7$ and $\sqrt{2}$ to six places.

14.2. Prove that if $\langle a \rangle$ converges, then every subsequence of $\langle a \rangle$ converges and has the same limit as a.

14.3. Suppose $\langle a \rangle, \langle b \rangle$ are sequences, with $b_n \neq 0$ for all n. Prove that if $a_n \to L$ and $b_n \to M \neq 0$, then $a_n/b_n \to L/M$. (Hint: show first that this holds if $a_n = 1$ for all n.)

14.4. Prove the following statements:
  a) If $b \leq c_n$ for all n, and $c_n \to L$, then $b \leq L$.
  b) If $b \leq L + \varepsilon$ for all $\varepsilon > 0$, then $b \leq L$.
  c) If $c_n \to L$ and $b_n - c_n \to 0$, then $b_n \to L$.

14.5. *The Nested Interval Property.* Suppose $\{I_n\}$ is a sequence of closed intervals, with I_n of length d_n, such that $I_{n+1} \subseteq I_n$ for all n and $d_n \to 0$. The Nested Interval Property states that for such a sequence, there is exactly one point that belongs to each I_n. Prove the following statements:
  a) The Completeness Axiom implies the Nested Interval Property.
  b) The Nested Interval Property implies the Completeness Axiom.

14.6. Suppose $a_n = p(n)/q(n)$, where p and q are polynomials and the degree of q is greater than the degree of p. Use properties of limits to prove that $a_n \to 0$.

14.7. If $a_1 = 1$ and $a_n = \sqrt{3a_{n-1} + 4}$ for $n > 1$, prove that $a_n < 4$ for all $n \in \mathbb{N}$.

14.8. Suppose $x_1 = 1$ and $2x_{n+1} = x_n + 3/x_n$ for $n \geq 1$. Prove that $\lim_{n \to \infty} x_n$ exists, and find the limit.

14.9. Suppose $x_1 = a$ and $x_{n+1} = \sqrt{1 + x_n}$ for $n \geq 1$. Prove that $\lim_{n \to \infty} x_n$ exists, and find the limit.

14.10. Suppose a sequence $\langle x \rangle$ satisfies the recurrence $x_{n+1} = x_n^2 - 4x_n + 6$.
  a) If $\lim_{n \to \infty} x_n$ exists and equals L, what possible values can L have?
  b) The behavior of x_n as $n \to \infty$ depends on the initial value x_0. For each $x_0 \in \mathbb{R}$, describe this behavior. (Hint: obtain a recurrence for the sequence $\{y_n\}$ defined by $y_n = x_n - 2$, and study its behavior.)

14.11. (+) *Generalization of Exercise 14.10.* Suppose $\langle x \rangle$ satisfies $x_n = f(x_{n-1})$ for $n \geq 1$, where $f(x) = x^2 + Ax + B$. Determine the possible values of $\lim_{n \to \infty} x_n$. Completely determine the limiting behavior of x_n in terms of x_0, A, and B.

14.12.   Suppose $f_1(x) = x$ for $x \in \mathbb{R}$ and $f_{n+1}(x) = (f_n(x))^2/2$ for $n \geq 1$. If $\lim_{n \to \infty} f_n(x)$ exists, what can the limit equal? For which x is the sequence $\{f_n(x)\}$, strictly increasing, constant, or strictly decreasing? Use this information to determine how $\lim_{n \to \infty} f_n(x)$ depends on x.

14.13.   Suppose that $a_{n+2} = (\alpha + \beta)a_{n+1} - \alpha\beta a_n$ with $\beta \neq \alpha$, and suppose that $a_0 = a_1 = 1$. Find the limit as $n \to \infty$ of a_{n+1}/a_n.

14.14.   (!) A runaway train is hurtling toward a brick wall at the speed of 100 miles per hour. When it is two miles from the wall, a fly begins to fly repeatedly between the train and the wall at the speed of 200 miles per hour. Determine how far the fly travels before it is smashed.

14.15.   (–) Suppose $\Sigma_{k=1}^{\infty} a_k$ and $\Sigma_{k=1}^{\infty} b_k$ converge to A and B, respectively. Prove that $\Sigma_{k=1}^{\infty}(a_k + b_k)$ converges and equals $A + B$.

14.16.   Find the expansion of $1/2$ in base 3, with proof. Determine the rational number that has ternary expansion $.121212\ldots$.

14.17.   (!) Consider $.247247247\cdots$, expressed as a decimal expansion. Write this as a rational number with numerator and denominator in base 10. Now suppose the expression $.247247247\cdots$ is an expansion in base 8. Write this as a rational number with numerator and denominator in base 10.

14.18.   Describe geometrically the set of numbers in the interval $[0, 1]$ whose ternary expansions contain no 1's. Prove that the set is uncountable.

14.19.   Exercise 13.9 requests a proof that the k-ary expansion of every rational number is "eventually periodic", meaning that after some initial portion the remainder repeats a finite list. Prove the converse: if the k-ary expansion of x is eventually periodic, then x is rational.

14.20.   *Alternative approach to the geometric series.*   Suppose that $y = 1/(1 - x)$, which is equivalent to $y = 1 + xy$. Suppose that $|x| < 1$ and that we have an initial guess y_0 for y. The equation $y = 1 + xy$ suggests two algorithms.

   a) Given y_0, define the sequence $\langle y \rangle$ by $y_{n+1} = 1 + xy_n$ for $n \geq 0$. Prove that $\langle y \rangle$ converges to $1/(1 - x)$.

   b) Given y_0, define $\langle y \rangle$ instead by $y_n = 1 + xy_{n+1}$, so $y_{n+1} = (y_n - 1)/x$. Why does this algorithm fail, even when $x \neq 0$?

14.21.   A *limit point* of a sequence $\langle a \rangle$ is a number L to which some subsequence of $\langle a \rangle$ converges. Construct a sequence with infinitely many limit points.

14.22.   (!) Suppose $a_n = p(n)x^n$, where p is a polynomial in n and $|x| < 1$. Prove that $a_n \to 0$. (Hint: consider the ratio a_{n+1}/a_n.) (Comment: For a fixed number x with $|x| < 1$, this exercise shows that x^n tends to zero so fast that multiplication by a polynomial in n does not affect the limit. Thus exponential decay dominates polynomial growth.)

14.23.   Suppose $\langle x \rangle$ is the sequence given by $x_1 = 1$ and $x_{n+1} = 1/(x_1 + \cdots + x_n)$ for $n \geq 1$. Prove that $\langle x \rangle$ converges, and obtain the limit.

14.24. Suppose a sequence of measurements are to be made. Each measurement involves some error but can be made to any specified accuracy. How can it be guaranteed that the total error is at most 1?

14.25. (+) *Measure zero.* A set $S \subset \mathbb{R}$ has *measure zero* if, for every $\varepsilon > 0$, there is a countable collection of intervals whose union contains S, such that the sum of the lengths of the intervals is less than ε. Prove that a union of countably many sets of measure zero also has measure zero. Conclude that the set of rational numbers has measure zero. (Hint: This uses what might be called the "ultimate" $\varepsilon/2$ argument; consider $\varepsilon/2^n$ for each n.)

14.26. Compute $\Sigma_{n=1}^{\infty} (\frac{x}{x+1})^n$. What assumptions must be made about x?

14.27. Compute $\Sigma_{n=1}^{\infty} \frac{1}{n(n+1)}$. Use this to obtain upper and lower bounds on $\Sigma_{n=1}^{\infty} \frac{1}{n^2}$. (Comment: the exact value of $\Sigma_{n=1}^{\infty} \frac{1}{n^2}$ is $\pi^2/6$.)

14.28. (!) Suppose the nth partial sum of a series equals $1/n$, for $n \geq 1$. Determine the nth term in the series.

14.29. Suppose $b_k = c_k - c_{k-1}$, where $\langle c \rangle$ is a sequence such that $c_0 = 1$ and $\lim_{k \to \infty} c_k = 0$. Use the definition of series to determine $\Sigma_{k=1}^{\infty} b_k$.

14.30. Change the Tennis Problem (Problem 14.2) so that the winner is the first one to reach four points. What is the server's probability of winning the game?

14.31. *Comparison test for divergence.* Suppose $\Sigma_{k=1}^{\infty} c_k$ diverges to ∞, and suppose $a_k \geq c_k$ for all k. Prove that $\Sigma_{k=1}^{\infty} a_k$ diverges to ∞.

14.32. Determine whether $1 + \frac{1}{3} + \frac{1}{5} + \frac{1}{7} + \cdots$ converges.

14.33. (!) *Convergence of alternating series.* Recall that if $\langle c \rangle$ is an increasing sequence, $\langle d \rangle$ is a decreasing sequence, and $d_n - c_n \to 0$, then $\langle c \rangle$ and $\langle d \rangle$ converge and have the same limit. Use this and the Squeeze Theorem to prove the following statement: If $\langle a \rangle$ is a sequence whose terms alternate in sign, converge to 0, and satisfy $|a_{k+1}| \leq |a_k|$ for all n, then the series $\Sigma_{k=0}^{\infty} a_k$ converges.

14.34. Consider $\Sigma_{k=1}^{\infty} \frac{(-1)^{k+1}}{k} = 1 - \frac{1}{2} + \frac{1}{3} - \frac{1}{4} + \cdots$. By Exercise 15.32, this series converges. (It converges to ln2, but that is not needed here.) Prove that the sum of the series is less than 5/6. The terms in the series can be summed in other orders. Prove that $1 + \frac{1}{3} - \frac{1}{2} + \frac{1}{5} + \frac{1}{7} - \frac{1}{4} + \frac{1}{9} + \frac{1}{11} - \frac{1}{6} + \cdots$ has sum greater than 5/6 (in fact, the sum exceeds 1). Find a reordering of the terms to obtain (with proof!) a convergent series whose sum exceeds 3/2.

14.35. Suppose that Σa_k converges, that $\Sigma |a_k|$ diverges, and that L is a real number. Prove that the terms of $\langle a \rangle$ can be reordered to obtain a series that converges to L. (Riemann)

14.36. Suppose $0 < a_n \leq a_{2n} + a_{2n+1}$ for all $n \geq 1$. Prove that $\Sigma_{n=1}^{\infty} a_n$ diverges.

14.37. *Ratio test for divergence.* Suppose $\langle a \rangle$ is a sequence such that $|a_{k+1}/a_k| \to \rho$ for some $\rho > 1$. Prove that $\Sigma_{k=1}^{\infty} a_k$ diverges.

14.38. In Example 12.33, we found the formula $f(x) = 1/(1 - x - x^2)$ for the generating function of the Fibonacci numbers. This implies that the series $\sum_{n=0}^{\infty} F_n x^n$ converges for all x such that $|x|$ is less than the smallest magnitude of a root of $1 - x - x^2$. Assuming this, use the ratio test to find $\lim \frac{F_{n+1}}{F_n}$. Compare this limit with the formula for the Fibonacci numbers in Solution 12.25.

14.39. *Limit comparison test.* Suppose $\langle a \rangle$ and $\langle b \rangle$ are sequences of positive numbers, and a_n/b_n converges to a nonzero real number L. Prove that $\sum_{k=1}^{\infty} b_k$ converges if and only if $\sum_{k=1}^{\infty} a_k$ converges.

14.40. (!) *Condensation test.*

a) Suppose $\langle a \rangle$ is a decreasing sequence of positive numbers. Prove that $\sum_{k=1}^{\infty} a_k$ converges if and only if $\sum_{j=0}^{\infty} 2^j a_{2^j}$ converges. (Hint: prove that partial sums of the first series are bounded by those in the second, and those in the second are bounded by twice those in the first.)

b) For $p \in \mathbb{R}$, prove by part (a) that $\sum_{k=1}^{\infty} k^{-p}$ converges if and only if $p > 1$.

14.41. Suppose that $\langle a \rangle$ and $\langle b \rangle$ are sequences of positive numbers such that $b_{k+1}/b_k \le a_{k+1}/a_k$ for all k in some tail. Prove that if $\sum_{k=1}^{\infty} a_k$ converges, then $\sum_{k=1}^{\infty} b_k$ converges.

14.42. *Raabe's test.* The ratio test for convergence of series is inconclusive when the ratio of successive terms converges to 1. This can be overcome if the convergence is slow enough. Suppose p is a fixed real number greater than 1.

a) (+) Prove that if $0 < x < 1$, then $(1 - px) < (1 - x)^p$.

b) Use part (a) and Exercise 14.41 with $a_k = 1/k^p$ to prove that $\sum_{k=1}^{\infty} b_k$ converges if $b_{k+1}/b_k \le 1 - p/k$ for all k in some tail (we are assuming $b_k > 0$ for all k).

14.43. (+) Use the divergence of $\sum_{k=1}^{\infty} 1/k$ to prove that every nonzero rational number is a finite sum of reciprocals of distinct integers. (Such an expression is known as an *Egyptian fraction*.)

14.44. Use the binomial theorem and the properties of limits to prove that $\exp(x + y) = \exp(x)\exp(y)$.

14.45. *Root test.* Suppose $\langle a \rangle$ is a sequence such that $|a_n|^{1/n} \to \rho$.

a) Prove that if $\rho < 1$, then $\sum_{k=1}^{\infty} a_k$ converges.

b) Prove that if $\rho > 1$, then $\sum_{k=1}^{\infty} a_k$ diverges.

c) Show by example that if $\rho = 1$, then $\sum_{k=1}^{\infty} a_k$ may converge or diverge.

Chapter 15

Continuity

Continuity is the precise mathematical formulation of an easy intuitive idea. For many phenomena, a small change in the input results in a small change in the output. Before defining continuous functions, we describe several problems where the idea arises.

15.1. Problem. *The Antipodal Point Problem.* Consider a circular wire. Suppose the temperature does not change abruptly from point to point. This implies that some pair of opposite points on the circle have the same temperature. Why? ■

15.2. Problem. *The Jewel Thieves Problem.* Two jewel thieves have stolen a circular necklace. The necklace has an even number of diamonds and an even number of rubies. The thieves want to split the necklace so that each keeps half of each type of jewel. Because the links are made of gold, they don't want to cut many links. Is it always possible, no matter how the jewels are arranged, to make two cuts so that each thief gets a segment containing half the jewels of each type? ■

15.3. Example. *The Butterfly Effect.* If a butterfly flaps its wings in Moscow, does the resulting wind current affect weather patterns in the USA? For years scientists believed the answer to be no, but recent studies suggest otherwise. Physical phenomena depend on many variables, resulting in possible "chaotic" behavior.[†] The flapping of the butterfly's wings produces a small change in one variable. Depending on the role of other variables, this small change may produce a significant change far away. See Exercise 15.5 for a related example. ■

[†]J. Gleick, *Chaos: Making a New Science*, Viking Press (New York, 1987), Chapter 1.

LIMITS AND CONTINUITY

To discuss continuity, we need a concept of limit for function values $f(x)$ as x approaches a. When computing limits of function values we don't consider what happens at a, but only at elements of the domain close to a. A *neighborhood* about a is an open interval containing a. We define a *deleted interval about* a or *deleted neighborhood of* a to be the set obtained by removing a from a neighborhood of a. For example, $\{x \in \mathbb{R}: |x - a| < \delta\}$ is a neighborhood of a and $\{x \in \mathbb{R}: 0 < |x - a| < \delta\}$ is a deleted neighborhood of a.

15.4. Definition. Suppose f is defined on a deleted interval around a. We write $\lim_{x \to a} f(x) = L$ if for every $\varepsilon > 0$ there exists $\delta > 0$ such that $0 < |x - a| < \delta$ implies $|f(x) - L| < \varepsilon$. We read $\lim_{x \to a} f(x) = L$ as "the *limit* of $f(x)$ as x approaches a is L". We also write this as "$f(x) \to L$ as $x \to a$", which we read as "$f(x)$ approaches L as x approaches a".

15.5. Example. *The role of epsilon and delta.* If $f(x) \to L$ as $x \to a$, then we can make $f(x)$ as close as we like to L by making x sufficiently close to a. For example, $\lim_{x \to 10} x^2 = 100$. If we want to guarantee that x^2 is within $\varepsilon = 1$ of 100, then we may choose $\delta = .04$. Choosing $\delta = .05$ does not suffice, because $10.05^2 = 101.0025$. ∎

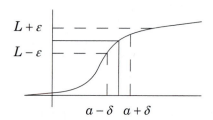

The definition of $\lim_{x \to a} f(x) = L$ says that given a desired "tolerance" ε, there is a real number $\delta > 0$ such that each input x within δ of a (except a itself) produces an output $f(x)$ within ε of L. By comparing this definition with that of convergence for sequences, we see that δ plays the same role here that N plays there:

$$(\forall \varepsilon > 0)(\exists \delta > 0)[(0 < |x - a| < \delta) \to (|f(x) - L| < \varepsilon)]$$
$$(\forall \varepsilon > 0)(\exists N \in \mathbb{N})[(n > N) \to (|a_n - L| < \varepsilon)]$$

Let us also consider what it means for $\lim_{x \to a} f(x) = L$ to be false. In this situation there exists some fixed $\varepsilon > 0$, say $\varepsilon = \varepsilon^*$, such that for every $\delta > 0$ (no matter how small), there is some x within δ of a such that $|f(x) - L| \geq \varepsilon^*$. In particular, taking $\delta = 1/n$, we find a number x_n such that $|x_n - a| < 1/n$ but $|f(x_n) - L| \geq \varepsilon^*$. The ability to construct such a sequence will be helpful in writing proofs by contradiction.

15.6. Example. Suppose c is a positive constant and $f(x) = cx\sin(1/x)$ for all $x \neq 0$. The values of the sine function are bounded by ± 1. Hence $|f(x)| \leq c|x|$. In proving that $\lim_{x \to 0} f(x) = 0$, we do not consider $x = 0$, where f is not defined. Given a particular $\varepsilon > 0$, we can choose δ (in terms of ε) to be ε/c. Then when $0 < |x - 0| < \delta$, we have $|x| < \varepsilon/c$, and hence $|f(x) - 0| \leq c|x| < \varepsilon$.

On the other hand, suppose $f(x)$ is defined to be the "sign" of x, meaning $+1$ when $x > 0$ and -1 when $x < 0$. We show that f has no limit at 0. No matter how we choose L, there is no deleted neighborhood of 0 on which $|f(x) - L| < 1$ for all x. If $L \geq 0$, the negative values of x are bad; if $L \leq 0$, the positive values of x are bad. ∎

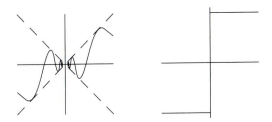

The definition of limit for function values parallels the definition of limit for sequences. Indeed, we can define limits for function values using sequences. We will prove that the two definitions are equivalent. The mapping that takes n to $f(x_n)$ is the composition of f with the sequence $\langle x \rangle$, and hence it defines a sequence $\langle y \rangle$. We write $f(x_n) \to L$ to mean that this sequence $\langle y \rangle$ converges to L.

15.7. Definition. A function f defined on a deleted neighborhood S of a *has sequential limit* L as x approaches a if $f(x_n) \to L$ for every sequence $\langle x \rangle$ converging to a in S.

15.8. Theorem. Given a real-valued function f defined on a deleted neighborhood S of a, the following two statements are equivalent:
A) $\lim_{x \to a} f(x) = L$.
B) f has sequential limit L as x approaches a.

Proof. We first prove A ⇒ B. Suppose $\lim_{x \to a} f(x) = L$. To prove that f has sequential limit L, we consider an arbitrary sequence $\langle x \rangle$ in S converging to a and prove that $f(x_n)$ converges to L. Given an arbitrary $\varepsilon > 0$, we need to find $N \in \mathbb{N}$ such that $n > N$ implies $|f(x_n) - L| < \varepsilon$. Using the definition of $\lim_{x \to a} f(x) = L$, we know for this particular ε that there exists a number $\delta > 0$ for which $0 < |x - a| < \delta$ implies $|f(x) - L| < \varepsilon$. Now we use the definition of $x_n \to a$; given $\delta > 0$, we can choose $N' \in \mathbb{N}$ such that $n > N'$ implies $0 < |x_n - a| < \delta$. If we set $N = N'$, then $n > N$ implies $|f(x_n) - L| < \varepsilon$.

We prove B $\Rightarrow$ A by proving $\neg A \Rightarrow \neg B$. If $\lim_{x \to a} f(x) = L$ does not hold, then there is some fixed $\varepsilon^* > 0$ such that for every $\delta > 0$, some x satisfies $0 < |x - a| < \delta$ and $|f(x) - L| \geq \varepsilon^*$. We consider this fixed ε^* and the sequence of choices for δ given by $\delta_n = 1/n$. The choice of ε^* guarantees that for each n we can find x_n such that $0 < |x_n - a| < 1/n$ and $|f(x_n) - L| \geq \varepsilon^*$. This constructs a sequence $\langle x \rangle$ such that $x_n \to a$, but the sequence of values $f(x_n)$ does not converge to L. Hence f does not have sequential limit L as x approaches a. ∎

When we prove that some hypothesis H about limits of functions implies a conclusion C about limits of functions, we use "ε" as we did for limits of sequences. The statements H and C have the same form; suppose C is $(\forall \varepsilon)(\exists \delta)(\forall x)P(x)$ and H is $(\forall \varepsilon')(\exists \delta')(\forall x)Q(x)$. Proving C requires proving a statement for every positive ε. When we invoke the hypothesis H, we may use the existence of a suitable δ' for any desired ε', since we have assumed H is true. We make an appropriate choice of ε' in terms of ε. We did this twice in the proof of A $\Rightarrow$ B above, using two different statements in the role of H. The ε' yields a δ', and we use that to construct the desired δ.

The equivalence of the two notions of limit implies that limits of function values have essentially the same properties as limits of sequences. Suppose $\square$ denotes a numerical binary operator, such as addition, subtraction, multiplication, division, etc. (not composition!). Recall that the function $h = f \square g$ is defined pointwise in terms of functions f and g by setting the value of h at x to be $f(x) \square g(x)$.

15.9. Lemma. Suppose $\square$ is a numerical binary operator such that $b_n \square c_n \to L \square M$ whenever $\langle b \rangle$, $\langle c \rangle$ are sequences such that $b_n \to L$ and $c_n \to M$. If f, g are functions such that $\lim_{x \to a} f(x) = L$ and $\lim_{x \to a} g(x) = M$, then $\lim_{x \to a}(f \square g)(x) = L \square M$.

Proof. Suppose $x_n \to a$, and let $b_n = f(x_n)$ and $c_n = g(x_n)$. The sequential version of convergence for f and g implies that $b_n \to L$ and $c_n \to M$. Hence $b_n \square c_n \to L \square M$. Since $(f \square g)(x_n) = b_n \square c_n$, we have $(f \square g)(x_n)$. Since this is true for each $\langle x \rangle$ converging to a, the function $f \square g$ has sequential limit $L \square M$ as x approaches a, and hence $\lim_{x \to a}(f \square g)(x) = L \square M$. ∎

The definition of "$\lim_{x \to a} f(x)$" omits the situation $|x - a| = 0$. If the value of f at a is the same as the limit, then the graph of f has no "gap" at $x = a$, and we think of f as being "continuous" at a:

15.10. Definition. A function f defined on an interval containing a is *continuous at* a if $\lim_{x \to a} f(x) = f(a)$. Equivalently, f is continuous at a if for every $\varepsilon > 0$ there is a $\delta > 0$ such that $|x - a| < \delta$ implies $|f(x) - f(a)| < \varepsilon$.

A function is *continuous on an open interval* (c, d) if it is continuous at every point of (c, d). A function is *continuous on a closed interval* $[c, d]$ if it is continuous on (c, d) and for every sequence $\langle x \rangle$ of numbers in $[c, d]$, $x_n \to c$ implies $f(x_n) \to f(c)$ and $x_n \to d$ implies $f(x_n) \to f(d)$.

Other phrasings of the definition of continuity at a include: 1) f is continuous at a if and only if $f(x_n) \to f(a)$ whenever $\langle x \rangle$ converges to a, 2) f is continuous at a if and only if for every $\varepsilon > 0$ there is some neighborhood of a in which the values of f are within ε of $f(a)$.

15.11. Corollary. If f and g are continuous at x, then $f + g$ and fg are continuous at x. If f is continuous and $f(x) \neq 0$, then $1/f$ is continuous at x. Every polynomial is continuous on $\mathbb{R}$. The ratio of two polynomials is continuous wherever the denominator is nonzero.

Proof. The first two statements follow immediately from Lemma 15.8. We leave the proofs for polynomials to Exercise 15.8. ∎

We use the squeeze theorem for sequences (Theorem 14.8) to prove a sufficient condition for continuity at a point.

15.12. Proposition. (Squeeze Theorem for Continuity) Suppose $A(x) \leq f(x) \leq C(x)$ for all x in an interval I containing α. If A and C are continuous at α and $A(\alpha) = C(\alpha)$, then f is continuous at α.

Proof. Let $\langle x \rangle$ be an arbitrary sequence in I converging to α. By the sequential definition of continuity, the sequences $A(x_n)$ and $C(x_n)$ converge to $A(\alpha) = L = C(\alpha)$. By the squeeze theorem, $f(x_n)$ also converges to L. This holds for every such sequence $\langle x \rangle$, so f has sequential limit L at α. Furthermore, $A(\alpha) = C(\alpha)$ implies $f(\alpha) = L$, and hence f is continuous at α. ∎

15.13. Example. If $|f(x)| \leq m|x|$ for some positive constant m and all x, then f is continuous at 0. Here $A(x) = -mx$ and $C(x) = mx$. ∎

When a function f is defined in a neighborhood of a, there are two ways for continuity to fail. One is that $\lim_{x \to a} f(x)$ exists but doesn't equal $f(a)$. This is called a "removable singularity"; by changing the definition of $f(a)$, we could make f continuous at a. The second type of failure is that $\lim_{x \to a} f(x)$ doesn't exist at all.

15.14. Example. *Failure of continuity.* Consider the functions defined by $f(x) = 1/x$, $g(x) = \sin(1/x)$, and $h(x) = \text{sign}(x)$ for $x \neq 0$. As x tends to 0, the first is unbounded, the second oscillates wildly, and the third has a "jump discontinuity". All three are discontinuous at 0. ∎

The sequential version of limit yields a simple proof that the composition of continuous functions is continuous:

15.15. Theorem. (Continuity of composite functions) If f is continuous at x and g is continuous at $f(x)$, then the composite function $h = g \circ f$ is continuous at x.

Proof. It suffices to prove that if $\langle x \rangle$ is an arbitrary sequence converging to x, then the sequence defined by $z_n = h(x_n)$ converges to $h(x)$. Because f is continuous at x, the sequence defined by $y_n = f(x_n)$ converges to $f(x)$. Because $y_n \to f(x)$ and g is continuous at $f(x)$, we conclude $z_n = h(x_n) = g(y_n) \to g(f(x))$. ∎

APPLICATIONS OF CONTINUITY

Sequential continuity makes some theorems about continuity particularly easy to prove. Nevertheless, the epsilons and deltas in the definition of continuity hold more than theoretical interest. The question of how small δ must be to keep error from exceeding ε can be a question of engineering.

15.16. Example. *Constructing a rectangle.* Consider a wooden board 12 feet long. We want to cut it into lengths of roughly 4 feet and 8 feet, then split the two pieces lengthwise to build a rectangle of area roughly 32 square feet. How close must we make the board lengths to 4 and 8 in order to make the area within ε of 32?

Ignoring losses due to sawdust, we suppose that our lengths are $(4 - x)$ and $(8 + x)$. The resulting area is $(4 - x)(8 + x) = 32 - 4x - x^2$, and we want to choose x so that $|4x + x^2| < \varepsilon$. For $0 < \varepsilon < 4$, this requires $-2 + 2\sqrt{1 - \varepsilon/4} < x < -2 + 2\sqrt{1 + \varepsilon/4}$ (Exercise 15.3). When $\varepsilon = 1$, the requirement is $-.268 < x < .236$ (asymmetric!), so we chose $\delta \le .236$. To keep the area within one square foot of 32 square feet, we must cut the board to within $.236$ feet (about 3 inches) of the desired length.

The alert reader will note that we omitted another region for x where the desired inequalities on the error hold. This is an interval around $x = -4$, which occurs when our error is so bad that the short side has become the long side and vice versa. ∎

With careful attention to the definition of continuity, we can prove statements about continuous functions that seem geometrically obvious. One such statement is the Intermediate Value Theorem (Theorem 15.18), suggested by the illustration below. When f is continuous on

$[a, b]$ and $f(a) < 0 < f(b)$, the graph of f must cross the horizontal axis between a and b, and this provides a solution to the equation $f(x) = 0$.

The validity of conclusions drawn from geometric reasoning may depend on aspects of the completeness property for $\mathbb{R}$ that are not evident in the picture. Consider what happens when the domain is $\mathbb{Q}$. Suppose $f \colon \mathbb{Q} \to \mathbb{R}$ by $f(x) = x^2 - 2$, as illustrated below. There is no x in the domain of f such that $f(x) = 0$. The graph looks the same whether the domain is $\mathbb{Q}$ or $\mathbb{R}$, but the Intermediate Value Theorem does not hold for polynomials defined on $\mathbb{Q}$.

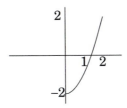

15.17. Lemma. If f is continuous on a neighborhood of a and $f(a) \neq 0$, then there exists some $\delta > 0$ such that $|x - a| < \delta$ implies that $f(x)$ is nonzero and has the same sign as $f(a)$.

Proof. Suppose $f(a) \neq 0$. Setting $\varepsilon = |f(a)|$ in the definition of continuity, we know there exists $\delta > 0$ such that $|x - a| < \delta$ implies $|f(x) - f(a)| < |f(a)|$. Hence for every x between $a - \delta$ and $a + \delta$, the distance from $f(a)$ to $f(x)$ is less than the distance from $f(a)$ to 0. This implies that $f(x)$ has the same sign as $f(a)$. ∎

We could also present this proof using the contrapositive. If $f(x)$ and $f(a)$ have opposite signs, then the vertical distance $|f(x) - f(a)|$ between them is $|f(x)| + |f(a)|$, which exceeds $|f(a)|$.

15.18. Theorem. (Intermediate Value Theorem) If f is a continuous real-valued function on $[a, b]$, and $f(a) < y < f(b)$, then there exists an $x \in (a, b)$ such that $f(x) = y$.

Proof. Let $S = \{x \in [a, b] \colon f(t) < y \text{ for all } t \in [a, x]\}$. Since $a \in S$, we know that S is non-empty. Also, b is an upper bound for S. By the least upper bound property (completeness axiom), we know that S has a least upper bound α. We claim that $f(\alpha) = y$.

If $f(\alpha) \neq y$, then by Lemma 15.17 there exists δ such that $f(x) - y$ has the same sign as $f(\alpha)$ when x is within δ of α. If $f(\alpha) < y$, then $f(x) - y$ is negative when $\alpha \leq x < \alpha + \delta$, and α is not an upper bound for S. If $f(\alpha) > y$, then $f(x) - y$ is positive when $\alpha - \delta < x \leq \alpha$, and α is not the least upper bound for S. Both cases are impossible, and we conclude that $f(\alpha) = y$. ∎

15.19. Example. The polynomial defined by $f(x) = x^5 - 12x - 13$ is continuous. Since $f(2) = -5$ and $f(2.8) = 125.5$, the Intermediate Value
Theorem implies that there exists $x \in (2, 2.8)$ such that $f(x) = 0$.

To obtain a better approximation, consider $f(2.4) = 37.8$. This implies that there is a solution between 2 and 2.4. Continuing this process, $f(2.2) = 12.1 > 0$, $f(2.1) = 2.6 > 0$, but $f(2.05) = -1.39 < 0$. Therefore, there is a solution between 2.05 and 2.1. We can continue this
"bisection" process to approximate the solution as accurately as desired.
Additional steps lead to 2.067916, accurate to six decimal places. ∎

The method of bisection provides an algorithm for approximating a
solution to the equation $f(x) = 0$. For equations in rational numbers, we
obtain exact answers; here we must be content with computing a solution of specified accuracy.

The algorithm also provides an alternative (constructive) proof of
the Intermediate Value Theorem. Suppose f is continuous and we have
numbers a_0 and c_0 such that $f(a_0)$ and $f(c_0)$ have opposite signs. By
successively replacing one of $\{a_n, c_n\}$ with their average (unless we find
an exact solution), we create two bounded monotone sequences $\langle a \rangle$ and
$\langle c \rangle$ such that $f(a_n) < 0$, $f(c_n) > 0$, and $\lim a_n = L = \lim c_n$. Since f is
sequentially continuous, we have $\lim f(a_n) = f(L) = \lim f(c_n)$. Since
$f(a_n) < 0$, we have $f(L) = \lim f(a_n) \leq 0$. Since $f(c_n) > 0$, we have
$f(L) = \lim f(c_n) \geq 0$. Therefore $f(L) = 0$. The method converges slowly;
in chapter 16 we will discuss a faster method that requires a stronger
hypothesis on the function f.

We next consider a continuous function f on $[0,1]$ such that
$0 \leq f(x) \leq 1$ for all x. Its graph crosses the line $y = x$, and hence the picture suggests that f has a fixed point (a point where $f(x) = x$). The
Intermediate Value Theorem provides the proof.

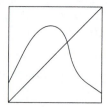

15.20. Corollary. A continuous function from $[0,1]$ to itself has a fixed
point.

Proof. The function f has a fixed point at x^* if and only if the function
g defined by $g(x) = x - f(x)$ is 0 at x^*. If f is continuous, then g is

continuous, since it is the difference of continuous functions (see Corollary 15.11). Since $0 \le f(x) \le 1$ for all x, we have $g(0) \le 0$ and $g(1) \ge 0$. If neither 0 nor 1 is a fixed point, then we can apply the Intermediate Value Theorem to conclude that there is a number $x^* \in (0, 1)$ such that $g(x^*) = 0$. Now x^* is the desired fixed point for f. ∎

15.21. Solution. *The Antipodal Point Problem.* We want to prove that some pair of opposite points on a circular wire have the same temperature. If the wire has circumference c, then we can represent the temperature as a continuous function f on the interval $[0, c]$ with $f(0) = f(c)$. We can extend the domain of f to $\mathbb{R}$ by setting $f(b) = f(a)$ whenever $b - a$ is an integer multiple of c. Then the temperature at the point opposite x is $f(x + c/2)$.

We seek an x^* such that $f(x^*) = f(x^* + c/2)$. Consider the function g defined by $g(x) = f(x) - f(x + c/2)$. If g is always zero, the temperature is constant, and our conclusion holds. Otherwise, since $g(x + c/2) = -g(x)$, the function g attains both positive and negative values. Since g is a difference of continuous functions, g is continuous. Applying the Intermediate Value Theorem to the interval between x and $x + c/2$, we obtain a number x^* such that $g(x^*) = 0$, and hence $f(x^*) = f(x^* + c/2)$. ∎

15.22. Solution. *The Jewel Thieves Problem.* Suppose our circular necklace has $2k$ diamonds and $2l$ rubies in some order. Splitting the necklace properly requires finding $k + l$ consecutive jewels in this order consisting of k diamonds and l rubies. This is a discrete problem, but the argument is analogous to that for the Antipodal Point Problem. We consider cuts that capture $k + l$ jewels for the first thief. When we shift the position of the cut counterclockwise by one jewel, we lose one jewel and pick up one jewel for the set captured by the first thief. This could leave the number of diamonds unchanged or change it by one. Since we always capture $k + l$ jewels, getting k diamonds ensures getting l rubies, so we focus only on the number of diamonds. Let $f(i)$ be the number of diamonds among the $k + l$ jewels starting with the ith jewel. We can extend this so that $f(i + 2k + 2l) = f(i)$.

When we move from starting at i to starting at $i + k + l$, one by one, we transform the set of beads captured for the first thief into the complementary set. If the first set has too many diamonds, then the second has too few. Technically, if $f(i) - k$ has one sign, then $f(i + k + l) - k$ has the opposite sign. Since f is integer-valued and its value changes by at most one when its argument changes by one, $f(i) - k$ cannot change sign without attaining the value 0 along the way. When f changes by at most one with each unit change in the argument, we have a discrete version of the Intermediate Value Theorem. ∎

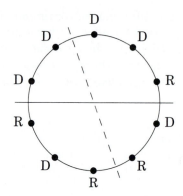

CONTINUITY AND CLOSED INTERVALS

We next show that continuous functions defined on closed bounded intervals are bounded. Recall that a real-valued function is *bounded* if its image is a bounded set.

15.23. Theorem. A continuous function on a closed and bounded interval is bounded.

Proof. Suppose f is continuous on $[a, b]$. If f is not bounded on $[a, b]$, then for every n there is an x_n in $[a, b]$ such that $|f(x_n)| > n$. The sequence $\langle x \rangle$ is bounded, since $a \le x_n \le b$. By the Bolzano-Weierstrass Theorem, $\langle x \rangle$ has a convergent subsequence $\{x_{n_k} : k \in \mathbb{N}\}$. Let c be the limit of this subsequence; note that $a \le x_{n_k} \le b$ implies $a \le c \le b$. Since f is continuous at c and $x_{n_k} \to c$, the sequential definition of continuity implies $\lim_{k \to \infty} f(x_{n_k}) = f(c)$. Hence $\{f(x_{n_k}) : k \in \mathbb{N}\}$ is a convergent sequence. On the other hand, our choice of $\langle x \rangle$ implies $|f(x_{n_k})| > n_k$, which means $\{|f(x_{n_k})|\}$ is unbounded and $f(x_{n_k})$ does not converge. The contradiction implies that f is bounded on $[a, b]$. ∎

15.24. Example. *The importance of closed intervals.* The function defined by $f(x) = 1/x$ is continuous on the open interval $(0, 1)$, but it is not bounded on this interval. Where does the proof fail? ∎

Suppose f is continuous on $[a, b]$. We now know that f takes on a bounded set S of values, which by the Completeness Axiom has a supremum and infimum. Must $\sup(S)$ belong to S? The answer is Yes; in other words, f attains its maximum value.

15.25. Theorem. (The Maximum-Minimum Theorem). A continuous function on a closed and bounded interval attains its maximum and minimum values.

Proof. Let f be continuous on $[a,b]$. By Theorem 15.23, the set $S = \{f(x): a \le x \le b\}$ is bounded, so we can set $\alpha = \inf(S)$ and $\beta = \sup(S)$. We prove that $\beta = f(x)$ for some $x \in [a,b]$. The definition of supremum guarantees a sequence $\langle y \rangle$ in S such that $y_n \to \beta$ (see Exercise 13.3). Since y_n belongs to the set S of images of f on the interval $[a,b]$, there is a number $x_n \in [a,b]$ such that $f(x_n) = y_n$. Since $a \le x_n \le b$ for each n, the sequence $\langle x \rangle$ is bounded, and the Bolzano-Weierstrass Theorem guarantees a convergent subsequence $\{x_{n_k}: k \in \mathbb{N}\}$. Let $c = \lim_{k \to \infty} x_{n_k}$; since $a \le x_n \le b$, we also have $a \le c \le b$.

We claim that $f(c) = \beta$. Since $\lim_{k \to \infty} x_{n_k} = c$ and f is continuous at c, we have $\lim_{k \to \infty} f(x_{n_k}) = f(c)$. This sequence of values is a subsequence of $\langle y \rangle$, and $y_n \to \beta$. Every subsequence of a convergent sequence converges to the same limit, so the value $f(c)$ to which $f(x_{n_k})$ converges must also be β.

To establish the attainment of the minimum, we can make a similar argument, or we can apply the statement for the maximum to the continuous function $-f$. ∎

15.26. Example. *The importance of bounded intervals.* The function defined by $f(x) = 1/(1 + x^2)$ is bounded for $x \ge 0$, since its values are between 0 and 1, but it does not attain its minimum value. ∎

We next introduce a property stronger than continuity; we will need it in Chapter 17 for a theorem about integration. We then prove that, for functions restricted to a closed and bounded interval, the property is equivalent to continuity.

15.27. Definition. A function f is *uniformly continuous* on an interval I if for every $\varepsilon > 0$, there exists a δ such that $y, x \in I$ and $|y - x| < \delta$ together imply $|f(y) - f(x)| < \varepsilon$.

This property is harder to satisfy than continuity at each point because we require more of δ. Instead of merely guaranteeing that $f(y)$ is within ε of $f(x)$ whenever y is within δ of a fixed number x, we want a single real number δ to work for every $x \in I$. When f is merely continuous at each x in I, the δ that we choose in order to make $|y - x| < \delta$ imply $|f(y) - f(x)| < \varepsilon$ can depend not only on ε *but also on* x. The order of the quantifiers in the definitions makes uniform continuity a stronger property than continuity at each point: continuity at each x is $(\forall \varepsilon, x)(\exists \delta)P$, while uniform continuity is $(\forall \varepsilon)(\exists \delta)(\forall x)P$, where P is the statement "$|y - x| < \delta$ implies $|f(y) - f(x)| < \varepsilon$".

15.28. Example. *Uniform continuity versus continuity at each point.* The function $f(x) = 1/x$ is continuous on the open interval $(0, 1)$, but it is not uniformly continuous on $(0, 1)$. If $x < y$, then

$$|f(x) - f(y)| = (1/x) - (1/y) = (y - x)/xy < (y - x)/x^2.$$

If $y - x$ is very small, then $|f(x) - f(y)|$ is very close to $(y - x)/x^2$ (by the continuity of f). To make $|f(x) - f(y)| < \varepsilon$, we need to choose δ smaller than $x^2\varepsilon$; $\varepsilon x^2/2$ suffices. We must choose a smaller δ as x becomes smaller. No single choice of δ works throughout the interval. ∎

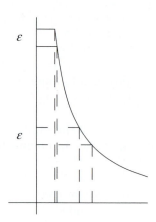

15.29. Theorem. If f is continuous on the closed and bounded interval $[a, b]$, then f is uniformly continuous on $[a, b]$.

Proof. The proof is by contradiction, using the Bolzano-Weierstrass Theorem and the sequential definition of pointwise continuity. Suppose f is not uniformly continuous on $[a, b]$. The negation of the definition of uniform continuity is this: there exists some fixed $\varepsilon^* > 0$ such that for every $\delta > 0$, there is some pair $x, y \in [a, b]$ satisfying $|y - x| < \delta$ and $|f(y) - f(x)| \geq \varepsilon^*$.

Since this holds for every $\delta > 0$, we may consider it for the sequence of numbers $\delta_n = 1/n$. Let (y_n, x_n) be the pair that results when we use $\delta = 1/n$ in the statement above about ε^*, so $|y_n - x_n| < 1/n$ and $|f(y_n) - f(x_n)| \geq \varepsilon^*$. The sequence $\langle x \rangle$ is a sequence of real numbers between a and b. By the Bolzano-Weierstrass Theorem, $\langle x \rangle$ has a convergent subsequence $\{x_{n_k} : k \in \mathbb{N}\}$ converging to some $c \in [a, b]$. Since $y_n - x_n \to 0$, we also have $y_{n_k} \to c$.

Since f is continuous at c, the sequential definition of continuity implies $f(x_{n_k}) \to f(c)$ and $f(y_{n_k}) \to f(c)$. Hence $\lim_{k \to \infty} f(x_{n_k}) - f(y_{n_k}) = 0$. This contradicts the choice of (y_n, x_n); thus $|f(x_{n_k}) - f(y_{n_k})| \geq \varepsilon^*$ for every k. ∎

We will use this theorem in Chapter 17 to obtain the area under the graph of a continuous function by integration.

EXERCISES

15.1. Using the "epsilon-delta" definition, prove that the absolute value function is continuous.

15.2. Suppose f is defined by $f(x) = 1/x$, and $a = .5$. How large can δ be if it is required that $f(x)$ is within $.1$ of $f(a)$ when x is within δ of a?

15.3. (!) When $f(x) = x^2 + 4x$, $\lim_{x \to 0} f(x) = 0$. How small must δ be so that $|x| < \delta$ implies that $|f(x)| < \varepsilon$? Express δ as a function of ε. Assume that $\varepsilon < 4$.

15.4. (−) Suppose $\lim_{x \to 0} f(x) = 0$. Prove that for all $n \in \mathbb{N}$, there exists x_n such that $|f(x_n)| < 1/n$.

15.5. Define a function of two variables by $f(a, n) = (1 + a)^n$, where a and n are positive.

   a) For fixed a, how does $f(a, n)$ behave as $n \to \infty$? For fixed n, how does $f(a, n)$ behave as $a \to 0$?

   b) Suppose L is a real number with $L \geq 1$. Prove that there exists a sequence $\langle a \rangle$ such that $a_n \to 0$ and $f(a_n, n) \to L$ as $n \to \infty$. In other words, depending on the rate at which a approaches 0, f may approach any value.

15.6. (!) *Often discontinuous functions.*

   a) Let $f \colon \mathbb{R} \to \mathbb{R}$ be the function defined by $f(x) = 0$ if $x \in \mathbb{Q}$ and $f(x) = 1$ if $x \notin \mathbb{Q}$. Prove that f is discontinuous at every real number.

   b) Let $g \colon \mathbb{R} \to \mathbb{R}$ be the function defined by $g(x) = 0$ if $x \in \mathbb{Q}$ and $g(x) = cx$ if $x \notin \mathbb{Q}$, where c is a nonzero real number. Prove that g is continuous at 0 and discontinuous at every other real number.

15.7. (!) Give two proofs that if $|f(x) - f(a)| \leq c|x - a|$ for some positive constant c and all x, then f is continuous at a. One proof should use the ε, δ definition, and the other should apply general results about continuity.

15.8. Prove that every polynomial is continuous on $\mathbb{R}$. Prove that the ratio of two polynomials is continuous at every point where the denominator is nonzero.

15.9. Suppose that f and g are continuous on the closed interval $[a, b]$. Suppose also that $f(a) > g(a)$ and that $f(b) < g(b)$. Prove that there exists $c \in [a, b]$ such that $f(c) = g(c)$.

15.10. (!) Prove that every polynomial of odd degree has at least one real root.

15.11. Given a fixed positive real number ε, prove that there is a positive real number c (depending on ε, but not on x or y) such that $|xy| \leq \varepsilon x^2 + cy^2$ for all $x, y \in \mathbb{R}$.

15.12. Write out the proof that a continuous function on $[a, b]$ has a lower bound, using the Bolzano-Weierstrass Theorem and definition of continuity.

15.13. Let S be the set of positive real numbers. Suppose $f \colon S \to S$ is continuous and injective.

   a) Prove that the inverse of f, defined on the image of f, is continuous.

   b) Suppose the sequence $\langle x \rangle$ satisfies $x_1 = c$ for some $c \in S$ and

$x_{n+1} = f(\Sigma_{j=1}^n x_j)$ for $n \geq 1$. Prove that if $\langle x \rangle$ converges, then its limit is 0. (Hint: prove that $\Sigma_{j=1}^n x_j$ converges.)

15.14. (!) Let $f_n(x) = (x^n + 1)^{1/n}$, defined on the set of positive real numbers. Determine the function g defined for positive x by $g(x) = \lim_{n \to \infty} f_n(x)$. More generally, find $\lim_{n \to \infty} (a^n + b^n)^{1/n}$. (Assume that for each $c \in \mathbb{R}$, x^c is continuous at each positive x.)

15.15. Use the method of bisection to compute $\sqrt{10}$ to four decimal places. Use the same method to find a solution of $x^7 - 5x^3 + 10 = 0$ to two decimal places.

15.16. Find a function f that is a counterexample to the following statement: If f is a real-valued function of two variables and all the limits described below exist, then

$$\lim_{y \to 0} \lim_{x \to 0} f(x, y) = \lim_{x \to 0} \lim_{y \to 0} f(x, y).$$

(When taking a limit in one variable, other variables are treated as constants.)

15.17. Suppose a car travels around a circular track. Along the circle are containers of gas of various sizes. The total amount of gas is exactly enough to enable the car to complete the trip. Prove that there is some place where the car can start and then complete the trip before running out of gas, assuming that the capacity of the car's gas tank is sufficiently large.

15.18. (+) Let n be a positive integer, and suppose f is continuous on $[0, 1]$ and $f(0) = f(1)$. Prove that the graph of f has a horizontal chord of length $1/n$. In other words, prove that there exists $x \in [0, (n-1)/n]$ such that $f(x + 1/n) = f(x)$. (Comment: surprisingly, for any α that is not the reciprocal of an integer, we can construct such a function f that has no horizontal chord of length α.)

15.19. (!) Suppose f is continuous on an interval I. For each $a \in I$ and $\varepsilon > 0$, let $m(a, \varepsilon) = \sup(\{\delta : |x - a| < \delta$ implies $|f(x) - f(a)| < \varepsilon\})$. What property must $m(a, \varepsilon)$ satisfy for f to be uniformly continuous on I?

15.20. *Continuous functions with constant multiplicity.*
   a) Construct a continuous function $f : \mathbb{R} \to \mathbb{R}$ such that every real number occurs as the image of exactly three numbers.
   b) (+) Suppose that $f : \mathbb{R} \to \mathbb{R}$ is continuous and that every real number occurs as the image of exactly k numbers. Prove that k must be odd. (Hint: Try to draw the graph of such a function with k even to see what goes wrong. Suppose k is even, z is a fixed real number, and $x_1, \ldots, x_k$ are the numbers such that $f(x) = z$. Use the Intermediate Value Theorem and the Maximum-Minimum Theorem to complete a proof by contradiction.)

Chapter 16

Differentiation

The famous calculus text by George Thomas begins, "Calculus is the mathematics of change and motion."[†] Studying how physical quantities change with time leads to the notions of continuity and differentiability. Our preparatory work with limits enables us to prove the basic theorems of differential calculus and to fully appreciate them.

16.1. Problem. *Approximation of function values.* The sides of a square with area 64 square inches are 8 inches long. Because the square root function is continuous, the sides of a square with area 65 square inches are approximately 8 inches long. How can we obtain a better estimate for $\sqrt{65}$? ■

16.2. Problem. *Solving equations iteratively.* In Algorithm 15.19, we used the bisection method to approximate the solution of an equation. Is there a faster algorithm? Using the tangent to the graph of a function, we can aim from a point on the graph to a "more educated" next guess for the solution. When does this process converge to a solution? ■

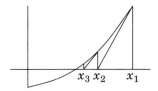

16.3. Problem. *Circle of Curvature.* Suppose a particle moves along a smooth curve in the plane. At each instant, its motion is closely approximated by motion along a circle. The center and radius of the best approximating circle change as the particle moves. The reciprocal of the radius measures the "curvature" of the motion. How can it be found? ■

[†]G. Thomas, *Calculus and Analytic Geometry*, 4th ed., Addison-Wesley, 1968, p1.

THE DERIVATIVE

The derivative is the precise formulation of the notion of rate of change. The ratio $(f(b) - f(a))/(b - a)$ is the average rate of change in $f(x)$ as x changes from a to b. As b approaches a, this ratio approaches the instantaneous rate of change at a. Perhaps the most familiar example of this is the relationship between the odometer and the speedometer in a car. The odometer measures distance traveled; the speedometer measures the speed, which is the instantaneous rate of change of distance traveled. Thus if we have traveled $f(a)$ miles at time a, and $f(b)$ miles at time b, then $(f(b) - f(a))/(b - a)$ represents our average speed in the time interval from a to b. The limit of this ratio as b approaches a equals the speed at time a.

The derivative also has a simple geometric interpretation. Given f, the slope $m_{a,b}$ of the line through $(a, f(a))$ and $(b, f(b))$ is $\frac{f(b) - f(a)}{b - a}$. As $b \to a$, this slope $m_{a,b}$ approaches the slope of the graph at $(a, f(a))$.

16.4. Definition.   The   function f is   *differentiable*   at x if $\lim_{h \to 0} \frac{f(x+h) - f(x)}{h}$ exists. When this limit exists, its value is the *derivative* of f at x. The derivative at x is written as $f'(x)$ or $\frac{df}{dx}(x)$. The ratio $\frac{f(x+h) - f(x)}{h}$ is the *difference quotient*.

The derivative of f at x equals the slope of the line tangent to the graph of f at the point $(x, f(x))$. If we move from $(x, f(x))$ along this line to $(x + h, y)$ for small h, then we expect y to be close to $f(x + h)$. This motivates the alternative definition of the derivative given below. The derivative of a function at x is the slope of the "best" linear approximation to the function at x, where a *linear approximation* to a function f at x is a linear function whose graph passes through $(x, f(x))$.

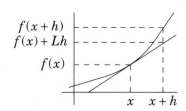

16.5. Definition.   An *error function* is a function e defined in a neighborhood of 0 such that $\lim_{h \to 0} e(h)/h = 0$. The function f is *differentiable* at $x \in \mathbb{R}$ if f is defined in a neighborhood of x and there exists a number L such that the function e_x defined by $f(x + h) = f(x) + Lh + e_x(h)$ is an error function. If so, the number L is called the *derivative* of f at x, and we write $L = f'(x)$.

16.6. Example. *Linearity.* If $f(x) = mx + b$, then $f(x + h) = f(x) + mh$. The error function e_x is identically zero, and $f'(x) = m$ for all x. ∎

16.7. Lemma. The two definitions of the derivative are equivalent.

Proof. For $h \neq 0$, the equation $f(x + h) - f(x) - Lh = e(h)$ is equivalent to $\dfrac{f(x+h)-f(x)}{h} = L + e(h)/h$. Thus $\lim_{h \to 0} e(h)/h = 0$ if and only if $\lim_{h \to 0} \dfrac{f(x+h)-f(x)}{h} = L$. ∎

In Definition 16.5, the error function e_x and value L depend on x. We write the derivative at x as $f'(x)$ to emphasize its dependence on x. Elementary calculus courses spend much time computing derivatives of particular functions. In principle, one can always compute a derivative by using either definition.

16.8. Example. If $f(x) = x^n$, then $f'(x) = nx^{n-1}$, where $n \in \mathbb{N} \cup \{0\}$. Using the binomial theorem, $(x + h)^n = \sum_{k=0}^{n} \binom{n}{k} x^{n-k} h^k$. We compute

$$\frac{f(x+h)-f(x)}{h} = \sum_{k=1}^{n} \binom{n}{k} x^{n-k} h^{k-1} = nx^{n-1} + h g_x(h),$$

where $g_x(h)$ is a polynomial in h. As $h \to 0$, $\dfrac{f(x+h)-f(x)}{h} \to nx^{n-1}$.

We can perform the same compuation using linear approximations. We expand $f(x + h) = (x + h)^n$ using the binomial theorem as before. We obtain $(x + h)^n = x^n + nx^{n-1} h + e_x(h)$, where $e_x(h)$ is a polynomial in h that is divisible by h^2. Hence $e_x(h)/h \to 0$. Since $e_x(h)$ equals $f(x + h) - f(x) - nx^{n-1}h$, we obtain $f'(x) = nx^{n-1}$. ∎

16.9. Example. If $f(x) = x^{1/2}$ for $x > 0$, then $f'(x) = \frac{1}{2} x^{-1/2}$. We compute

$$\frac{f(x+h)-f(x)}{h} = \frac{\sqrt{x+h}-\sqrt{x}}{h} = \frac{\sqrt{x+h}-\sqrt{x}}{h} \frac{\sqrt{x+h}+\sqrt{x}}{\sqrt{x+h}+\sqrt{x}} = \frac{x+h-x}{h \cdot (\sqrt{x+h}+\sqrt{x})}.$$

Simplifying yields $1/(2\sqrt{x})$ for the limit of the difference quotient. ∎

16.10. Solution. *Approximating the square root* (Problem 16.1). Using Example 16.9 and the linear approximation, we have $\sqrt{x + h} = \sqrt{x} + \dfrac{1}{2\sqrt{x}} h + e(h)$. Hence $\sqrt{65}$ is approximately $8 + \dfrac{1}{2\sqrt{64}} 1 = 8 + 1/16 = 8.0625$. The correct value is 8.06226 to five decimal places. ∎

16.11. Remark. Many standard approximations use the linear approximation definition of the derivative. For example, when h is small, we approximate $1/(1 - h)$ by $1 + h$, $(1 + h)^\alpha$ by $1 + \alpha h$, $\sin h$ by h, and e^h by $1 + h$. This requires knowing the values of the derivatives at $x = 0$ of $1/(1 - x)$, $(1 + x)^\alpha$, $\sin x$, and e^x, which are $1, \alpha, 1, 1$, respectively. (We cannot compute the last two until we give definitions for $\sin x$ and e^x.) ∎

Linear approximations and properties of error functions make it easy to prove elementary rules for differentiation. We first discuss simple properties of error functions.

16.12. Lemma. Suppose that e, e_1, e_2 are error functions.
a) $e(h) \to 0$ as $h \to 0$.
b) The sum $e_1 + e_2$ is an error function.
c) If $c \in \mathbb{R}$ and u is a bounded function defined on a neighborhood of 0, then the products ce and ue are error functions.

Proof. We use the elementary properties of limits for sums and products (Lemma 15.9). (a) We have $e(h) = h \cdot [e(h)/h] \to 0 \cdot 0 = 0$. (b) We have $(e_1 + e_2)(h)/h = e_1(h)/h + e_2(h)/h \to 0 + 0 = 0$. (c) When $c \in \mathbb{R}$, we have $ce(h)/h \to c \cdot 0 = 0$. For the general statement, let c be an upper bound on $|u|$ in a neighborhood of 0. Since $0 \le |u(h)e(h)/h| \le c|e(h)/h|$, the result for ue follows from the Squeeze Theorem. ∎

16.13. Lemma. Suppose that e is an error function.
a) If $c \in \mathbb{R}$, then $\dfrac{1}{1+ch+e(h)} - (1-ch)$ defines an error function.
b) If $s(h) \to 0$, then the composition $e \circ s$ is an error function.

Proof. (a) Computing the difference yields
$$\frac{1}{1+ch+e(h)} - (1-ch) = \frac{1-(1-ch)(1+ch+e(h))}{1+ch+e(h)} = \frac{e(h)(1-ch)+c^2h^2}{1+ch+e(h)}.$$
We now divide by h and let $h \to 0$. Since $e(h)/h \to 0$, $1 - ch \to 1$, $c^2h \to 0$, and $1 - ch + e(h) \to 1$, the displayed esspression tends to 0. This yields the desired conclusion.

(b) Since e is an error function, for each $\varepsilon > 0$ there is a $\delta > 0$ such that $|t| < \delta$ implies $|e(t)| \le |t|\varepsilon$. Therefore, $|e(s(h))| \le |s(h)|\varepsilon$ for $|s(h)| < \delta$. Since $s(h) \to 0$, we can choose δ' such that $|h| < \delta'$ implies $|s(h)| < \delta$, and hence $|h| < \delta'$ implies $|e(s(h))/h| < \varepsilon$. This proves that $(e \circ s)(h)/h \to 0$. ∎

16.14. Theorem. If f and g are differentiable at x and c is a constant, then $f + g$, cf, $f \cdot g$, and f/g (if $g(x) \ne 0$) are differentiable at x. The derivatives are
a) $(f + g)'(x) = f'(x) + g'(x)$,
b) $(cf)'(x) = c \cdot f'(x)$,
c) $(fg)'(x) = f(x)g'(x) + f'(x)g(x)$ (product rule),
d) $(f/g)'(x) = \dfrac{g(x)f'(x) - f(x)g'(x)}{[g(x)]^2}$ (quotient rule).

Proof. We obtain the derivatives as linear approximations. In these computations, x, $f(x)$, $g(x)$, $f'(x)$, $g'(x)$ do not change as h changes.

(a) Recall that $(f + g)(x + h) = f(x + h) + g(x + h)$. Using this and the linear approximations for f and g, we compute

$$(f + g)(x + h) = f(x) + f'(x)h + e_1(h) + g(x) + g'(x)h + e_2(h)$$
$$= (f + g)(x) + (f'(x) + g'(x))h + (e_1 + e_2)(h).$$

By Lemma 16.12b, $e_1 + e_2$ is an error function, and hence $(f + g)'(x) = f'(x) + g'(x)$.

(b) Recall that $(cf)(x + h) = c \cdot f(x + h)$. We compute

$$(cf)(x + h) = c[f(x) + f'(x)h + e(h)] = cf(x) + cf'(x)h + ce(h).$$

By Lemma 16.12c, ce is an error function, and hence $(cf)'(x) = c \cdot f'(x)$.

(c) Recall that $(fg)(x + h) = f(x + h) \cdot g(x + h)$. Again we compute

$$(fg)(x + h) = [f(x) + f'(x)h + e_1(h)] \cdot [g(x) + g'(x)h + e_2(h)]$$
$$= f(x)g(x) + [f'(x)g(x) + f(x)g'(x)]h$$
$$+ e_1(h)[g(x) + g'(x)h] + e_2(h)[f(x) + f'(x)h] + e_1(h)e_2(h).$$

By Lemma 16.12c, each of the last three terms defines an error function. By Lemma 16.12b, their sum is an error function, and hence $(fg)'(x) = f'(x)g(x) + f(x)g'(x)$.

(d) The formula for $(f/g)'(x)$ follows from (c) and the case where f is identically 1 (Exercise 16.3). To differentiate $1/g$, we compute

$$\frac{1}{g(x + h)} = \frac{1}{g(x) + g'(x)h + e(h)} = \frac{1}{g(x)} \frac{1}{1 + g'(x)h/g(x) + e(h)/g(x)}$$

By Lemma 16.12c with $c = 1/g(x)$, $e(h)/g(x)$ is an error function. By Lemma 16.13a with $c = g'(x)/g(x)$, we can rewrite $\dfrac{1}{1 + g'(x)h/g(x) + e(h)/g(x)}$ as $1 - g'(x)h/g(x) + e_3(x)$, where e_3 is an error function. Hence

$$\frac{1}{g(x + h)} = \frac{1}{g(x)} [1 - \frac{g'(x)}{g(x)} h + e_3(h)] = \frac{1}{g(x)} - \frac{g'(x)}{[g(x)]^2} h + e_4(h),$$

where e_4 is an error function. Therefore, $(1/g)'(x) = g'(x)/[g(x)]^2$. ∎

16.15. Corollary. A polynomial is differentiable at every point. More generally, the ratio of two polynomials is differentiable at every point where the denominator is nonzero.

Proof. The first statement follows from Theorem 16.14 and the differentiability of x^n. The second statement then follows immediately from Theorem 16.14. ∎

Differentiability is a stronger condition than continuity.

16.16. Theorem. If f is differentiable at x, then f is continuous at x.

Proof. Using linear approximations, $f(x + h) = f(x) + f'(x)h + e(h)$, where e is an error function. Since $\lim_{h \to 0} e(h) = 0$ (Lemma 16.12a), we have $\lim_{h \to 0} f(x) + f'(x)h + e(h) = f(x)$, and hence f is continuous at x.

Alternatively, to prove that f is continuous at x, we can prove $f(x+h) - f(x) \to 0$ by computing $\frac{f(x+h)-f(x)}{h} \cdot h \to f'(x) \cdot 0 = 0$. ∎

The converse does not hold.

16.17. Example. *Continuous but not differentiable.* The absolute value function is continuous but not differentiable at 0. The difference quotient is $\frac{|h|}{h}$, which has no limit as $h \to 0$.

More generally, if g is bounded, but not continuous at 0, then the function f defined by $f(x) = xg(x)$ for $x \neq 0$ and $f(0) = 0$ is continuous at 0 but not differentiable at 0. The difference quotient is $g(h)$, which has no limit as $h \to 0$. In Example 16.64 and Example 16.65, we present continuous functions that are not differentiable anywhere! ∎

We next give a sufficient condition for differentiability at a point that is an Analogue of the Squeeze Theorem for continuous functions (Proposition 15.12). When a function f is squeezed between differentiable functions having equal values and equal derivatives at a point, f must also have that derivative there (see Example 16.39).

16.18. Theorem. (Squeeze Theorem for Differentiability) Let A and C be functions differentiable at x, with $A(x) = C(x)$ and $A'(x) = C'(x) = L$. If $A(t) \leq f(t) \leq C(t)$ for t in a neighborhood of x, then f is differentiable at x, and $f'(x) = L$.

Proof. The hypotheses imply that $A(x) = f(x) = C(x)$. Hence $A(x+h) - A(x) \leq f(x+h) - f(x) \leq C(x+h) - C(x)$. Using error functions, this becomes $A'(x)h + e_1(h) \leq f(x+h) - f(x) \leq C'(x)h + e_2(h)$. Subtracting Lh yields $e_1(h) \leq f(x+h) - f(x) - Lh \leq e_2(h)$. Since a function squeezed between two error functions is itself an error function (Exercise 16.7), f is differentiable at x with derivative L. ∎

16.19. Corollary. If $|g(t) - g(x)| \leq c|t - x|^{1+\alpha}$ for all t, where c and α are positive constants, then g is differentiable at x, and $g'(x) = 0$.

Proof. This is the special case of Theorem 16.18 with $f(t) = g(t) - g(x)$, $A(t) = -c|t - x|^{1+\alpha}$, and $C(t) = c|t - x|^{1+\alpha}$. ∎

We have studied how composition of functions behaves under various operations. Consider the composition of two linear functions. If $f(x) = ax + b$ and $g(x) = cx + d$, then $(g \circ f)(x) = c(ax + b) + d = acx + (bc + d)$. The composition is also linear, and its derivative is the product of the derivatives of f at x and g at $f(x)$.

The useful formula for differentiation of composite functions is called the "chain rule". Our proof of the chain rule amounts to showing that if f and g are differentiable, then $g \circ f$ is differentiable, and its linear approximation is the composition of the linear approximations to g and f. Therefore, the derivatives multiply as they do in the linear case.

16.20. Theorem. (Chain Rule) If f is differentiable at x and g is differentiable at $f(x)$, then the composite function $\phi = g \circ f$ is differentiable at x, and $\phi'(x) = g'(f(x))f'(x)$.

Proof. Let $y = f(x)$, and write $L = f'(x)$ and $M = g'(y)$. Differentiability of f and g allows us to write $f(x+h) = f(x) + Lh + e_1(h)$ and $g(y+k) = g(y) + Mk + e_2(k)$, where e_1 and e_2 are error functions. To prove that ϕ is differentiable at x, we seek a real number N such that $\phi(x+h) = \phi(x) + Nh + e(h)$, where $e(h)/h \to 0$.

To do this, we evaluate $\phi(x+h)$. We have

$$g(f(x+h)) = g(f(x) + Lh + e_1(h)) = g(f(x)) + M[Lh + e_1(h)] + e_2(Lh + e_1(h))$$

$$= g(f(x)) + MLh + [Me_1(h) + e_2(Lh + e_1(h))].$$

Let $e(h) = Me_1(h) + e_2(Lh + e_1(h))$. If we can prove that $e(h)/h \to 0$, then $\phi'(x)$ exists and equals $ML = g'(f(x))f'(x)$.

By Lemma 16.12, $Me_1(h)$ is an error function. Because the sum of error functions is an error function, showing that $e_2(Lh + e_1(h))/h \to 0$ will imply that e is an error function and will complete the proof.

Let $s(h) = Lh + e_1(h)$. We have $e_2(s(h)) = 0$ when $s(h) = 0$, and otherwise we write $\left| \frac{e_2(s(h))}{h} \right| = \left| \frac{e_2(s(h))}{s(h)} \right| \left| \frac{s(h)}{h} \right|$. Since $e_1(h)/h \to 0$, we have $s(h)/h \to L$. Since $s(h) \to 0$ and e_2 is an error function, we have $e_2(s(h))/s(h) \to 0$, by Lemma 16.13b. ∎

16.21. Example. Given $m, n \in \mathbb{N}$, let $f(x) = (x^n + 1)^m$. Then $f'(x) = m(x^n + 1)^{m-1}nx^{n-1}$. ∎

The definition of the derivative via linear approximation and our proof of the chain rule extend to functions of several variables. All the formal rules of differentiation (such as the product rule, quotient rule, etc.) follow as corollaries of the general chain rule. This approach makes Theorem 16.14, for example, trivial.

APPLICATIONS OF THE DERIVATIVE

Differential calculus provides a method for finding the maximum and minimum values of a function on an interval. We know from the Maximum-Minimum Theorem that a continuous function on a closed and bounded interval attains its maximum and its minimum values. The proof did not provide a method for computing them. The points where the maximum and minimum values of a differentiable function are attained must be points where the derivative is zero or endpoints of the interval. Thus it suffices to compare the values at these points.

16.22. Definition. A *local maximum* for the function f occurs at x if $f(t) \le f(x)$ for all t in some neighborhood of x. Similarly, a *local minimum* occurs at x if $f(t) \ge f(x)$ for all t in some neighborhood of x. A *local extremum* occurs where a local maximum or a local minimum occurs.

The next theorem gives a necessary (but not sufficient) condition for a local extremum of a differentiable function to occur at x; the derivative must be 0 at x. To find the extreme values of f on the interval $[a, b]$, we need only check the endpoints and the places where the derivative is 0.

16.23. Theorem. If f is differentiable at x and a local extremum of f occurs at x, then $f'(x) = 0$.

Proof. We first suppose that the local extremum is a local maximum. Thus there exists $\delta > 0$ such that $|h| < \delta$ implies $f(x + h) \le f(x)$. Since f is differentiable at x, the limit of the difference quotient $[f(x + h) - f(x)]/h$ exists as $h \to 0$. This ratio is nonnegative when $-\delta < h < 0$ and nonpositive when $0 < h < \delta$. Hence the limit L must satisfy $L \ge 0$ and $L \le 0$, and we conclude that $L = 0$.

If a local minimum occurs at x, then we can apply the analogous argument with all inequalities reversed, or we can apply the result about local maximums to the differentiable function $-f$. ∎

16.24. Example. *Maximum area in a pen.* A farmer plans to use n feet of wire fencing to form three sides of a rectangular pen against the side of a barn. The pen will stand out x feet from the wall, for some x. How should x be chosen to maximize the area of the rectangle?

The dimensions of the rectangle are x by $n - 2x$; the area is $f(x) = x(n - 2x)$, with $0 \le x \le n/2$. The value at the endpoints of the interval on which f is defined is 0. The derivative is $n - 4x$, which is 0 at $x = n/4$. The maximum area is $n^2/8$, achieved when $x = n/4$. We can also minimize any quadratic polynomial without using calculus (see Exercise 1.13). ∎

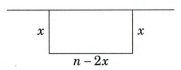

16.25. Example. *Necessary but not sufficient.* The condition $f'(x) = 0$ is necessary but not sufficient for a local extremum at x. If $f(x) = x^3$, then $f'(0) = 0$, but this f has no local extremum. ∎

A differentiable function must have a local extremum between two numbers where it has the same value. If the function has a larger value somewhere on this interval, then it has a local maximum, and otherwise it has a local minimum. In other words, "what goes up and comes down must turn around." The next theorem makes this statement precise.

16.26. Theorem. (Rolle's Theorem) If f is differentiable on (a, b), is continuous on $[a, b]$, and $f(a) = f(b) = 0$, then there exists a $c \in (a, b)$ such that $f'(c) = 0$.

Proof. Suppose first that $f(x) = 0$ for all $x \in (a, b)$; in this case, $f'(x) = 0$ for all $x \in (a, b)$. Hence we may eliminate this case. If $f(x) \le 0$ for all $x \in (a, b)$, then we can consider $-f$ instead. Hence we may assume that there exists $x \in (a, b)$ with $f(x) > 0$. Since f is continuous on $[a, b]$, we know from the Maximum-Minimum Theorem that f achieves its maximum value on the interval $[a, b]$. The maximum cannot occur at a or b, since $f(a) = f(b) = 0 < f(x)$. Hence it occurs at some $c \in (a, b)$, and thus $f(c)$ must be a local maximum. The necessary condition for local extrema (Theorem 16.23) now yields $f'(c) = 0$. ∎

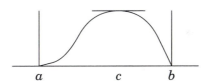

$$a \qquad c \qquad b$$

16.27. Example. *Necessity of continuity.* The hypothesis of continuity on $[a, b]$ is necessary in Rolle's Theorem. Consider the function f defined by $f(x) = x$ for $0 \le x \le 1$, but $f(1) = 0$. Then $f(0) = f(1) = 0$ and f is differentiable on $(0, 1)$, but $f'(x) = 1$ for all $x \in (0, 1)$. ∎

Rolle's Theorem leads to the Mean Value Theorem. When $a \ne b$, the equation of the line through (a, A) and (b, B) is $y = \dfrac{b - x}{b - a} A + \dfrac{x - a}{b - a} B$. The slope of this line is $m_{a,b} = \dfrac{f(b) - f(a)}{b - a}$ when $A = f(a)$ and $B = f(b)$.

16.28. Theorem. (Mean Value Theorem) If f is differentiable on (a, b) and continuous on $[a, b]$, then there exists $c \in (a, b)$ such that $f'(c) = \dfrac{f(b) - f(a)}{b - a}$.

Proof. By subtracting a linear function from f, we obtain a function to which Rolle's Theorem applies. The linear function defined by $g(x) =$

$\frac{b-x}{b-a} f(a) + \frac{x-a}{b-a} f(b)$ satisfies $g(a) = f(a)$ and $g(b) = f(b)$. Letting $h(x) = f(x) - g(x)$, we have $h(a) = h(b) = 0$. Also h is differentiable on (a, b) and continuous on $[a, b]$, since it is the difference of two functions that have those properties. Hence Rolle's Theorem applies to h and yields some $c \in (a, b)$ such that $h'(c) = 0$. But $h'(x) = f'(x) - g'(x) = f'(x) - \frac{f(b) - f(a)}{b - a}$. Hence $f'(c) = \frac{f(b) - f(a)}{b - a}$, as desired. ∎

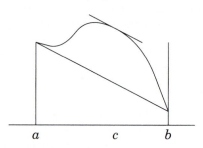

16.29. Corollary. If f is differentiable with derivative 0 on an open interval, then f is constant on that interval.

Proof. If there are two numbers a, b in the interval at which f has different values, then the Mean Value Theorem says that the derivative is nonzero somewhere between them. ∎

16.30. Example. *The Pennsylvania Turnpike.* Each driver entering the Pennsylvania Turnpike receives a card noting the entry point and time, so the correct toll can be assessed when exiting. A driver entered at Pittsburgh and exited four hours later at Philadelphia, 300 miles away. He received a speeding ticket! When a driver travels 300 miles in four hours, the Mean Value Theorem implies that the speed was 75 miles per hour at some point along the way. ∎

The Mean Value Theorem has many generalization and applications. We present a generalization due to Cauchy and apply it to compute limits of ratios.

16.31. Theorem. (Cauchy Mean Value Theorem). If f and g are differentiable on (a, b) and continuous on $[a, b]$, then there exists $c \in (a, b)$ such that $[f(b) - f(a)]g'(c) = [g(b) - g(a)]f'(c)$.

Proof. Define F on $[a, b]$ by

$$F(x) = [f(b) - f(a)] \cdot [g(x) - g(a)] - [g(b) - g(a)] \cdot [f(x) - f(a)].$$

Since this function satisfies the hypotheses of Rolle's Theorem, there exists $c \in (a, b)$ such that $F'(c) = 0$. By differentiating F with respect to x, we see that c has the desired property. ∎

The method for computing limits of ratios, called l'Hôpital's Rule, is named for Marquis de l'Hôpital (1661-1704). It was discovered by John Bernoulli (1667-1748) and given to l'Hôpital in return for salary.

We have seen that the limit of a quotient f/g is the quotient of the limits of f and g when both limits exist and $\lim g$ is nonzero. When both limits equal 0, the limit of the quotient may still exist, and we may be able to evaluate it using derivatives.

16.32. Theorem. (l'Hôpital's Rule) Suppose $\lim_{x \to a} f(x) = 0$ and $\lim_{x \to a} g(x) = 0$. If f and g are differentiable in an interval containing a and $\lim_{x \to a} \frac{f'(x)}{g'(x)}$ exists, then $\lim_{x \to a} \frac{f(x)}{g(x)} = \lim_{x \to a} \frac{f'(x)}{g'(x)}$.

Proof. Since f and g are differentiable at a, they are continuous at a, and hence $f(a) = g(a) = 0$. We can thus write $\frac{f(x)}{g(x)}$ as $\frac{f(x) - f(a)}{g(x) - g(a)}$. By the Cauchy Mean Value Theorem, there is a c between a and x such that $[f(x) - f(a)]g'(c) = [g(x) - g(a)]f'(c)$, and hence $f(x)g'(c) = g(x)f'(c)$. We treat $c = c(x)$ as a function of x. The existence of $\lim_{x \to a} \frac{f'(x)}{g'(x)}$ requires that g' be nonzero in some deleted neighborhood of a, and therefore $\frac{f(x)}{g(x)} = \frac{f'(c(x))}{g'(c(x))}$ for all x near a. For $\langle x \rangle$ converging to a from above, the Squeeze Theorem implies $c(x_n) \to a$. The same argument applies to sequences converging to a from below. Using sequential limits, we conclude that $\lim_{x \to a} \frac{f(x) - f(a)}{g(x) - g(a)} = \lim_{c \to a} \frac{f'(c)}{g'(c)}$, if the latter limit exists. ∎

16.33. Example. Because the numerator and denominator are both 0 when $x = 2$, we have $\lim_{x \to 2} \frac{x^3 - 2x^2 + x - 2}{x^2 - 7x + 10} = \lim_{x \to 2} \frac{3x^2 - 4x + 1}{2x - 7} = -\frac{5}{3}$. ∎

NEWTON'S METHOD

We return to the problem of solving equations. A *zero* of the function f is a solution to $f(x) = 0$. Sir Isaac Newton (1642-1727) invented an algorithm that often produces a sequence $\langle x \rangle$ converging to a zero of a differentiable function f. Given a guess x_n for a solution, Problem 16.2 suggests using the linear approximation to f at x_n to make a better guess. The graph of the linear approximation is the line tangent to the graph of f at $(x_n, f(x_n))$. We move to the point where this line intersects the horizontal axis. This yields the next guess $x_{n+1} = x_n - \frac{f(x_n)}{f'(x_n)}$.

16.34. Algorithm. *Newton's Method.* Given an initial guess x_0 for a zero of a differentiable function f, Newton's method generates a sequence of guesses via the recurrence $x_{n+1} = x_n - \frac{f(x_n)}{f'(x_n)}$ for $n \geq 0$. ∎

16.35. Proposition. Suppose that f is differentiable and that f' is continuous. If $\langle x \rangle$ is a sequence defined by $x_{n+1} = x_n - \dfrac{f(x_n)}{f'(x_n)}$ for $n \geq 0$ and $x_n \to L$, then $f(L) = 0$.

Proof. If $\langle x \rangle$ converges, then $x_{n+1} - x_n \to 0$. Hence $f(x_n)/f'(x_n) \to 0$. Since f is continuously differentiable, $f'(x_n) \to f'(L)$. Hence $\lim f(x_n) = \lim \dfrac{f(x_n)}{f'(x_n)} f'(x_n) = 0 \cdot f'(L) = 0$. Since every differentiable function is continous, $f(L) = \lim f(x_n) = 0$. ∎

Newton's method does not always work; it fails when $f'(x_n) = 0$. Even when $f'(x_n)$ is never zero, the process may not converge. The set of initial guesses producing convergence may be hard to describe. Also, it may be impossible to reach some solutions by Newton's method without guessing them exactly as x_0. We now explore favorable cases.

16.36. Example. *Newton's method for pth roots of real numbers.* For the recurrence $x_{n+1} = \frac{1}{2}(x_n + 2/x_n)$, we proved in Example 14.10 that $\langle x \rangle$ converges to $\sqrt{2}$ if $x_0 > 0$. This is also the recurrence produced by applying Newton's method to the function f defined by $f(x) = x^2 - 2$.

Suppose we seek the pth root of a positive real number a. This is a zero of the function f defined by $f(x) = x^p - a$. Because $f'(x) = px^{p-1}$, Newton's method yields the recurrence $x_{n+1} = (1 - 1/p)x_n + (1/p)(a/x_n^{p-1})$. Proposition 16.35 implies that if the resulting sequence converges, then the limit is a pth root of a. The limit does exist; this follows by generalizing Example 14.10 (Exercise 16.40) or by applying Theorem 16.46. ∎

When Newton's method does converge, it generally converges faster than the method of bisection.

16.37. Example. *Speed of convergence.* Given a sequence $\langle x \rangle$ converging to L, we may study how fast the differences $x_n - L$ tend to 0. Consider the recurrence $x_{n+1} = \frac{1}{2}(x_n + 2/x_n)$ arising from Newton's method for $x^2 - 2 = 0$. Let $y_n = x_n - \sqrt{2}$. In the recurrence for $\langle x \rangle$, this yields

$$y_{n+1} + \sqrt{2} = \frac{1}{2}\left(y_n + \sqrt{2} + \frac{2}{y_n + \sqrt{2}}\right) = \frac{1}{\sqrt{2}}\left(\frac{y_n}{\sqrt{2}} + 1 + \frac{1}{1 + y_n/\sqrt{2}}\right).$$

Subtracting $\sqrt{2}$ from both sides leads to

$$y_{n+1} = \frac{1}{\sqrt{2}}\left(-1 + \frac{y_n}{\sqrt{2}} + \frac{1}{1 + y_n/\sqrt{2}}\right).$$

When y_n is small, the geometric series expands $1/(1 + y_n/\sqrt{2})$ as $1 - y_n/\sqrt{2} + y_n^2/2 - \cdots$. Substituting this into the recurrence yields $y_{n+1} = y_n^2/(2\sqrt{2}) + \cdots$, where the error involves higher powers of y_n. Convergence is therefore rapid once we get close to the limit. Each iteration nearly doubles the number of accurate places in the binary expansion. The method of bisection adds only one accurate place per iteration. ∎

The conclusion of Proposition 16.35 need not hold if f' is not continuous (Exercise 16.42). When studying differentiable functions, we often need the stronger hypothesis that f' is continuous. In the next section, we need the yet stronger hypothesis that f' is differentiable.

16.38. Definition. A function f is *continuously differentiable* on an open interval if f is differentiable there and f' is continuous there. When f' is differentiable, we write its derivative as f'' and say that f is *twice differentiable*. For $k \geq 2$, we define the *kth derivative* $f^{(k)}$ of f to be the derivative of $f^{(k-1)}$, when it exists. A function f is *smooth* or *infinitely differentiable* if for each $k \in \mathbb{N}$, the kth derivative exists.

16.39. Example. *Differentiable but not continuously differentiable.* Consider the function $f(x)$ defined by $f(x) = x^2 \sin(1/x)$ if $x \neq 0$ and $f(x) = 0$ if $x = 0$. Since $|\sin y| \leq 1$ for all y, we have $|f(x)| \leq x^2$ for all $x \neq 0$. This inequality also holds at $x = 0$, by the definition of f. Hence Corollary 16.19 implies that f is differentiable at 0 and that $f'(0) = 0$. To compute $f'(x)$ when $x \neq 0$, we use the chain rule and the product rule. By the chain rule, the derivative of $\sin(1/x)$ is $-x^{-2}\cos(1/x)$. By the product rule, this yields $f'(x) = 2x\sin(1/x) - \cos(1/x)$ for $x \neq 0$. This function has no limit as $x \to 0$, so f' is not continuous at 0. ∎

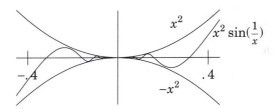

CONVEXITY AND CURVATURE

Geometric considerations yield an important class of functions where Newton's method works, including those in Example 16.36.

16.40. Definition. A function f is *convex* on an interval I if, for all x, z, y in I with $x < z < y$, the point $(z, f(z))$ lies at or below the line segment joining $(x, f(x))$ and $(y, f(y))$. Equivalently, for all $t \in [0, 1]$, we have the *convexity inequality*

$$f((1-t)x + ty) \leq (1-t)f(x) + tf(y).$$

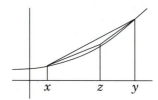

The equivalence of the two statements in the definition follows from setting $z = (1-t)x + ty$. The expression $(1-t)x + ty$ is called a "convex combination" of x and y. A convex combination is a weighted average. We can interpret the convexity inequality as saying that every weighted average of two function values is larger than the function value at the corresponding weighted average of the arguments. To show that a continuous function is convex, it suffices to know that the convexity inequality holds for each pair x, y when $t = 1/2$ (Exercise 16.38). The convexity inequality has a simple interpretation in terms of slopes.

16.41. Lemma. Suppose $f \colon [x, y] \to \mathbb{R}$. If $0 < t < 1$ and $z = (1-t)x + ty$, then the following are equivalent:
A) $f(z) \le (1-t)f(x) + tf(y)$ (the convexity inequality).
B) $m_{x,z} \le m_{x,y}$.
C) $m_{x,y} \le m_{z,y}$.

Proof. A $\Leftrightarrow$ B. We rewrite A as $f(z) - f(x) \le t[f(y) - f(y)]$. Since $t = (z - x)/(y - x)$, we have $f(z) - f(x) \le \frac{z-x}{y-x}[f(y) - f(x)]$. Since $z - x > 0$, we divide by it to obtain $m_{x,z} \le m_{x,y}$. These steps are reversible.

A $\Leftrightarrow$ C. We rewrite A as $f(z) - f(y) \le (1-t)[f(x) - f(y)]$. Since $1 - t = (y - z)/(y - x)$, we obtain $f(z) - f(y) \le \frac{y-z}{y-x}[f(x) - f(y)]$. Multiplying this inequality by -1 and dividing by $y - z$ yields $m_{z,y} \ge m_{x,y}$. These steps are reversible. ∎

The geometric definition of convexity does not mention differentiation, but Lemma 16.41 and the Mean Value Theorem combine to characterize convex differentiable functions.

16.42. Theorem. A differentiable function f is convex on an interval I if and only if f' is nondecreasing on I.

Proof. Suppose first that f is convex, and choose $x, y \in I$ with $x < y$. For all z between x and y, Lemma 16.41 yields $m_{x,z} \le m_{x,y} \le m_{z,y}$. Letting z decrease to x in the first inequality yields $f'(x) \le m_{x,y}$. Letting z increase to y in the second inequality yields $f'(y) \ge m_{x,y}$. Therefore, $f'(x) \le f'(y)$ and f' is nondecreasing on I.

If f is not convex on I, then Lemma 16.41 yields points x, z, y in I with $x < z < y$ such that $m_{x,z} > m_{x,y}$ and $m_{x,y} > m_{z,y}$. Since f is differentiable, the Mean Value Theorem yields points c, d with $x < c < z$ and $z < d < y$ such that $f'(c) = m_{x,z}$ and $f'(d) = m_{z,y}$. Now $f'(c) > f'(d)$, and f' is not nondecreasing. ∎

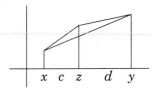

16.43. Corollary. If f is twice differentiable, then f is convex if and only if f'' is nonnegative.

Proof. When f is twice differentiable, each condition is equivalent to f' being nondecreasing. This uses Theorem 16.42 and the application of Exercise 16.22 to f' and f''. ∎

16.44. Example. *pth roots.* Suppose $f(x) = x^p - a$, so $f'(x) = px^{p-1}$ (see Exercise 16.6). If $p > 0$ and $x > 0$, then $f'(x) > 0$, and hence f is injective for $x > 0$. If $p \geq 1$, then $f''(x) = p(p-1)x^{p-2} > 0$, so f is convex for $x > 0$. We will prove that Newton's method converges for convex differentiable functions; hence we can use it to compute the unique positive pth root of a to any desired accuracy. ∎

16.45. Lemma. If differentiable functions f and g are equal at a and satisfy $f'(x) > g'(x) > 0$ for $x > a$, then $f(x) > g(x)$ for $x > a$.

Proof. By the Cauchy Mean Value Theorem, there exists $c \in (a, x)$ such that $[f(x) - f(a)]g'(c) = [g(x) - g(a)]f'(c)$. Since $f'(c) > g'(c) > 0$, we have $f(x) - f(a) > g(x) - g(a)$. Since $f(a) = g(a)$, we conclude that $f(x) > g(x)$. ∎

16.46. Theorem. Suppose f is convex and differentiable and has a zero. If $f'(x_0) \neq 0$, then Newton's method starting at x_0 converges to a zero of f. Furthermore, all zeros of f arise in this way.

Proof. Consider four sets of real numbers:
$$S = \{x: f(x) > 0, f'(x) < 0\} \qquad U = \{x: f(x) < 0, f'(x) > 0\}$$
$$T = \{x: f(x) < 0, f'(x) < 0\} \qquad V = \{x: f(x) > 0, f'(x) > 0\}$$

By Theorem 16.42, f' is nondecreasing. Hence every element of $S \cup T$ is less than every element of $U \cup V$. Because a function with positive derivative is increasing, every element of U is less than every element of V, and similarly every element of S is less than every element of T. Thus the four sets appear as illustrated, except that any of S, $T \cup U$, and V may be empty.

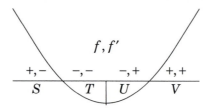

We claim that Newton's method converges to a zero of f between S and T if $x_0 \in S \cup T$ and to a zero of f between U and V if $x_0 \in U \cup V$. By symmetry, we may assume $x_0 \in U \cup V$. The picture suggests that $x_0 \in U$ implies $x_1 \in V$. To see this, we let g be the linear approximation to f at x_0 and use Lemma 16.45 to conclude that $f(x_1) > g(x_1) = 0$.

Now suppose $x_n \in V$. The assumption that f has a zero guarantees that V has a lower bound. Since f and f' are both positive on V, we have $x_{n+1} < x_n$. Since $(x_{n+1}, 0)$ is on the tangent to the graph at $(x, f(x))$, we again obtain $f(x_{n+1}) > g(x_{n+1}) = 0$, and thus $x_{n+1} \in V$.

Since the sequence decreases and remains in V, the Monotone Convergence Theorem implies that it converges to a number at least $\inf V$. We know from Proposition 16.35 that the sequence can only converge to a zero of f. If $T \cup U$ is non-empty and f has two zeros, then we can find them by starting the sequence in $U \cup V$ and in $S \cup T$. ∎

16.47. Example. For the function defined by $f(x) = x^2 - 2$, Newton's method converges to $\sqrt{2}$ when $x_0 > 0$ and to $-\sqrt{2}$ when $x_0 < 0$. When $x_0 = 0$, the sequence is undefined. ∎

We close this section with a discussion of curvature.

16.48. Solution. *Circle of curvature (optional).* Suppose that the function g defined on an open interval I containing x_0 is twice differentiable. The subset $\{(x, g(x)): x \in I\}$ of the plane is a curve γ. We determine the radius of curvature at the point (x_0, y_0) on γ.

The circle of curvature C is the set $\{(x, y): (x - a)^2 + (y - b)^2 = r^2\}$, where the pair (a, b) is the center of the circle and r is its radius. To determine the three unknowns a, b, r, we need to specify three pieces of information. We want the function y describing C to agree with g, g', g'' at x_0, so we require

$y(x_0) = g(x_0)$ (the circle intersects γ at (x_0, y_0))
$y'(x_0) = g'(x_0)$ (the circle is tangent to γ at (x_0, y_0))
$y''(x_0) = g''(x_0)$ (the circle has the same curvature as γ at (x_0, y_0))

The equation for the circle determines (two choices of) y implicitly as a function of x. Our calculations do not require making this choice at the start. We differentiate both sides of $(x - a)^2 + (y - b)^2 = r^2$ with

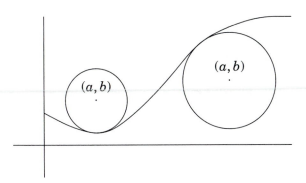

respect to x to obtain $2(x-a)+2(y-b)y'(x)=0$. This yields $y'(x)=-\frac{x-a}{y-b}$. Differentiating this by using the quotient rule and the chain rule yields $y''(x)=-\frac{(y-b)-(x-a)y'(x)}{(y-b)^2}=-\frac{(y-b)^2+(x-a)^2}{(y-b)^3}$.

We evaluate these expressions at x_0 to write the required equations in terms of the parameters a,b,r.

1) $(x_0-a)^2+(y_0-b)^2=r^2$

2) $-\frac{x_0-a}{y_0-b}=g'(x_0)$

3) $-\frac{(y_0-b)^2+(x_0-a)^2}{(y_0-b)^3}=g''(x_0).$

From (2) and (1), we obtain $1+(g'(x_0))^2=\frac{(y_0-b)^2+(x_0-a)^2}{(y_0-b)^2}=\frac{r^2}{(y_0-b)^2}$. We rewrite (3) as $g''(x_0)=-\frac{r^2}{(y_0-b)^3}$. By eliminating (y_0-b), we obtain an expression for r^2:

$$r^2=\left[\frac{r^2}{(y_0-b)^2}\right]^3\left[-\frac{r^2}{(y_0-b)^3}\right]^{-2}=\frac{[1+g'(x_0)^2]^3}{g''(x_0)^2}.$$

When $g''(x_0)\neq 0$, the radius of curvature is $r=\frac{[1+g'(x_0)^2]^{3/2}}{|g''(x_0)|}$.

When $g''(x_0)=0$, the circle degenerates to a straight line (the radius becomes infinite). The *curvature* of γ at the point $(x,g(x))$ is defined to be the reciprocal of the radius of curvature. This reflects the intuition that curvature should be larger when the graph is more curved. (Some authors define curvature by measuring the rate at which the line tangent to the curve is changing.) Because the numerator in the formula for r is never zero, curvature is well-defined, and it equals 0 when $g''(x_0)=0$.

When $g''(x_0)$ is positive, the circle lies above the curve γ, and $y-b$ is the negative square root of $r^2-(x_0-a)^2$. When $g''(x_0)$ is negative, $y-b$ is the positive square root. Computing with y^2 instead of y enabled us to consider both cases simultaneously. ∎

SERIES OF FUNCTIONS

In the remainder of this chapter, we study sequences and series in which each term is a function. After proving the basic theorems about convergence of series of functions, we use such series to construct examples of continuous but nowhere differentiable functions. Constructing functions via series (as described in the next example) also has many scientific applications.

16.49. Example. *Power series and Fourier series.* A *power series* is a series of the form $\Sigma_{n=0}^{\infty} a_n(x-p)^n$. We view this as the sum of the functions f_n defined by $f_n(x) = a_n(x-p)^n$. Perhaps the most important power series is given by the exponential function $\exp(x) = \Sigma_{n=0}^{\infty} \dfrac{x^n}{n!}$. In Chapter 14, we thought of x as fixed when writing this series. Now we treat the summands as functions of x. Convergence of power series is fairly easy to understand.

Fourier series, named for Joseph Fourier (1768-1830), are series of the form $\Sigma_{n=0}^{\infty} (a_n \sin(nx) + b_n \cos(nx))$. Convergence is a delicate matter for Fourier series. Physicists and engineers often use them, because they can represent rather general functions by superposition (summation) of waves. Convergence of Fourier series is too delicate a matter to be discussed in this book, but we mention a simple example the reader might enjoy. Consider $\Sigma_{n=1}^{\infty} \dfrac{\sin(nx)}{n}$. Graph the first few partial sums to see what is happening; to what function does the series seem to converge? Both Fourier series and power series are best understood using complex numbers, which we introduce in Chapter 18. ∎

We begin by defining uniform convergence for sequences of functions. The limit of such a sequence will itself be a function. We use the notation $\{f_n\}$ rather than $\langle f \rangle$ to name the sequence in order to use f to name the limit function. The definition of pointwise convergence is the application at each point x of the definition of convergence for the sequence of numbers $f_n(x)$.

16.50. Definition. Let $\{f_n\}$ be a sequence of functions defined on an interval I. The sequence $\{f_n\}$ *converges pointwise* to f on I if for every $x \in I$ and every $\varepsilon > 0$, there exists $N \in \mathbb{N}$ such that $n > N$ implies $|f_n(x) - f(x)| < \varepsilon$. The sequence $\{f_n\}$ *converges uniformly* to f on I if for every $\varepsilon > 0$, there exists $N \in \mathbb{N}$ such that, for all choices of $n > N$ and $x \in I$, we have $|f_n(x) - f(x)| < \varepsilon$. The sequence $\{f_n\}$ is *uniformly Cauchy* on I if for every $\varepsilon > 0$, there exists $N \in \mathbb{N}$ such that, for all choices of $n, m > N$ and $x \in I$, we have $|f_n(x) - f_m(x)| < \varepsilon$.

16.51. Remark. *Testing uniform convergence.* Uniform convergence asks us to prove that many sequences (one for each x) converge "at the same rate". To do so, we must control the worst case. In particular, $\{f_n\}$ converges uniformly to f on I if and only if $\sup_{x \in I} |f_n(x) - f(x)|$ converges to 0 as n tends to infinity. ∎

This remark is particularly easy to apply when we can bound $\sup_{x \in I} |f_n(x) - f(x)|$ easily, as in the next example and in Exercise 16.52.

16.52. Example. We contrast two sequences of functions defined on the interval $[0, 1]$. Let $f_n(x) = x$ for $0 \le x \le 1/n$ and $f_n(x) = 0$ for $1/n < x \le 1$. Let $g_n(x) = nx$ for $0 \le x \le 1/n$ and $g_n(x) = 0$ for $1/n < x \le 1$. Both $\{f_n\}$ and $\{g_n\}$ converge pointwise to the zero function. Since $\max_{x \in [0,1]} |f_n(x)| = 1/n$, the sequence $\{f_n\}$ converges uniformly. On the other hand, $\max_{x \in [0,1]} |g_n(x)| = 1$, so $\{g_n\}$ does not converge uniformly. ∎

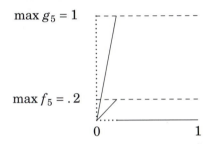

The condition of uniform convergence is stronger than the condition of pointwise convergence, because it requires a single natural number N chosen in terms of ε that works simultaneously for the tail of the sequence at every $x \in I$. Similarly, uniformly Cauchy is stronger than each sequence $f_n(x)$ being a Cauchy sequence, because the same natural number must work simultaneously at all x.

16.53. Lemma. Suppose $\{f_n\}$ is a sequence of bounded functions on an interval I. Then $\{f_n\}$ converges uniformly to some function on I if and only if $\{f_n\}$ is uniformly Cauchy on I.

Proof. Suppose $\{f_n\}$ converges uniformly to f. Consider an arbitrary $\varepsilon > 0$. From the definition of uniform convergence, we obtain a natural number N such that $n > N$ and $x \in I$ imply $|f_n(x) - f(x)| < \varepsilon/2$. Using this N, choose $n, m > N$. Now for each $x \in I$ we have

$$|f_n(x) - f_m(x)| \le |f_n(x) - f(x)| + |f(x) - f_m(x)| < \varepsilon/2 + \varepsilon/2 = \varepsilon.$$

We have proved that $\{f_n\}$ is uniformly Cauchy.

Conversely, suppose $\{f_n\}$ is uniformly Cauchy. For each fixed x, the numbers $f_n(x)$ form a Cauchy sequence indexed by n. By the

Cauchy Convergence Criterion, this sequence has a limit; call it $f(x)$. Doing this for each $x \in I$ defines a function f on I; we claim that $\{f_n\}$ converges uniformly to f. Consider an arbitrary $\varepsilon > 0$, and choose ε' such that $0 < \varepsilon' < \varepsilon$. From the definition of uniformly Cauchy, we obtain a natural number N such that $n, m > N$ and $x \in I$ imply $|f_n(x) - f_m(x)| < \varepsilon'$. Keeping n fixed, let $m \to \infty$. Using the definition of f, the continuity of the absolute value function, and the preservation of inequalities under limits (Lemma 14.13), we have

$$|f_n(x) - f(x)| = |f_n(x) - \lim_{m \to \infty} f_m(x)| = \lim_{m \to \infty} |f_n(x) - f_m(x)| \le \varepsilon' < \varepsilon.$$

Hence this choice of N in terms of ε verifies that $\{f_n\}$ converges uniformly to f on I. ∎

16.54. Definition. Suppose $\{g_n\}$ is a sequence of bounded functions on an interval I. If for each $x \in I$, the series $\Sigma_{n=1}^{\infty} g_n(x)$ converges, then setting $g(x) = \Sigma_{n=1}^{\infty} g_n(x)$ defines a function from I to R; we say that the *series* of functions $\Sigma_{n=1}^{\infty} g_n$ *converges pointwise* to g. Given such a sequence $\{g_n\}$, let $f_n = \Sigma_{k=1}^{n} g_k$. Then the series $\Sigma_{n=1}^{\infty} g_n$ *converges uniformly* to g on I if the sequence $\{f_n\}$ converges uniformly to g on I.

16.55. Corollary. (Weierstrass M-test) Suppose $\{g_n\}$ is a sequence of bounded functions on an interval I, with $|g_n(x)| \le M_n$ for $x \in I$. Suppose also that $\Sigma_{n=1}^{\infty} M_n$ converges. Then $\Sigma_{n=1}^{\infty} g_n$ converges uniformly on I.

Proof. By Lemma 16.53, it suffices to show that the sequence $\{f_n\}$ of partial sums is uniformly Cauchy. Consider an arbitrary $\varepsilon > 0$. We have $f_n - f_m = \Sigma_{k=m+1}^{n} g_k$. Hence $x \in I$ implies that $|f_n(x) - f_m(x)| \le \Sigma_{k=m+1}^{n} |g_k(x)| \le \Sigma_{k=m+1}^{n} M_k$. This last summation equals $s_n - s_m$, where $\langle s \rangle$ is the sequence of partial sums. Since $\Sigma_{k=1}^{\infty} M_k$ converges, $\langle s \rangle$ is a Cauchy sequence. Hence we can choose $N \in \mathbb{N}$ in terms of our given ε such that $|s_n - s_m| < \varepsilon$. With this N, we have $|f_n(x) - f_m(x)| < \varepsilon$ whenever $n, m > N$ and $x \in I$, and $\{f_n\}$ is uniformly convergent. ∎

We use this notation because M_n "majorizes" g_n. The corollary provides an easy way to prove uniform convergence of series of functions.

16.56. Example. *Applying the M-test.* Let $g_n(x) = x^n / n^2$, and let $M_n = 1/n^2$. Because $\Sigma_{n=1}^{\infty} 1/n^2$ converges (Exercise 14.27 or 14.40) and $|g_n(x)| \le M_n$, we conclude that $\Sigma_{n=1}^{\infty} g_n$ converges uniformly on $[-1, 1]$. ∎

We next obtain a sufficient condition for continuity of the limit of a convergent sequence of functions.

16.57. Theorem. If $\{f_n\}$ are continuous on an interval I, and $\{f_n\}$ converges uniformly to f on I, then f is continuous at every x in I.

Proof. Given $\varepsilon > 0$, uniform convergence allows us to choose a natural number N such that $n > N$ and $x \in I$ imply $|f_n(x) - f(x)| < \varepsilon/3$. Now, let a be an arbitrary element of I; we prove that f is continuous at a. For all $x \in I$ and $n > N$, we have

$$|f(x) - f(a)| \le |f(x) - f_n(x)| + |f_n(x) - f_n(a)| + |f_n(a) - f(a)|$$

$$\le \varepsilon/3 + |f_n(x) - f_n(a)| + \varepsilon/3.$$

Since f_n is continuous, we can choose $\delta > 0$ such that $|x - a| < \delta$ implies $|f_n(x) - f_n(a)| < \varepsilon/3$. With this choice of δ in terms of ε, we have $|f(x) - f(a)| < \varepsilon$ when $|x - a| < \delta$. Hence f is continuous at a. ∎

16.58. Example. The hypothesis of uniform convergence is needed in Theorem 16.57. Suppose f_n is defined on $[0, 1]$ by $f_n(x) = x^n$. The sequence $\{f_n\}$ converges pointwise to the function f defined by $f(x) = 0$ for $0 \le x < 1$ and $f(1) = 1$. This function is not continuous. ∎

16.59. Corollary. If $\{g_n\}$ are continuous functions on an interval I, and $\sum_{n=1}^{\infty} g_n$ converges uniformly to g on I, then g is continuous on I.

Proof. The sum of n continuous functions is a continuous function. Hence $f_n = \sum_{k=1}^{n} g_k$ is continuous, and the theorem applies. ∎

By combining Corollaries 16.55 and 16.59, we obtain the following statement: If $\{g_n\}$ is a sequence of bounded continuous functions, with bounds $|g_n(x)| \le a_n$ such that $\sum_{n=1}^{\infty} a_n$ converges, then $g = \sum_{n=1}^{\infty} g_n$ is a continuous function.

16.60. Example. Define $g: \mathbb{R} \to \mathbb{R}$ by $g(x) = \sum_{n=1}^{\infty} \exp(-nx^2)/n^2$. Here each g_n is continuous (see Exercise 16.43 for continuity of the exponential function). Also $\exp(-nx^2) \le 1$ for all n and x, and $\sum_{n=1}^{\infty} 1/n^2$ converges. We conclude that g is a continuous function on all of $\mathbb{R}$. ∎

These results about uniform convergence allow us to interchange certain limits involving infinite series.

16.61. Corollary. If $\{f_n\}$ is a uniformly convergent sequence of continuous functions on I and $a \in I$, then

$$\lim_{n \to \infty} \lim_{x \to a} f_n(x) = \lim_{x \to a} \lim_{n \to \infty} f_n(x).$$

Proof. Suppose $\{f_n\}$ converges to f. By Theorem 16.57, f is continuous. Therefore, $\lim_{x \to a} f(x) = f(a) = \lim_{n \to \infty} f_n(a)$. Because each f_n is continuous, the last expression equals $\lim_{n \to \infty} \lim_{x \to a} f_n(x)$. Because $f_n \to f$, the first expression equals $\lim_{x \to a} \lim_{n \to \infty} f_n(x)$. ∎

16.62. Corollary. (Interchange of limit and uniformly convergent sum) Suppose $\{g_n\}$ are continuous functions on I. If $\Sigma_{n=1}^{\infty}g_n$ converges uniformly on I, and $a \in I$, then

$$\lim_{x \to a} \Sigma_{n=1}^{\infty}g_n(x) = \Sigma_{n=1}^{\infty}g_n(a) = \Sigma_{n=1}^{\infty}\lim_{x \to a}g_n(x).$$

Proof. By Corollary 16.59, $\{g_n\}$ converges to a function g that is continuous on I. Continuity of g implies $\lim_{x \to a} g(x) = g(a)$, which is the desired equality. ∎

Since an infinite sum is a limit, Corollaries 16.61 and 16.62 both state conditions under which we can interchange limits. This does not hold in general, as illustrated by Exercise 15.16 and the next example.

16.63. Example. *Failure of interchange of limits.* To see that the order of taking limits can matter, consider summing the entries in this infinite matrix:

$$\begin{pmatrix} 0 & 1 & 0 & 0 & \cdots \\ -1 & 0 & 1 & 0 & \cdots \\ 0 & -1 & 0 & 1 & \cdots \\ 0 & 0 & -1 & 0 & \cdots \\ \vdots & \vdots & \vdots & \vdots & \cdots \end{pmatrix} \qquad a_{i,j} = \begin{cases} +1 & \text{if } j = i+1 \\ -1 & \text{if } j = i-1 \\ 0 & \text{if otherwise} \end{cases}$$

One way to do this is to evaluate $\Sigma_{i=1}^{k}\Sigma_{j=1}^{k}a_{i,j}$ and let k tend to infinity. This corresponds to considering k by k matrices in the upper left corner. For each k, the entries sum to 0, so the limit is 0.

Alternatively, we can think of the sum as taking the limit of $\Sigma_{i=1}^{m}\Sigma_{j=1}^{n}a_{i,j}$ as n and m approach infinity. The result depends on the order in which we perform the two limit operations. For fixed m, n, the sum is 1 if $n > m$ (more columns than rows), and the sum is -1 if $m > n$ (more rows than columns). Therefore, $\lim_{m \to \infty}\lim_{n \to \infty}\Sigma_{i=1}^{m}\Sigma_{j=1}^{n}a_{i,j} = 1$ and $\lim_{n \to \infty}\lim_{m \to \infty}\Sigma_{i=1}^{m}\Sigma_{j=1}^{n}a_{i,j} = -1$. ∎

Uniform convergence helps in constructing functions that are continuous on an interval but not differentiable anywhere. We describe two classical examples; the first we only sketch.

16.64. Example. *Everywhere continuous and nowhere differentiable.* We define a sequence of functions $\{f_n\}$, each mapping $[0, 1]$ to $[0, 1]$. We start with the function $f_0(x) = x$. It is easiest to define each successive function by considering their graphs. To obtain f_1, break the graph of f_0 into thirds, reflect the middle third of the graph through its average height, and connect the ends of this new segment to the ends of the interval. To obtain f_{n+1} from f_n, do this to each segment in the graph of f_n; we illustrate the first two steps.

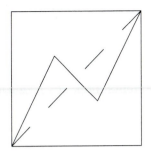 

It is fairly easy to show that $\{f_n\}$ is uniformly Cauchy and converges pointwise to a continuous function f (Exercise 16.53). It is much more difficult to prove that f is nowhere differentiable. ∎

16.65. Example. *Everywhere continuous and nowhere differentiable.* Given $x \in \mathbb{R}$, let $d(x)$ denote the distance from x to the nearest integer. Define g by $g(x) = \sum_{n=0}^{\infty} d(10^n x)/10^n$. We show that g is continuous on all of $\mathbb{R}$ but is not differentiable anywhere. (This construction is due to van der Waerden.)

The continuity of g is easy to prove. Since $d(x) = \inf_{n \in \mathbb{Z}} |x - n|$, and since the absolute value function is continuous, the function d is continuous as well. Hence the function g_n defined by $g_n(x) = d(10^n x)/10^n$ is continuous. Since $0 \le d(x) \le \frac{1}{2}$ for all x, we have $|g_n(x)| \le \frac{1}{2} \cdot 10^{-n}$ for all x. Since $\sum_{n=0}^{\infty} \frac{1}{2} \cdot 10^{-n} = 10/18$ (geometric series), we know by the Weierstrass M-test that the series $\sum_{n=0}^{\infty} g_n$ converges uniformly on $\mathbb{R}$. Corollary 16.59 then implies that g is continuous everywhere.

To prove that g is nowhere differentiable, we choose an arbitrary $x \in \mathbb{R}$ and consider the decimal expansion of its fractional part. Suppose that this decimal expansion is $.a_1 a_2 a_3 \cdots$. We define a sequence $\langle h \rangle$ as follows: When a_m is not 4 or 9, we put $h_m = 10^{-m}$. When a_m equals 4 or 9, we put $h_m = -10^{-m}$. This choice makes $g(x + h_m) - g(x)$ easy to compute. If a_m does not equal 4 or 9, then from each of $10^n x + 10^{n-m}$ and $10^n x$ we move up to the nearest integer, or from each we move down. If a_m equals 4 or 9, then the same phenomenon occurs for $10^n x - 10^{n-m}$ and $10^n x$. In each case, we are comparing $d(10^n(x + h_m))$ with $d(10^n x)$ (see Exercise 16.54 for an example).

In order to obtain a formula for the difference quotient $[g(x + h_m) - g(x)]/h_m$, we let $\alpha_n = +1$ if $a_n \in \{0, 1, 2, 3, 4\}$ and $\alpha_n = -1$ if $a_n \in \{5, 6, 7, 8, 9\}$. We now compute

$$\frac{d(10^n(x + h_m)) - d(10^n x)}{h_m} = \begin{cases} \alpha_n 10^n & \text{if } n < m \\ 0 & \text{if } n \ge m \end{cases}$$

From this, we have

$$\frac{g(x + h_m) - g(x)}{h_m} = \sum_{n=0}^{\infty} \frac{d(10^n x + 10^n h_m) - d(10^n x)}{10^n h_m} = \sum_{n=0}^{m-1} \alpha_n.$$

The infinite sum in the expression for g collapses to a finite sum in the difference quotient. Furthermore, the difference quotient for h_m is the sum of the first m terms of some sequence of $+1$'s and -1's. The existence of a limit for the difference quotients $[g(x + h_m) - g(x)]/h_m$ is equivalent to the convergence of the series $\sum_{n=0}^{\infty} \alpha_n$. A necessary condition for convergence of a series is that the terms approach 0; this fails here, since each α_n is 1 or -1. Hence this sequence of difference quotients does not converge, and therefore $\lim_{h \to 0}[g(x + h) - g(x)]/h$ does not exist. ∎

Although it is difficult to explicitly describe functions that are continuous but nowhere differentiable, more advanced considerations imply that "almost all" continuous functions are nowhere differentiable.

EXERCISES

16.1. Suppose $f(x) = \Pi_{j=1}^{n}(x + a_j)$, where $a_1, \ldots, a_n \in \mathbb{R}$. Compute $f'(x)$.

16.2. Derive the product rule for differentiation using difference quotients. (Hint: add and subtract an appropriate quantity to the numerator.)

16.3. Suppose f and g are differentiable at x and $g(x) \neq 0$. Using the product rule and the formula for $(1/g)'(x)$, prove the quotient rule

$$(f/g)'(x) = \frac{g(x)f'(x) - f(x)g'(x)}{[g(x)]^2}.$$

16.4. (!) Compute the derivative of the cube root function using either definition. (Hint: Use the factorization $a^3 - b^3 = (a - b)(a^2 + ab + b^2)$ to simplify the difference of cube roots.)

16.5. The following inductive argument fails to prove that $(d/dx)x^n = nx^{n-1}$ for nonnegative integers n. "Basis step $(n = 0)$: $\lim_{h \to 0}(1 - 1)/h = 0$. Inductive step $(n > 0)$: Using the induction hypothesis for $n - 1$ and the product rule for differentiation, we compute $(d/dx)x^n = (d/dx)xx^{n-1} = x \cdot (n - 1)x^{n-2} + 1 \cdot x^{n-1} = nx^{n-1}$." explain the error and correct the proof.

16.6. Suppose $r = p/q$, where $p \in \mathbb{Z}$ and $q \in \mathbb{N}$. We define x^r to be $(x^p)^{1/q}$. Determine $f'(x)$, where $f(x) = x^r$. (Hint: We have determined this already for $r \in \mathbb{N}$. Derive the formula first for $p = 1$ and then for $r \in \mathbb{Q}$. Comment: When $r \in \mathbb{R}$, the same formula holds for f'. The proof uses properties of the exponential function and appears in Exercise 17.16.)

16.7. Suppose that e_1 and e_2 are error functions and that $e_1(h) \leq e(h) \leq e_2(h)$ for all h in a neighborhood of 0. Prove that e is an error function.

16.8. (!) Suppose $f(x) = x + x^2$ if x is rational, and $f(x) = x$ if x is irrational. Prove that f is differentiable at $x = 0$.

16.9. (–) *Sufficient conditions for differentiability at a point.*
  a) Suppose that $|f(x)| \le x^2 + x^4$ for all x. Prove that $f'(0)$ exists.
  b) Suppose that $|f(x)| \le g(x)$, where $g(x) \ge 0$ for all x and $g'(0) = g(0) = 0$. Prove that $f'(0)$ exists.
  c) Suppose that g is a bounded function and that $f(x) = (x-a)^2 g(x)$ for all x. Prove that $f'(a)$ exists.

16.10. Suppose that $|f(x) - f(y)| \le |g(x) - g(y)|$ for all $x, y \in \mathbb{R}$, and suppose g is differentiable at a with $g'(a) = 0$. Prove using difference quotients that f is differentiable at a and that $f'(a) = 0$.

16.11. *Squeeze theorem for differentiability.* Suppose f, g, h are functions defined on a neighborhood S of x, and $A(t) \le f(t) \le C(t)$ for all $t \in S$. Suppose that $A(x) = C(x)$, that A and C are differentiable at x, and that $C'(x) = C'(x)$. Use difference quotients to prove that f is differentiable at x and $A'(x) = C'(x)$.

16.12. Suppose f is differentiable, with $f(0) = 0$. Let $g(x) = f(x)/x$ for $x \ne 0$.
  a) How should $g(0)$ be defined to make g continuous at 0?
  b) If $g(0)$ is defined so that g is continuous at 0, does it follow that g is differentiable at 0? Give a proof or a counterexample.

16.13. (–) The volume of a ball of radius r is $\frac{4}{3}\pi r^3$. Suppose that air is escaping from a ball at the rate of 36 cubic inches per second. How fast is the radius of the ball decreasing at the moment when the radius is 6 inches?

16.14. (–) What real number most exceeds its square?

16.15. Suppose $f(x) = ax^2 + bx + c$ with $a > 0$. Determine the minimum value of f on $\mathbb{R}$. What condition on a, b, c is necessary and sufficient for the minimum value to be positive?

16.16. A company wishes to set the price of its new liquid to maximize profit. A marketing analysis indicates that if the price is set at x dollars per gallon, then the number of gallons sold per day will be $g(x) = 1000/(5 + x)$. The government, wishing to stimulate production, will also pay the company (per day) \$50 times $\sqrt{g(x)}$. Determine the maximum and minimum values of the company's daily profit and the prices that yield these values.

16.17. (!) Suppose that $m_1, \ldots, m_k$ are nonnegative real numbers with sum n.
  a) Using calculus and induction, prove that $\Sigma_{i<j} m_i m_j \le (1 - \frac{1}{k})\frac{n^2}{2}$, with equality only when $m_1 = \cdots = m_k$.
  b) In the case where $m_1, \ldots, m_k$ are integers, give a combinatorial proof that $\Sigma_{i<j} m_i m_j$ is maximized when each m_i is $\lfloor n/k \rfloor$ or $\lceil n/k \rceil$.

16.18. (!) Prove that two differentiable functions on an interval (a, b) have the same derivative if and only if they differ by a constant.

16.19. Derive the Mean Value Theorem from the Cauchy Mean Value Theorem.

16.20. Suppose f is differentiable on $[a, b]$ and $f'(a) < y < f'(b)$. Prove that there exists $c \in (a, b)$ such that $f'(c) = y$. (Comment: This is the Intermediate Value Property for the function f'. It does not require that f' be continuous.)

16.21. (!) Suppose f is differentiable, and $f'(x) < 1$ for all x. Prove that f has at most one fixed point. (Recall that x is a fixed point of f if $f(x) = x$.)

16.22. (–) Suppose f is differentiable. Prove that f' is nonnegative everywhere if and only if f is nondecreasing.

16.23. Suppose f is differentiable and $f'(0) > 0$. Suppose also that f is not monotone in any neighborhood of 0. Explain why f' must be discontinuous at 0. Construct an example of such a function f. (Hint: modify Example 16.17).

16.24. Suppose f is differentiable, and suppose that f and f' are positive on $\mathbb{R}$. Prove that the function $g = f/(1 + f)$ is bounded and increasing.

16.25. Suppose $f(a) = f(b) = 0$ and f is differentiable on $[a, b]$. Determine f under the condition that $f'(x) \geq 0$ for all $x \in [a, b]$.

16.26. A function whose domain and target are themselves sets of functions is sometimes called an *operator*. Differentiation, for example, is an operator whose domain is the set of differentiable functions. In this exercise we consider another operator A defined on this set. The image of the function f under the operator A is defined to be the function Af whose value at x is $\lim_{t \to 1} \frac{f(tx) - f(x)}{tf(x) - f(x)}$. (If $f(x) = 0$, then Af is not defined at x.)

    a) Suppose the function f is continuously differentiable. Use L'Hôpital's Rule to compute Af.

    b) Use part (a) to compute $(Af)(x)$ when $f(x) = x^n$ and when $f(x) = e^x$.

    c) If f' is not continuous, the computation using L'Hôpital's Rule is not valid. Give a direct proof of the formula in part (a) that is valid even when f' is not continuous. (Hint: Replace t by $1 + h$. When $x \neq 0$, replace hx by u.)

16.27. Suppose $f \colon \mathbb{R} \to \mathbb{R}$. In some contexts, difference is more appropriate than differentiation. The *first forward difference* of a function f is the function Δf defined by $\Delta f(x) = f(x + 1) - f(x)$. The *$k$th forward difference* of f is defined by $\Delta^k f(x) = \Delta^{k-1} f(x + 1) - \Delta^{k-1} f(x)$.

    a) Prove that $\Delta^k f(x) = \sum_{j=0}^{k} (-1)^j \binom{k}{j} f(x + j)$.

    b) Prove that $f^{(k)}(x) = \lim_{h \to 0} \frac{1}{h} \sum_{j=0}^{k} (-1)^j \binom{k}{j} f(x + jh)]$ when the limit exists.

16.28. Suppose f is smooth. Prove that f is a polynomial of degree at most k if and only if $f^{(k+1)}(x) = 0$ for all x.

16.29. (+) Suppose f is smooth, $f(0) = 0$, and f has a local minimum at 0. If $f^{(j)}(0) \neq 0$ for some natural number j, let k be the smallest such number. Prove that k is even. Give an example of a smooth function f such that $f(x) = 0$ if and only if $x = 0$, and $f^{(j)}(0) = 0$ for all $j \in \mathbb{N}$.

16.30. Suppose f and g are smooth (Definition 16.38). Compute the kth derivative of $f \circ g$, for $1 \leq k \leq 5$. Describe the form of the expression for general k. (Comment: The integer coefficients arising here are the *Stirling numbers of the second kind* and also arise in combinatorics.)

16.31. (–) Using an initial guess of 1 for the solution, apply Newton's method to seek a solution to the equation $x^5 = 33$ and compute the first four iterations. Repeat this with an initial guess of 2. (Use a calculator.)

16.32. Find a quadratic function f for which the recurrence generated by Newton's method is $x_{n+1} = \frac{1}{2}(x_n - 1/x_n)$. Use the graph of the function to explain the behavior of the recurrence as $n \to \infty$.

16.33. Find a differentiable function f and a sequence $\langle x \rangle$ such that $x_n \to 0$, $f'(x_n) \to \infty$, and $f(x_n) = 1$ for every n. Determine $\lim[x_n - f(x_n)/f'(x_n)]$. What does this exercise say about Proposition 16.35?

16.34. (!) Suppose f and g are convex and $c \in \mathbb{R}$. Which of the three functions $f + g$, $c \cdot f$, and $f \cdot g$ must be convex? (Give proofs or counterexamples.)

16.35. Suppose f is convex on the interval $[a, b]$. Prove that the maximum of f on $[a, b]$ is $f(a)$ or $f(b)$. (Comment: convex functions need not be differentiable.)

16.36. Suppose f is twice differentiable and f'' is nonnegative everywhere. Given that $f(a) = A$ and $f(b) = B$, what is the maximum possible value of $f((a+b)/2)$? For what function is this bound attained?

16.37. Which polynomials of odd degree are convex on $\mathbb{R}$?

16.38. (!) Characterize the fourth degree polynomials that are convex on $\mathbb{R}$ by giving a necessary and sufficient condition on the coefficients.

16.39. Suppose Y takes on n possible values $y_1, \ldots, y_n$ with given probabilities p_i, where $-1 \le y_i \le 1$ for all i. Suppose also that the expectation of Y is 0 and that f is convex. Prove that the expectation of $f(Y)$ is at most $[f(1) + f(-1)]/2$.

16.40. (+) Suppose f is continuous and $f(\frac{x+y}{2}) \le \frac{f(x)+f(y)}{2}$ for all $x, y \in \mathbb{R}$. Prove that f is convex. (Hint: first prove that the inequality of Definition 16.40 holds when t is a fraction whose denominator is a power of 2, then apply the continuity of f.)

16.41. (+) Starting with $x_0 = a$, define a sequence $\langle x \rangle$ by $x_{n+1} = (1 - 1/p)x_n + (1/p)(a/x_n^{p-1})$ for $n \ge 0$. By using the convexity of x^p as a function of x, but without differentiability or Newton's method, prove that $x_n \to a^{1/p}$.

16.42. (+) Consider the polynomial f defined by $f(x) = (x - a)(x - b)(x - c)(x - d)$ with $a < b < c < d$. Describe the set of starting points x_0 such that Newton's method converges to a root. (Hint: draw very careful pictures. The set of starting points x_0 that fail is an uncountable set.)

16.43. Define $f_n \colon \mathbb{R} \to \mathbb{R}$ by $f_n(x) = n^2/(x^2 + n^2)$, and define $f = \lim f_n$. Determine f. Does $\{f_n\}$ converge uniformly to f?

16.44. Recall that $\exp(x) = \Sigma_{n=0}^{\infty} x^n/n!$. Define g_n by $g_n(x) = x^n/n!$.
    a) Prove that $\Sigma_{n=0}^{\infty} g_n$ converges uniformly to $\exp(x)$ on any bounded interval I (and hence $\exp(x)$ is continuous).
    b) Prove that $\exp(x + y) = \exp(x)\exp(y)$.
    c) Determine $\lim_{h \to 0}(\exp(h) - 1)/h$. (Comment: l'Hôpital's rule cannot be applied here, since we do not yet know that $\exp(x)$ is differentiable. The explicit series definition of $\exp(h)$ must be used.)
    d) Use (b) and (c) to prove that the derivative of $\exp(x)$ is $\exp(x)$.

16.45. Suppose $a > 0$, and define f by $f(x) = \exp(-ax^2)$. Determine where f is convex. Sketch the graph of f.

16.46. (!) Find an explicit formula for $\sum_{n=0}^{\infty} n^2 x^n$ when $|x| < 1$. Using derivatives of the geometric series or Theorem 9.42, describe $\sum_{n=0}^{\infty} q(n)x^n$, where q is a polynomial. (Hint: the answer is a polynomial in $1/(1-x)$.)

16.47. (+) Consider two kinds of baseball players. One hits singles with probability p; the other hits home runs with probability $p/4$. The players otherwise strike out. Assume that singles advance each runner by two bases. Compare a team composed of the home run hitters with a team composed of the singles hitters. How many runs does each team expect to score per inning?

16.48. (!) Express $\sum_{k=0}^{n} kx^k$ as a ratio of two polynomials in x.

16.49. Prove that if q is a polynomial, then $\sum_{k=0}^{n} q(k)x^k$ is the ratio of two polynomials in x.

16.50. Suppose $0 < p < 1$. Let X be a random variable such that $\text{Prob}(X = n) = p(1 - p)^n$, for each nonnegative integer n.
  a) The probability generating function for X is $\phi(t) = \sum_{n=0}^{\infty} \text{Prob}(X = n)t^n$. Find an explicit formula for $\phi(t)$ by evaluating the sum. Use this to verify that these probabilities sum to 1.
  b) Compute $E(X)$.
  c) Obtain a simple formula for $\text{Prob}(X \le 20)$.

16.51. Suppose $y(x) = x^n$, where $n \ge 2$.
  a) Find the curvature at the point $(x, y(x))$.
  b) Find an equation for the value of x where the curvature is maximized.
  c) Solve the equation in (b) to find the value of x where the graph of $y(x) = x^3$ has greatest curvature.

16.52. (!) *Critical exponent for uniform convergence.*
  a) Suppose that $f_n(x) = x^n(1 - x)$. Prove that $f_n(x) \to 0$ uniformly on $[0, 1]$.
  b) Suppose that $f_n(x) = n^2 x^n(1 - x)$. Prove that $f_n(x)$ converges to 0 pointwise but not uniformly on $[0, 1]$.
  c) Suppose that $f_n(x) = n^\alpha x^n(1 - x)$, where $\alpha \ge 0$. Prove that $f_n(x) \to 0$ uniformly on $[0, 1]$ if and only if $\alpha < 1$. (Assume that $(1 - 1/n)^n \to e^{-1}$.)

16.53. Prove that the sequence defined in Example 16.64 is uniformly Cauchy.

16.54. Consider the proof that the function g in Example 16.65 is nowhere differentiable. Compute explicitly the difference quotient $[g(x + h_m) - g(x)]/h_m$ for all m in the following two cases: $x = 0$ and $x = .1496$.

16.55. Define $f: \mathbb{R} \to \mathbb{R}$ by $f(x) = \sum_{n=0}^{\infty} \frac{\sin(3^n x)}{2^n}$. Prove that f is continuous on $\mathbb{R}$ and is not differentiable at 0. (Comment: In fact, f is nowhere differentiable.)

16.56. Given a continuous and nowhere differentiable function on $\mathbb{R}$, construct from it a continuous function that is differentiable at exactly one point.

Chapter 17

Integration

Integration is the mathematical process that enables us to calculate areas and volumes; it is the continuous analogue of summation. In this chapter, we present the theory of integration and its relationship to differentiation and infinite series. The Fundamental Theorem of Calculus shows that integration is in a sense the inverse operation to differentiation. We use integration to define the logarithm function and define the exponential function to be its inverse. This definition yields the same function as the series definition in Example 14.33. We also define the trigonometric functions by infinite series. This leads to a definition of π and a proof, using integration, that the area inside the unit circle is π. Integration also arises in discussion of probability and expectation and in physical considerations such as work and center of mass, since the average value of a function over a set can be expressed as an integral.

17.1. Problem. *The Rainfall Problem.* Suppose rain is falling uniformly over a square region. What fraction of the raindrops do we expect to fall within an inscribed circle? (A circle inscribed in a square has diameter equal to the side length of the square.) ∎

17.2. Problem. *Continuous Compounding.* Suppose that the interest rate on an account is x (per year). If the interest is compounded once per year (simple interest), then after m years the total amount is $(1 + x)^m$ times the original amount. Most banks compound interest more often, such as quarterly or daily. If the interest is compounded n times per year, then the total amount after m years is $(1 + x/n)^{nm}$. The more often interest is compounded, the higher the "yield", which is the effective interest rate. For example, 5% interest ($x = .05$) compounded daily has a yield of 5.13%. The yield under continuous compounding is $\lim_{n \to \infty}(1 + x/n)^n - 1$. What is the value of this limit? ∎

DEFINITION OF THE INTEGRAL

What do we mean by the "area" of a region in the plane? This question leads to deep mathematical issues, requiring clear understanding of subtle properties of geometry and limits. The area of a square is defined to be the square of the lengths of its sides. Using squares, we obtain upper and lower bounds for the area of a bounded region R in the plane. We lay a fine grid over the plane. Let S_1 be the union of all the squares in the grid that are contained in R, and let S_2 be the union of all the squares in the grid that contain points of R. Because R is bounded, S_1 and S_2 consist of finitely many squares of equal size; we count the squares and multiply by the area of each square to find the areas of S_1 and S_2. We obtain $Area(S_1) \leq Area(R) \leq Area(S_2)$. By using a finer grid, we hope to obtain better bounds.

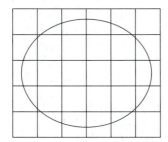
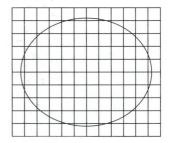

In the first picture above, using unit squares, we bound the area of the ellipse by $8 \leq Area(R) \leq 26$. With the finer grid, we obtain $13 \leq Area(R) \leq 23$. Let $\mathbf{U}$ be the set of upper bounds we can obtain using grids, and let $\mathbf{L}$ be the set of lower bounds. If $\sup \mathbf{L} = \inf \mathbf{U} = a$, then we believe that $Area(R) = a$.

We use this idea of upper and lower approximation to define integrals. Suppose f is a continuous (hence bounded) positive-valued function on the interval $[a, b]$. Let R be the region defined by $\{(x, y) \in \mathbb{R}^2 : a \leq x \leq b \text{ and } 0 \leq y \leq f(x)\}$.

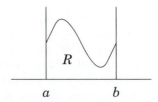

It is efficient to use rectangles instead of squares to obtain upper and lower bounds on $Area(R)$. We break the interval $[a, b]$ into subintervals to obtain the bases of the rectangles, and we use the values of f on these subintervals to determine the heights of the rectangles.

17.3. Definition. A *partition* of $[a, b]$ is a collection of n subintervals $\{[x_{i-1}, x_i]\}$ such that $a = x_0 \leq \cdots \leq x_n = b$. We specify a partition P by its list of "breakpoints" $x_0, \ldots, x_n$. A partition Q is a *refinement* of a partition P if each breakpoint for P is also a breakpoint for Q. The *least common refinement* of two partitions is the partition whose set of breakpoints is the union of their sets of breakpoints.

17.4. Definition. Suppose $f: [a, b] \to \mathbb{R}$ is a bounded function. Let P be a partition of $[a, b]$ specified by $x_0, \ldots, x_n$. Let
$$l_i = \inf\{f(x): x \in [x_{i-1}, x_i]\} \quad \text{and} \quad u_i = \sup\{f(x): x \in [x_{i-1}, x_i]\}.$$
The *lower sum* of f corresponding to P is the sum $L(f, P) = \sum_{i=1}^{n}(x_i - x_{i-1})l_i$. The *upper sum* of f on P corresponding to P is the sum $U(f, P) = \sum_{i=1}^{n}(x_i - x_{i-1})u_i$.

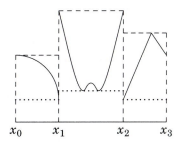

$$x_0 \qquad x_1 \qquad\qquad x_2 \qquad x_3$$

17.5. Example. The figure shows an interval partitioned into three subintervals. We have graphed a function f that is neither continuous nor monotone. The dashed horizontal line above the interval $[x_{i-1}, x_i]$ is at height u_i above the axis, and the dotted horizontal line is at height l_i. The number $(x_i - x_{i-1})u_i$ is the area of the taller rectangle above $[x_{i-1}, x_i]$, and the sum of these numbers, $U(f, P)$, is an upper bound on the area under the graph of f. Similarly, $L(f, P)$ is a lower bound on this area. ∎

Our results about upper and lower sums require understanding how they behave when we refine a partition. For the infimum and supremum of the values of a function f over a set S, we write $\inf_S f$ and $\sup_S f$. We need the simple observation that if $S \subseteq T$, then $\inf_T f \leq \inf_S f \leq \sup_S f \leq \sup_T f$.

17.6. Lemma. If P is a partition of the interval $[a, b]$, and Q is a refinement of P, then $L(f, P) \leq L(f, Q) \leq U(f, Q) \leq U(f, P)$.

Proof. (Exercise 17.3, as illustrated below). ∎

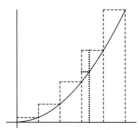

If f is a constant function, then $L(f, P) = U(f, P)$ for every partition P. Otherwise, $L(f, P) < U(f, P)$ for some partition. We hope to make these numbers arbitrarily close by refining the partition. The quantity squeezed between them is what we will call the integral of f on $[a, b]$. There are three equivalent ways to make this notion precise.

17.7. Proposition. Given a bounded function f defined on the interval $[a, b]$, the following three statements are equivalent.

a) For every $\varepsilon > 0$, there exists a partition R of $[a, b]$ such that $U(f, R) - L(f, R) < \varepsilon$.

b) $\sup_P L(f, P) = \inf_Q U(f, Q)$.

c) There is a sequence $\langle R \rangle$ of partitions such that $\lim_{n \to \infty} L(f, R_n)$ $= \lim_{n \to \infty} U(f, R_n)$.

Proof. By Lemma 17.6, $L(f, P) \le U(f, Q)$ for partitions P, Q of $[a, b]$. Hence the set of lower sums is bounded above by each upper sum, and the set of upper sums is bounded below by each lower sum. By the completeness axiom, $l = \sup_P L(f, P)$ and $u = \inf_Q U(f, Q)$ exist, and $l \le u$.

If (a) holds, then for each n we have a partition R_n such that $U(f, R_n) - L(f, R_n) < 1/n$. Since $L(f, R_n) \le l \le u \le U(f, R_n)$, we have $|u - l| < 1/n$. Since n is arbitrary, l must equal u. Thus (a) implies (b).

Suppose (b) holds. By the elementary properties of infs and sups (Exercise 13.3), we have partitions P_n and Q_n such that $l - L(f, P_n) < 1/(2n)$ and $U(f, Q_n) - u < 1/(2n)$. Since $l = u$, summing the inequalities yields $U(f, Q_n) - L(f, P_n) < 1/n$. Let R_n be the least common refinement of P_n and Q_n. Lemma 17.6 now implies that $U(f, R_n) - L(f, R_n) \le U(f, Q_n) - L(f, P_n) < 1/n$. Thus (b) implies (c).

Suppose (c) holds. When two sequences have the same limit, their difference converges to 0. Hence we can use the definition of a limit and the partitions guaranteed by (c) to obtain (a). ∎

17.8. Definition. The (bounded) function f is *integrable on* $[a, b]$ if the equivalent conditions in Proposition 17.7 hold. When f is integrable, the *integral of* f from a to b, written as $\int_a^b f(x)dx$ or $\int_a^b f$, is the common value of $\sup_P L(f, P)$ and $\inf_Q U(f, Q)$.

17.9. Remark. When f is integrable on $[a,b]$ and P is a partition of $[a,b]$, the definitions of lower sum and upper sum imply that $L(f,P) \le \int_a^b f(x)dx \le U(f,P)$. In other words, each of $U(f,P)$ and $L(f,P)$ approximates the integral, and $\int_a^b f(x)dx$ is squeezed between them. When f also is nonnegative on $[a,b]$, we define the "area under the graph of f" to be $\int_a^b f(x)dx$. ∎

Comments. A careful explanation of the notation "dx" is beyond the scope of this book; here we can view it as specifying the variable of integration. Informally, we also think of dx as the base of an infinitesimal rectangle at x with height $f(x)$. The integral sign is a stylized "S" representing the sum of the infinitesimal areas $f(x)dx$.

To integrate continuous or monotone functions, we can use partitions whose intervals have equal size. Similar methods apply also to functions that are piecewise continuous or piecewise monotone (continuous or monotone on each subinterval in some partition of $[a,b]$).

17.10. Example. $\int_0^1 x^p dx = 1/(p+1)$. Let P_n be the partition of $[0,1]$ into n equal parts. Then $U(f,P_n) = \Sigma_{i=1}^n \frac{1}{n}(\frac{i}{n})^p = n^{-(p+1)}\Sigma_{i=1}^n i^p$, and $L(f,P_n) = \Sigma_{i=1}^n \frac{1}{n}(\frac{i-1}{n})^p = \Sigma_{i=0}^{n-1} \frac{1}{n}(\frac{i}{n})^p = n^{-(p+1)}\Sigma_{i=0}^{n-1} i^p$. By Theorem 9.12, $\Sigma_{i=1}^n i^p$ is a polynomial in n with leading term $n^{p+1}/(p+1)$. Hence $\lim U(f,P_n) = 1/(p+1)$. Since $L(f,P_n) = U(f,P_n) - 1/n$, we also have $\lim L(f,P_n) = 1/(p+1)$. Since the limits are equal, the integral equals $1/(p+1)$. ∎

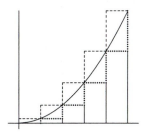

The definition we have given for $\int_a^b f(x)dx$ assumes $a \le b$. If $a > b$, we define $\int_a^b f(x)dx = -\int_b^a f(x)dx$, if the latter exists. In our understanding of how integration corresponds to computing area, this negation corresponds to the notion that when integrating from a larger x to a smaller x we are "erasing" area under the curve. This definition for $a > b$ is consistent with the observation that the definition for $a \le b$ implies $\int_a^a f(x)dx = 0$.

We are now prepared to develop the theory of integration. First we relate infimum and supremum to sums.

17.11. Lemma. When f and g are bounded real-valued functions on S, $\inf_S(f + g) \geq \inf_S f + \inf_S g$ and $\sup_S(f + g) \leq \sup_S f + \sup_S g$.

Proof. Let $B = S \times S$, and let $A = \{(x, x): x \in S\}$. Since $A \subseteq B$ implies $\inf_A h \geq \inf_B h$, we have

$$\inf_S(f + g) = \inf_{(x,x) \in A}(f(x) + g(x)) \geq \inf_{(x,y) \in B}(f(x) + g(y)) = \inf_S f + \inf_S g.$$

This proves the first inequality; the second is similar (Exercise 17.1). ∎

17.12. Proposition. (Linearity of integration) If f, g are integrable on $[a, b]$, and $c \in \mathbb{R}$, then $f + g$ and cf are integrable on $[a, b]$, and the following formulas hold:

a) $\int_a^b (f + g)(x)dx = \int_a^b f(x)dx + \int_a^b g(x)dx$,

b) $\int_a^b cf(x)dx = c\int_a^b f(x)dx$.

Proof. We first consider (a). Since f and g are integrable, Proposition 17.7 yields sequences $\langle P \rangle$ and $\langle Q \rangle$ of partitions of $[a, b]$ such that $\lim L(f, P_n) = \lim U(f, P_n)$ and $\lim L(g, Q_n) = \lim U(g, Q_n)$. Let R_n be the least common refinement of P_n and Q_n. By Lemma 17.6,

$$L(f, P_n) \leq L(f, R_n) \leq U(f, R_n) \leq U(f, P_n).$$

Hence $\lim L(f, R_n) = \lim U(f, R_n)$; also $\lim L(g, R_n) = \lim U(g, R_n)$.

The crucial point is to squeeze the lower and upper sums for $f + g$ between $L(f, R_n) + L(g, R_n)$ and $U(f, R_n) + U(g, R_n)$. For each subinterval J_i of R_n, Lemma 17.11 yields $\inf_{J_i} f + \inf_{J_i} g \leq \inf_{J_i}(f + g)$. Multiplying by the length of J_i, applying the distributive law, and summing over i yields $L(f, R_n) + L(g, R_n) \leq L(f + g, R_n)$. Similarly, we obtain $U(f, R_n) + U(g, R_n) \geq U(f + g, R_n)$. Also $L(f + g, R_n) \leq U(f + g, R_n)$, by Lemma 17.6. Together these yield

$$L(f, R_n) + L(g, R_n) \leq L(f + g, R_n) \leq U(f + g, R_n) \leq U(f, R_n) + U(g, R_n).$$

Here the leftmost and rightmost expressions converge to $\int_a^b f + \int_a^b g$. Hence the middle terms, squeezed between them, have the same limit. By Proposition 17.7, $f + g$ is integrable, and $\int_a^b (f + g) = \int_a^b f + \int_a^b g$.

We prove (b) for $c \geq 0$ and leave the case $c = -1$ to Exercise 17.4; the case $c < 0$ follows from these. Since f is integrable, there is a sequence $\langle P \rangle$ such that $L(f, P_n) \to \int_a^b f$ and $U(f, P_n) \to \int_a^b f$. Since $c \geq 0$, we have $\inf_J cf = c\inf_J f$ and $\sup_J cf = c\sup_J f$. Hence $L(cf, P_n) = cL(f, P_n) \to c\int_a^b f$ and $U(cf, P_n) = cU(f, P_n) \to c\int_a^b f$. By Proposition 17.7, cf is thus integrable, and $\int_a^b cf = c\int_a^b f$. ∎

17.13. Proposition. If f is integrable on $[a, b]$, then the following properties hold:

a) If $c \in [a, b]$, then $\int_a^b f(x)dx = \int_a^c f(x)dx + \int_c^b f(x)dx$.

b) $|f|$ is integrable on $[a, b]$, and $|\int_a^b f| \le \int_a^b |f| \le (b-a)\sup_{[a,b]}|f|$.

Proof. We leave the proof of (b) to Exercise 17.7. For (a), since f is integrable on $[a, b]$, f also is integrable on $[a, c]$ and on $[c, b]$ (Exercise 17.5). Hence we can find a sequence $\langle P \rangle$ of partitions of $[a, c]$ and a sequence $\langle Q \rangle$ of partitions of $[c, b]$ such that

$$\lim L(f, P_n) = \lim U(f, P_n) = \int_a^c f \quad \text{and} \quad \lim L(f, Q_n) = \lim U(f, Q_n) = \int_c^b f.$$

Let R_n be the partition of $[a, b]$ whose set of breakpoints is the union of the sets of breakpoints of P_n and Q_n. We have

$$L(f, R_n) = L(f, P_n) + L(f, Q_n) \quad \text{and} \quad U(f, R_n) = U(f, P_n) + U(f, Q_n).$$

Hence both $L(f, R_n)$ and $U(f, R_n)$ converge to $\int_a^c f + \int_c^b f$, proving (a). ∎

Propositions 17.12 and 17.13 justify defining area using integrals; they yield some of the properties we believe about area. Propositions 17.12a and 17.13a express the area of a region as the sum of the areas of two regions composing it. Proposition 17.12b explains how area behaves under a vertical change of scale (along with a change of orientation if $c < 0$). The statement of how area behaves under a horizontal change of scale is a special case of the change of variables formula (Theorem 17.20). Proposition 17.13b generalizes the discrete triangle inequality $|\Sigma x_i| \le \Sigma |x_i|$ to integrals ("continuous sums").

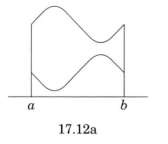
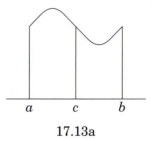

17.12a	17.13a

Not all bounded functions are integrable. For example, the function that is 1 on the rationals and 0 on the irrationals is not integrable on any interval (Exercise 17.6). We next prove that every continuous function is integrable. Every monotone function is integrable, regardless of whether it is continuous (Exercise 17.9).

17.14. Theorem. If f is continuous on the interval $[a, b]$, then f is integrable on $[a, b]$.

Proof. By Theorem 15.23 and Theorem 15.29, every function that is continuous on $[a, b]$ is bounded and uniformly continuous on $[a, b]$. Given $\varepsilon > 0$, we seek a partition P such that $U(f, P) - L(f, P) < \varepsilon$. Uniform continuity on $[a, b]$ yields a number $\delta > 0$ such that $t', t \in [a, b]$ with $|t' - t| < \delta$ implies $|f(t') - f(t)| < \dfrac{\varepsilon}{b-a}$. For $n > (b - a)/\delta$, we let P be the partition of $[a, b]$ into n subintervals $J_1, \ldots, J_n$ of equal length. Since each subinterval has length $(b - a)/n$, the elements of J_i differ by less than δ. Hence $\sup_{J_i} f - \inf_{J_i} f < \dfrac{\varepsilon}{b-a}$. By summing the contributions from each J_i (length times the maximum difference in heights), we have $U(f, P) - L(f, P) < \dfrac{b-a}{n} \sum_{i=1}^{n} \dfrac{\varepsilon}{b-a} = \varepsilon$. Hence P has the desired property, which proves that f is integrable on $[a, b]$. ∎

More advanced analysis considers other types of integrals. Our definition is the *Riemann integral*, due to G.F.B. Riemann (1826-1866). This definition applies only to bounded intervals and to bounded functions. "Improper integrals" sometimes overcome these limitations.

17.15. Definition. *Improper integrals.* We define $\int_a^\infty f(x)dx$ to be $\lim_{b \to \infty} \int_a^b f(x)dx$, if this limit exists. When f is unbounded at a, we define $\int_a^b f(x)dx$ to be $\lim_{\varepsilon \to 0} \int_{a+\varepsilon}^b f(x)dx$, if this limit (through positive values of ε) exists.

The definition of improper integral for unbounded intervals is analogous to the definition of infinite series using partial sums. Many important functions in mathematics can be expressed as improper integrals. For example, after we have defined the exponential function, we can show that $n! = \int_0^\infty e^{-x}x^n dx$ (Exercise 17.32).

An example of the second type of improper integral, when $-1 < \alpha < 0$, is $\int_0^1 x^\alpha dx$. We compute $\int_\varepsilon^1 x^\alpha dx = \dfrac{1}{\alpha + 1} - \dfrac{\varepsilon^{\alpha+1}}{\alpha + 1}$. As $\varepsilon \to 0$, this approaches $1/(\alpha + 1)$. The evaluation of $\int_\varepsilon^1 x^\alpha dx$ uses the Fundamental Theorem of Calculus, which we develop next.

THE FUNDAMENTAL THEOREM OF CALCULUS

The Fundamental Theorem of Calculus states precisely the sense in which differentiation and integration are inverse operations. This is the basis for the method of antiderivatives used to find indefinite integrals. From the Fundamental Theorem we obtain several techniques of integration, such as change of variables and integration by parts.

We treat $\int_a^x f(t)dt$ as a function of x, calling it $F(x)$. Thus $F(x)$ is the area under the graph of f from a to x. Intuitively, this area changes at the instantaneous rate $f(x)$. The first form of the Fundamental Theorem of Calculus makes this precise, stating that a continuous function is the derivative of its integral. In Theorem 17.18 we prove a second form, stating that a continuously differentiable function is the integral of its derivative.

17.16. Theorem. (Fundamental Theorem of Calculus) Suppose f is integrable on $[a,b]$, and let $F(x) = \int_a^x f(t)dt$ for $a < x < b$. If f is continuous at x, then F is differentiable at x, and $F'(x) = f(x)$.

Proof. Because f is integrable on $[a,b]$, it is also integrable on $[a,x]$ (Exercise 17.4), so $F(x)$ exists for each x in (a,b). To prove that $F'(x) = f(x)$, we show that $F(x+h) = F(x) + hf(x) + e(h)$, where e is an error function. Using Proposition 17.12a, we see that $e(h) = F(x+h) - F(x) - hf(x) = \int_x^{x+h} f(t)dt - hf(x)$. Since $f(x)$ is independent of t, we have $hf(x) = h \frac{1}{h} \int_x^{x+h} f(x)dt$. Thus $e(h) = \int_x^{x+h}[f(t) - f(x)]dt$.

To show that e is an error function, we prove that $e(h)/h \to 0$. Let J be the interval $[x, x+h]$ if $h > 0$ and $[x+h, x]$ if $h < 0$; observe that the length of J is $|h|$. Using Proposition 17.13b, we see that $|e(h)/h| = \frac{1}{|h|}|\int_x^{x+h}(f(t) - f(x))dt| \leq \sup_{t \in J}|f(t) - f(x)|$. Since f is continuous at x, this converges to 0 as $h \to 0$. ∎

17.17. Example. *Necessity of continuity.* In the Fundamental Theorem of Calculus, we cannot drop the hypothesis that f is continuous at x. Suppose $f(x) = 1$ for $0 \leq x \leq 1$ and $f(x) = -1$ for $-1 \leq x < 0$. Then f is integrable on $[-1, 1]$; let $F(x) = \int_{-1}^x f(t)dt = |x| - 1$. The function f is not continuous at 0, and F is not differentiable at 0. ∎

We define the *average value* of a function f integrable on $[a, b]$ to be $\frac{1}{b-a}\int_a^b f$. Theorem 17.16 states that the slope $m_{x,x+h}$ for the function F converges to $f(x)$ as $h \to 0$. The proof is similar to our proof about the sequence of averages in Example 14.20. To study the sequence $(1/n)\Sigma_{i=1}^n a_i$ formed from a convergent sequence $\langle a \rangle$, we defined $c_n = (1/n)\Sigma_{i=1}^n(a_i - a_n)$. In both cases, we subtracted the average.

The second form of the Fundamental Theorem states that the slope $m_{a,x}$ for F equals the average value of F' on the interval $[a, x]$.

17.18. Theorem. (Fundamental Theorem of Calculus - second form) If F is continuously differentiable on an open interval containing $[a, b]$, then $\int_a^x F'(t)dt = F(x) - F(a)$ for all $x \in [a, b]$.

Proof. Let $G(x) = -F(x) + \int_a^x F'(t)dt$, which is well-defined since F' is integrable. By the first version of the Fundamental Theorem, G is differentiable, and $G'(x) = -F'(x) + F'(x) = 0$ for all $x \in [a, b]$. By Corollary 16.29, G is a constant. Setting $x = a$ in the definition of G yields $G(a) = -F(a)$. Since $G(x) = G(a)$ for all x, we conclude that $-F(a) = -F(x) + \int_a^x F'(t)dt$. ∎

The Fundamental Theorem of Calculus enables us to evaluate the limit of sums that is $\int_a^x f$ by finding a function F whose derivative is f.

17.19. Example. *Integral of the pure powers.* Since we know that $(d/dx)x^{p+1}/(p+1) = x^p$, the Fundamental Theorem of Calculus allows us to compute $\int_0^1 x^p$ by taking the difference between the values at $x = 1$ and $x = 0$ of $x^{p+1}/(p+1)$. This avoids the limit computation of Example 17.10. As before, the result is $1/(p+1)$. ∎

It is convenient to write $F(b) - F(a) = F(x)|_a^b$, which we read as "$F(x)$ evaluated from $x = a$ to $x = b$." Since functions with the same derivative differ by a constant, we sometimes introduce an additive constant C and informally say "the *indefinite* integral of $f(x)$ is $F(x) + C$." The notation $F(x) + C$ represents an equivalence class of functions under the relation where two functions are equivalent when they differ by a constant. Each representative of the appropriate equivalence class leads to the same answer when we compute a definite integral of f.

The Fundamental Theorem of Calculus implies the change of variables formula for a definite integral.

17.20. Theorem. Suppose that f is continuous on $[a, b]$ and that g is a continuously differentiable bijection from the interval $[a, b]$ to the interval $[g(a), g(b)]$. For every $x \in [a, b]$,
$$\int_a^x f(g(z))g'(z)dz = \int_{g(a)}^{g(x)} f(t)dt.$$

Proof. Both sides depend on a and x. Fixing a, we view them as functions of x. By the Fundamental Theorem of Calculus, both these functions are differentiable (using the continuity of $(f \circ g) \cdot g'$). We show they are equal by showing that they have the same derivative and that they agree at $x = a$ (equality then follows from Corollary 16.29).

Both functions are 0 at $x = a$. The derivative of the function on the left is $f(g(x))g'(x)$, by the Fundamental Theorem. The derivative of the function on the right is also $f(g(x))g'(x)$, by the Fundamental Theorem and the chain rule. ∎

The product rule for differentiation combines with the Fundamental Theorem of Calculus to yield an important technique of integration.

17.21. Theorem. (Integration by Parts) If u and v are continuously differentiable, then

$$\int_a^b v \cdot u' = (uv)\big|_a^b - \int_a^b u \cdot v',$$

where $(uv)\big|_a^b = u(b)v(b) - u(a)v(a)$.

Proof. Let $F = uv$. By the product rule, $F' = u'v + uv'$, which is continuous. By the Fundamental Theorem of Calculus, $\int_a^b [u'v + uv'] = F(b) - F(a)$. Subtracting $\int_a^b u \cdot v'$ from both sides completes the proof. ∎

Integration by parts has a particularly nice geometric interpretation for monotone functions. A monotone continuous function is a bijection from the domain interval to its image, and therefore it has an inverse function.

17.22. Theorem. If f is a monotone increasing function, with inverse function f^{-1}, then $\int_a^b f(x)dx = yf^{-1}(y)\big|_s^t - \int_s^t f^{-1}(y)dy$, where $s = f(a)$ and $t = f(b)$.

Proof. By translating the origin if necessary, we may reduce the computation to the case where a and s are positive; similarly we may assume $b \geq a$ and $t \geq s$ (Exercise 17.19). It suffices to prove that $\int_a^b f(x)dx + \int_s^t f^{-1}(y)dy = yf^{-1}(y)\big|_s^t$. Both sides compute the same area, as suggested by the picture below. The right side is the area of the difference between the two rectangles. The left side computes this as the sum of two pieces, separated by the graph of f.

If $xf'(x)$ is integrable as a function of x, then we can also prove this using integration by parts:

$$\int_a^b f(x)dx = xf(x)\big|_a^b - \int_a^b xf'(x)dx = f^{-1}(y)y\big|_s^t - \int_s^t f^{-1}(y)dy. \quad ∎$$

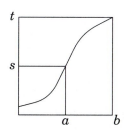

EXPONENTIALS AND LOGARITHMS

We are now prepared to define the logarithm and with it the expo-nential function. We use the Fundamental Theorem of Calculus to derive their basic properties.

17.23. Definition. For $x > 0$, the *natural logarithm* of x (written $\ln x$) equals $\int_1^x \frac{1}{t}\,dt$.

17.24. Theorem. The natural logarithm is a strictly increasing func-tion having the property that $\ln(xy) = \ln x + \ln y$.

Proof. Since $1/t$ is continuous for $t > 0$, the integral $\int_1^x \frac{1}{t}\,dt$ exists as a function of x. By the Fundamental Theorem of Calculus, its derivative is $1/x$, which is positive, and hence $\ln x$ is increasing.

To prove $\ln(xy) = \ln x + \ln y$, we consider $\ln(xy) - \ln x$ as a function of x for fixed y and show it has the constant value $\ln y$. To show first that $\ln(xy) - \ln x$ is a constant, we compute its derivative with respect to x. Using the chain rule, we know that $(d/du)\ln u = 1/u$ by the definition of the logarithm and the Fundamental Theorem of Calculus, and we have $(d/dx)(xy) = y$. Hence $(d/dx)\ln(xy) = (1/xy)y = 1/x$, and $(d/dx)\ln x = 1/x$, so $(d/dx)\ln(xy) - \ln x = 0$. A function with 0 derivative is a constant, so we have $\ln(xy) - \ln x = c$. Since $\ln 1 = 0$, setting $x = 1$ in $\ln(xy) - \ln x = c$ yields $\ln y = c$. ∎

The property $\ln(xy) = \ln x + \ln y$ implies that $\ln x$ is unbounded above as x increases and unbounded below as x approaches 0 (Exercise 17.14). Since the function also is strictly increasing, it is a bijection from the set of positive real numbers to $\mathbb{R}$. We give a clean definition of exponentiation by using the inverse of the logarithm function.

17.25. Definition. The *exponential function* is the bijection from $\mathbb{R}$ to the set of positive real numbers that is the inverse of the loga-rithm function. The value of the exponential function at x, written e^x, is the number $y > 0$ such that $x = \ln y$. For $a > 0$, we define a^x to be $e^{x \ln a}$.

In particular, the real number we call "e" is the unique y such that $\int_1^y (1/t)\,dt = 1$. The notation for the exponential function is motivated by the elementary property $e^{a+b} = e^a e^b$, which follows from Theorem 17.24 (see Exercise 17.16).

Many applied problems involve a "feedback" mechanism in which the growth rate of a time-dependent function is proportional to its value. These include compounding of interest, analysis of current in electrical circuits, and problems of radioactive decay. The exponential function arises in such problems because it is its own derivative.

17.26. Proposition. If f is differentiable and strictly monotone, then f^{-1} exists and is differentiable, and the derivative of f^{-1} at y is $1/f'(f^{-1}(y))$.

Proof. Since f is strictly monotone function, the inverse function exists (Proposition 3.34). The chain rule allows us to differentiate both sides of $y = f(f^{-1}(y))$. Exercise 17.20 requests the details. ∎

17.27. Corollary. If $g(y) = e^y$, then $g'(y) = g(y)$. Furthermore, if $a > 0$ and $h(y) = a^y$, then $h'(y) = a^y \ln a = h(y)\ln(a)$.

Proof. The function g is defined to be the inverse of ln, which is strictly increasing. By the Fundamental Theorem of Calculus, $(d/dx)\ln x = 1/x$. Letting $f(x) = \ln x$, we have $g = f^{-1}$, and $g'(y) = 1/f'(g(y)) = e^y$. The second statement follows from the definition of a^y, the first statement, and the chain rule. ∎

The differential equation $g'(x) = cg(x)$ is analogous to the recurrence $a_{n+1} - a_n = ca_n$. In each case, the change is proportional to the function value. The recurrence has solution $a_n = A(1 + c)^n$, where $A = a_0$ (Theorem 12.14), and the differential equation has solution $g(x) = Ae^{cx} = A(e^c)^x$, where $A = g(0)$. The analogy extends to higher-order constant coefficient differential equations. The constant coefficient second-order differential equation $g''(x) - (\alpha + \beta)g'(x) + \alpha\beta g(x) = 0$ has solution $g(x) = A_1 e^{\alpha x} + A_2 e^{\beta x}$ when $\alpha \neq \beta$. Here again we need two initial conditions to determine A_1 and A_2. We will not explore the vast subject of differential equations.

One simple application of the exponential function is its role in the compounding of interest. This application requires the evaluation of a certain limit, which is sometimes given as the definition of e^x.

17.28. Theorem. If $x \in \mathbb{R}$ and $a_n = (1 + x/n)^n$, then $a_n \to e^x$.

Proof. Letting $t = 1/n$, we consider $(1 + xt)^{1/t}$. For $t \neq 0$, this is a continuous function of t. By the sequential definition of convergence, $\lim_{n \to \infty} a_n$ must be the limit of this function as $t \to 0$, if it exists. Let $f(t) = \ln((1 + tx)^{1/t}) = \frac{\ln(1 + tx)}{t}$. We can use l'Hôpital's rule (Theorem 16.32) to evaluate $\lim_{t \to 0} f(t) = \lim_{t \to 0} \frac{x}{1 + tx} = x$. Thus $\ln(a_n) \to x$. Since the exponential function is continuous, $a_n \to e^x$. ∎

17.29. Solution. *Continuous Compounding.* The problem of determining the yield under continuous compounding of interest is solved by determining the limit of $(1 + x/n)^n$. This is the factor by which the original amount ("principal") in an account with interest rate x is multiplied after one year under continuous compounding. The limit is e^x. The factor is 1 plus the yield, so the yield is $e^x - 1$. For small x, the yield is approximately $1 + x + x^2/2$. ∎

The alert reader will recall that we defined $\exp(x)$ in Example 14.33 using series. We prove in Corollary 17.48 that this is the same function as the exponential function defined in Definition 17.24. To do so, we develop machinery allowing us to differentiate a power series term-by-term. The same technique yields the definitions and properties of the trigonometric functions sine and cosine, which we discuss before presenting the technical details of differentiating term-by-term.

TRIGONOMETRIC FUNCTIONS AND π

The fundamental constant π and the sine and cosine functions arise throughout mathematics and science; π is the area inside the unit circle, and the trigonometric functions describe the relationship between the sides and the angles of a right triangle with hypotenuse 1. Because geometric reasoning is difficult to make precise, we define π and these functions using series. This leads to a proof that π is the area of the unit circle. It is possible to define the sine and cosine functions rigorously using geometric reasoning; this also would allow us to prove Proposition 17.31 and from it the series expansions that we use as the definitions in 17.30. Our approach is direct but not geometric.

17.30. Definition. The *sine* and *cosine* functions are defined on $\mathbb{R}$ by

$$\sin x = \Sigma_{n=0}^{\infty} \frac{(-1)^n}{(2n+1)!} x^{2n+1}$$

$$\cos x = \Sigma_{n=0}^{\infty} \frac{(-1)^n}{(2n)!} x^{2n}$$

17.31. Proposition. The sine and cosine functions are defined and differentiable on all of $\mathbb{R}$. The derivatives are $(d/dx)\sin x = \cos x$ and $(d/dx)\cos x = -\sin x$.

Proof. These functions are defined for all x because the ratio test gives an infinite radius of convergence for the series. For a fixed x, the absolute value of ratio of successive terms in the series is $\dfrac{x^2}{(2n+3)(2n+2)}$ for $\sin x$ and $\dfrac{x^2}{(2n+2)(2n+1)}$ for $\cos x$; both converge to 0. To show that sine and cosine are differentiable, we would like to interchange the order of differentiation and summation so that we can differentiate the series term-by-term. Theorem 17.47 allows us to do this. When we differentiate term-by-term, the series for $\sin x$ becomes the series for $\cos x$, and the series for $\cos x$ becomes the series for $-\sin x$. ∎

17.32. Proposition. $\sin^2 x + \cos^2 x = 1$ for all $x \in \mathbb{R}$.

Proof. From the series definition, we obtain $\sin 0 = 0$ and $\cos 0 = 1$. Now let $f(x) = \sin^2 x + \cos^2 x$. Using the chain rule, we compute $f'(x) = 2 \sin x \cos x - 2 \cos x \sin x = 0$. By Corollary 16.29, f is constant, so $f(x) = f(0) = 1$ for all x. ∎

17.33. Corollary. The sine and cosine functions are bounded, with $|\sin x| \le 1$ and $|\cos x| \le 1$ for all x.

Proof. Immediate from Proposition 17.32. ∎

17.34. Proposition. There is a point $x_0 > 0$ such that $\cos x_0 = 0$.

Proof. Since $\cos 0 = 1 > 0$ and differentiability implies continuity, it suffices by the Intermediate Value Theorem to show that there exists a positive x such that $\cos x < 0$. We show that $\cos 2$ is negative. From the definition, $\cos 2 = 1 - \frac{2^2}{2} + \frac{2^4}{24} + \Sigma_{n=3}^{\infty} \frac{(-1)^n}{(2n)!} x^{2n}$. The first three terms sum to $-1/3$. We consider the remaining terms in pairs. The two terms with $n = 2k - 1$ and $n = 2k$ sum to

$$\frac{(-1)^{2k-1}}{(4k-2)!} 2^{4k-2} + \frac{(-1)^{2k}}{(4k)!} 2^{4k} = -\frac{2^{4k-2}}{(4k-2)!} \left(1 - \frac{2^2}{4k(4k-1)}\right).$$

This is negative when $k(4k - 1) > 1$. Hence each successive pair of the remaining terms is negative, and $\cos 2 < -1/3$. ∎

Thus the set $S = \{x > 0 : \cos x = 0\}$ is non-empty. Since S is bounded below by 0, it has an nonnegative infimum α. The set S contains a sequence $\langle x \rangle$ converging to α. Since the cosine function is continuous, $\cos \alpha = \cos(\lim x_n) = \lim(\cos x_n) = 0$. Since $\cos 0 = 1 \ne 0$, we have $\alpha > 0$.

17.35. Definition. The number π is defined to be 2α, where α is the smallest positive zero of $\cos x$.

Since $\cos 2 < 0$, we have shown that $\pi < 4$. We will soon obtain more accurate estimates for π. First we relate π to the area of a circle.

17.36. Lemma. $\int_0^{\pi/2} \sin^2 x \, dx = \int_0^{\pi/2} \cos^2 x \, dx = \pi/4$

Proof. By Proposition 17.32, we have $\int_0^{\pi/2} \sin^2 x \, dx = \int_0^{\pi/2}(1 - \cos^2 x) \, dx = \pi/2 - \int_0^{\pi/2} \cos^2 x \, dx$. Hence the two desired integrals sum to $\pi/2$.

Using integration by parts (Theorem 17.21), we have

$$\int_0^{\pi/2} \cos x (\cos x \, dx) = \sin x \cos x \big|_0^{\pi/2} + \int_0^{\pi/2} \sin^2 x \, dx.$$

Since $\sin 0 = 0$ and $\cos(\pi/2) = 0$, the first term is 0. Hence the two integrals in the statement of the lemma are equal. Since they sum to $\pi/2$, each equals $\pi/4$. ∎

17.37. Proposition. The functions mapping x to $\sin x$ and x to $\cos x$ are bijections from $[0, \pi/2]$ to $[0, 1]$.

Proof. By the definition of π, the sine function is increasing and the cosine function is decreasing on the interval $[0, \pi/2]$; hence they are injective. Since the values are 0 and 1 at the endpoints of $[0, \pi/2]$, the Intermediate Value Theorem implies that they are surjective. ∎

17.38. Theorem. The area inside a circle of radius 1 is π.

Proof. Consider a circle with center at the origin. The circle is defined to be $\{(u, v) \in \mathbb{R}^2 : u^2 + v^2 = 1\}$. By symmetry, the area is four times the area inside the quarter circle in the first quadrant. This is the area enclosed by the axes and by the curve $v = \sqrt{1 - u^2}$. Because $\sin x$ defines a bijection from $[0, \pi/2]$ to $[0, 1]$, we can let $u = \sin x$ and use the change of variables formula (Theorem 17.20) and Lemma 17.36 to compute $\int_0^1 \sqrt{1 - u^2} \, du = \int_0^{\pi/2} \cos^2 x \, dx = \pi/4$. ∎

17.39. Solution. *The Rainfall Problem.* If the rain falls uniformly, the fraction of the rain that falls inside the circle is the ratio of the area of the circle to the area of the square. The sides of the square are twice as long as the radius of the circle. By our understanding of scale factors in area, the ratio of the two areas remains the same no matter what the radius is. When the radius is 1, the area of the circle is π and the area of the square is 4. Hence the answer is $\pi/4$. ∎

The value of π has been computed to millions of decimal places. To ten places it is 3.1415926535. The methods we have developed yield some crude estimates for π. First consider the formula we have using the integral: $\pi = 4 \int_0^1 \sqrt{1 - x^2} \, dx$. We can use the definition of integration to approximate π. Let P be the partition of $[0, 1]$ into 100 equal pieces. For $f(x) = \sqrt{1 - x^2}$, we have $L(f, P) = (1/100^2) \Sigma_{k=1}^{100} (100^2 - k^2)^{1/2}$ and $U(f, P) = (1/100^2) \Sigma_{k=0}^{99} (100^2 - k^2)^{1/2}$. Using a calculator, we obtain $4L(f, P) = 3.12042$ and $4U(f, P) = 3.16042$. The convergence is slow.

We can also obtain a series that converges to π. Using the change of variables $x = \sin y / \cos y$, we have $\int_0^1 (1 + x^2)^{-1} dx = \int_0^{\pi/4} dy = \pi/4$. Since $|1/(1 + x^2)| < 1$ for $x > 0$, we can expand $|1/(1 + x^2)|$ using a geometric series to obtain $\Sigma_{n=0}^{\infty}(-x^2)^n$ as the integrand. Theorem 17.40 justifies the interchange of the summation and integration operations (applied with f_n equal to the nth partial sum of the series). Integrating term-by-term, we obtain $\pi/4 = \Sigma_{n=0}^{\infty}(-1)^n \int_0^1 x^{2n}dx = 1 - \frac{1}{3} + \frac{1}{5} - \frac{1}{7} + \cdots$. Taking 100 terms in the sum and multiplying by 4 yields the approximation 3.15149, which differs from π by about .01. Again convergence is slow.

A more geometric method is to inscribe regular n-gons in a circle of radius one and find their areas. Taking the limit as $n \to \infty$ yields the area of the circle, π. This approach is due to Pythagoras and also converges slowly. There are many methods for computing the decimal expansion of π, some of which converge much more rapidly than these.

A RETURN TO INFINITE SERIES

Differentiation, integration, and evaluation of series all involve limits. In general, one cannot interchange the order of limiting operations. We will use uniform convergence to prove sufficient conditions for interchanging the order of limit operations. That sometimes permits us to differentiate and integrate series term-by-term.

17.40. Theorem. Suppose $\{f_n\}$ is a sequence of continuous functions on an interval $[a,b]$, and suppose $x \in [a,b]$. If $\{f_n\}$ converges uniformly to f on $[a,b]$, then $\int_a^x f_n(t)dt \to \int_a^x f(t)dt$.

Proof. We prove that $\int_a^x f_n(t)dt - \int_a^x f(t)dt$ converges to 0. Consider an arbitrary $\varepsilon > 0$. Applying Propositions 17.12 and 17.13b, we have

$$|\int_a^x f_n(t)dt - \int_a^x f(t)dt| \le \int_a^x |f_n(t) - f(t)|dt.$$

Using uniform convergence, we can choose N such that $n > N$ and $t \in [a,b]$ implies $|f_n(t) - f(t)| < \varepsilon/(b-a)$. We conclude that $n > N$ implies $\int_a^x |f_n(t) - f(t)|dt < \varepsilon$, as desired. ■

17.41. Theorem. Suppose $\{F_n\}$ is a sequence of continuously differentiable functions on an interval $[a,b]$, and suppose $F_n(a)$ converges. If $\{F_n'\}$ converges uniformly to f on $[a,b]$, then $\{F_n\}$ converges to a continuously differentiable function F whose derivative is f.

Proof. Using the Fundamental Theorem of Calculus, we may write $F_n(x) - F_n(a) = \int_a^x F_n'(t)dt$. By Theorem 17.40, the right side converges to $\int_a^x f(t)dt$. By the Fundamental Theorem, this is differentiable, and its derivative is f. Furthermore, since the sequence $\{F_n'\}$ of continuous functions converges uniformly to f, f is continuous. Hence we have shown that $F_n(x) - F_n(a)$ converges to a continuously differentiable function whose derivative is f. Since by hypothesis $F_n(a)$ converges, we conclude that $F_n(x)$ also converges. ∎

17.42. Lemma. Let $g_n(x) = a_n x^n$. If $\Sigma_{n=0}^{\infty} a_n r^n$ converges and $0 < p < r$, then $\Sigma_{n=0}^{\infty} g_n$ converges uniformly on the interval $[-p, p]$.

Proof. Since $\Sigma_{n=0}^{\infty} a_n r^n$ converges, $a_n r^n \to 0$. Hence there is a constant C such that $|a_n r^n| \le C$ for all n. To prove that $\Sigma_{n=0}^{\infty} g_n$ converges uniformly, it suffices to show that the sequence of partial sums is uniformly Cauchy (by Lemma 16.53).

Let $f_n = \Sigma_{k=0}^{n} g_k$. Given n, m, x with $n > m$ and $|x| \le p$, we have

$$|f_n(x) - f_m(x)| = |\Sigma_{k=m+1}^{n} a_k x^k| \le \Sigma_{k=m+1}^{n} |a_k| |x^k|$$

$$= \Sigma_{k=m+1}^{n} |a_k| |x/r|^k |r|^k \le C |p/r|^{m+1} \Sigma_{j=0}^{\infty} |p/r|^j.$$

Since $|p/r| < 1$, the geometric series converges to a constant D. Given $\varepsilon > 0$, we can choose N so that $|p/r|^N < \varepsilon/(CD)$, because $|p/r|^N \to 0$ as $N \to \infty$ when $|p| < |r|$. With this choice of N, we have shown that $n, m > N$ implies $|f_n(x) - f_m(x)| < \varepsilon$ for all $x \in [-p, p]$, as desired. ∎

We extend the concept of limit to sequences of real numbers that do not converge. Given a bounded sequence $\langle b \rangle$, let $L_m = \sup_{n>m} b_n$. By the Completeness Axiom, each L_m is defined. Each successive supremum is taken over a more restricted set, so $\{L_m\}$ is a decreasing sequence bounded below by $\inf\{b_n\}$. By the Monotone Convergence Theorem, $\{L_m\}$ converges.

17.43. Definition. The lim sup of a bounded sequence $\langle b \rangle$ is the number $L = \lim_{m \to \infty} \sup_{n>m} b_n$. We abbreviate this to $L = \limsup b_n$. When $\langle b \rangle$ has no upper bound, we say that $\limsup b_n$ is infinite.

17.44. Example. If $a_n = (-1)^n + 1/n$, then $\lim a_n$ does not exist, but $\limsup a_n = 1$. ∎

Observe that if $\lim b_n = \rho$, then also $\limsup b_n = \rho$. More generally, $\limsup b_n$ is the supremum of the limits of all convergent subsequences of $\langle b \rangle$. Thus the lim sup of a bounded sequence always exists. This enables us to extend the root test for convergence of a series.

17.45. Theorem. (Root test) Suppose $\langle a \rangle$ is a sequence such that $L = \limsup |a_n|^{1/n}$. If $L < 1$, then $\Sigma_{k=1}^{\infty} a_k$ converges. If $L > 1$ (or L is infinite), then $\Sigma_{k=1}^{\infty} a_k$ diverges.

Proof. The proof is analogous to the proof of the root test for $\lim |a_n|^{1/n}$ (Exercise 17.31). ∎

We apply the test to discuss convergence of power series. The quantity $1/L$ in this discussion is often written as R and called the *radius of convergence*.

17.46. Corollary. Suppose F is a power series defined by $F(x) = \sum_{n=0}^{\infty} c_n x^n$. Let $L = \limsup |c_n|^{1/n}$. If $L = 0$, then $F(x)$ converges for all x; if L is infinite, then $F(x)$ converges only for $x = 0$. Suppose L is a positive real number. If $|x| < 1/L$, then $F(x)$ converges. If $|x| > 1/L$, then $F(x)$ diverges.

Proof. Let $a_n = c_n x^n$, so $|a_n|^{1/n} = |c_n|^{1/n}|x|$, and $\limsup |a_n|^{1/n} = L|x|$. The claim now follows by applying the root test above to test the convergence of $\sum_{n=0}^{\infty} a_n$. ∎

17.47. Theorem. Suppose $F(x) = \sum_{n=0}^{\infty} c_n x^n$ is a power series in x. If $F(x)$ converges whenever $|x| \le r$, then F is differentiable at all x with $|x| < r$, and $F'(x) = \sum_{n=1}^{\infty} c_n n x^{n-1}$.

Proof. Let $F_n(x) = \sum_{k=0}^{n} c_k x^k$, and consider the interval $I = [-p, p]$ for some $0 < p < r$. Since $F(x)$ converges when $|x| \le r$, we know that F_n converges pointwise to F; in particular, $F_n(-p)$ converges. Each F_n is a finite sum of continuously differentiable functions and hence is continuously differentiable. Since $F_n \to F$, it suffices by Theorem 17.41 to show that $\{F_n'\}$ converges uniformly to f on I, where $f(x) = \sum_{k=1}^{\infty} c_k k x^{k-1}$.

Let $a_k = k c_k$ for $k \ge 0$. Since $F_n'(x) = \sum_{k=1}^{n} a_k x^{k-1}$, it suffices by Lemma 17.42 to show that the series $\sum_{k=0}^{\infty} k c_k x^{k-1}$ converges for $|x| < r$. By Theorem 17.45 and our hypothesis about the convergence of $F(x)$, $\limsup |c_k|^{1/k} < 1/r$. Since $\limsup |k c_k|^{1/k} = \limsup |c_k|^{1/k}$, Corollary 17.46 guarantees that $f(x)$ also converges for $|x| < r$. ∎

We are now ready to prove that the two definitions of the exponential function agree.

17.48. Corollary. The function defined by $\exp(x) = \sum_{k=0}^{\infty} x^k/k!$ is the inverse of the natural logarithm.

Proof. In Example 14.33, we proved that $f(x) = \sum_{k=0}^{\infty} x^k/k!$ converges for all x. Since $(1/k!)k x^{k-1} = x^{k-1}/(k-1)!$, Theorem 17.47 tells us that $f(x)$ is differentiable and $f'(x) = f(x)$.

We first show that $f(x) > 0$ for all x. This holds for $x > 0$ because then the coefficients of the power series are positive. We also know that $f(0) = 1$. From this and $\exp(a + b) = \exp(a)\exp(b)$ (Exercise 14.44), we have $1 = f(0) = f(x - x) = f(x)f(-x)$. Hence $f(x) = 1/f(-x) > 0$ if $x < 0$.

To prove that f is the inverse of the natural logarithm, we prove that the composition $g = \ln \circ f$ is the identity function. Because f is positive, g is well-defined. Because f is differentiable, g is also differentiable. Using the chain rule, we have $g'(x) = (1/f(x))f'(x) = 1$. Also, $f(0) = 1$ implies $g(0) = 0$. By the Fundamental Theorem of Calculus, $g(x) = \int_0^x g'(t)dt = \int_0^x 1\,dt = x$ for all x. Hence g is the identity function, as desired. ∎

EXERCISES

17.1. (–) Suppose f and g are bounded real-valued functions on a set S. Prove that $\sup_S(f + g) \leq \sup_S f + \sup_S g$.

17.2. (–) Suppose $f(x) = x^2$ and $[a, b] = [1, 3]$. Let P_n be the partition of $[1, 3]$ into n subintervals of equal length. Compute $L(f, P_n)$ and $U(f, P_n)$ explicitly as functions of n. Verify that these functions have the same limit. Determine how large n must be to ensure that $U(f, P_n)$ is within $.01$ of $\int_1^3 f(x)dx$.

17.3. Suppose $f: [a, b] \to \mathbb{R}$. For partitions P, Q, R of $[a, b]$, prove that
a) $L(f, P) \leq L(f, R) \leq U(f, R) \leq U(f, P)$ when R is a refinement of P.
b) $L(f, P) \leq U(f, Q)$. (Hint: consider their least common refinement.)

17.4. Prove that if f is integrable on $[a, b]$, then $-f$ is integrable on $[a, b]$, with $\int_a^b(-f) = -\int_a^b f$. What is the geometric interpretation of $\int_a^b(f - g) = \int_a^b f - \int_a^b g$?

17.5. Suppose $a < c < b$, and suppose f is integrable on $[a, b]$. Prove that f is integrable on $[a, c]$ and on $[c, b]$.

17.6. Define $f: [0, 1] \to [0, 1]$ by $f(x) = 1$ if x is rational and $f(x) = 0$ if x is irrational. Prove that f is not integrable.

17.7. *Integrability of f and $|f|$.*
a) Prove that if f is integrable on $[a, b]$, then $|f|$ is integrable on $[a, b]$, and $|\int_a^b f| \leq \int_a^b |f| \leq (b - a)\sup_{[a,b]}|f|$. (Comment: For finite sums, the analogous inequality is $|\Sigma_{i=1}^n x_i| \leq \Sigma_{i=1}^n |x_i|$, which follows from the triangle inequality by induction.)
b) Give an example of a function f such that $|f|$ is integrable on $[0, 1]$ but f is not integrable on $[0, 1]$.

17.8. (!) *(Mean Value Theorem for integrals).* Suppose f is continuous on $[a, b]$. Prove that there is a $c \in [a, b]$ such that $f(c) = \frac{1}{b - a}\int_a^b f$. (Hint: first prove the special case when $\int_a^b f = 0$. Consider the function $f - \frac{1}{b - a}\int_a^b f$ to reduce the general statement to this case.)

17.9. (!) *Integration of monotone functions.* Suppose f is increasing on the interval $[a, b]$, and let P_n be the partition of $[a, b]$ into n intervals of equal length. Obtain a formula for $U(f, P_n) - L(f, P_n)$. Use this to show that f is integrable on $[a, b]$.

17.10. Suppose that f is a continuous function on the interval $[a, b]$.

a) Prove that if $f(x) \geq 0$ for $x \in [a, b]$ and f is not everywhere zero on $[a, b]$, then $\int_a^b f(x)dx > 0$.

b) Prove that if $\int_a^b f(t)g(t)dt = 0$ for every continuous function g on $[a, b]$, then $f(x) = 0$ for $a \leq x \leq b$.

17.11. Let $g(x) = \int_0^x (1+t^2)^{-1}dt + \int_0^{1/x}(1+t^2)^{-1}dt$. Prove that $g(x)$ is a constant. (Comment: We have shown that $\int_0^1 (1+t^2)^{-1}dt = \pi/4$, so if $g(x)$ is a constant, then the constant must be $\pi/2$.)

17.12. Suppose $f: [0, 1] \to [0, 1]$ is a bijection with $f(0) = 1$ and $f(1) = 0$. Prove that $\int_0^1 f(x)dx = \int_0^1 f^{-1}(y)dy$. Conclude that $\int_0^1 (1 - x^a)^{1/b} dx = \int_0^1 (1 - x^b)^{1/a} dx$. Evaluate the integral when a and b are reciprocals of natural numbers.

17.13. Use Theorem 17.21 to find indefinite integrals of $\ln(x)$ and $\tan^{-1}(x)$.

17.14. Use the property $\ln(xy) = \ln x + \ln y$ to prove that the logarithm function is unbounded above and below.

17.15. Suppose $f(x) = \int_0^x f(t)dt + c$. Determine f.

17.16. *Properties of exponentiation.*

a) Use the properties of the logarithm to prove that $e^{x+y} = e^x e^y$.

b) Suppose α is a fixed positive real number. Compute $(d/dx)x^\alpha$. (Hint: use Definition 17.25.)

c) Fill in the details of the proof of Corollary 17.27 to prove that $(d/dx)a^x = a^x \ln a$.

17.17. Compute the sum of the finite geometric series $\Sigma_{k=0}^n e^{kx}$. Differentiate both sides p times to prove that $\Sigma_{k=0}^n k^p$ is the value at $x = 0$ of $(\frac{d}{dx})^p \frac{1 - e^{(n+1)x}}{1 - e^x}$. (Comment: compare with Solution 9.10.)

17.18. (!) Define a sequence $\{f_n\}$ of functions by $f_n(x) = ae^{-anx} - be^{-bnx}$, where a, b are real constants with $0 < a < b$. Compute $\Sigma_{n=1}^\infty \int_0^\infty f_n(x)dx$, and compute $\int_0^\infty \Sigma_{n=1}^\infty f_n(x)dx$. (Hint: these are not equal!)

17.19. Suppose the formula $\int_a^b f(x)dx = yf^{-1}(y)|_u^v - \int_u^v f^{-1}(y)dy$ holds for every monotone increasing function f if $0 \leq a \leq b$ and $0 \leq u \leq v$, where $u = f(a)$ and $v = f(b)$. (This was proved in Theorem 17.22.) Extend the validity of the formula to all situations with $a \leq b$ and $u \leq v$ by using a translation of the origin to reduce the proof to this case.

17.20. Suppose f is a monotone differentiable function on an interval S. Prove that the inverse function f^{-1} is differentiable, with $(d/dy)f^{-1}(y) = 1/f'(f^{-1}(y))$.

316 Chapter 17: Integration

17.21. Because the exponential function is its own derivative, the integral $\int_0^1 e^x dx$ equals $e - 1$, by the Fundamental Theorem of Calculus. The steps below evaluate the integral as a limit of sums.

a) Write down the lower sum $L(f, P_n)$, where $f(x) = e^x$ and P_n is the partition of $[0, 1]$ into n equal parts.

b) Use a finite geometric sum to evaluate the sum in part (a).

c) Verify directly that $\lim_{n \to \infty} L(f, P_n) = e - 1$. (What properties of the exponential function does this use?)

17.22. (–) Determine the yields on bank accounts paying 6% simple interest, 6% interest compounded daily, and 6% compounded continuously.

17.23. (–) How many years does it take to double the value of a bank account paying 4% simple interest? How many years if the interest rate is $p\%$?

17.24. (!) Suppose $f(x) = x/\ln x$ for $x > 1$. Find the minimum value of f. Use this information to determine which is larger: π^e or e^π.

17.25. (+) Suppose that $h: \mathbb{R} \to \mathbb{R}$ and that $h(x^n) = h(x)$ for all $n \in \mathbb{N}$ and $x \in \mathbb{R}$.

a) Prove that if h is continuous at $x = 1$, then h is constant.

b) Show that without this assumption h need not be constant.

c) Suppose that $f(x^n) = nx^{n-1} f(x)$ for all $x > 0$ and all $n \in \mathbb{N}$. Suppose also that $\lim_{x \to 1} f(x)/\ln(x)$ exists. What does this imply about f?

17.26. Suppose f and g are differentiable. Compute the derivative of f^g.

17.27. *AGM Inequality.*

a) Prove that $y^a z^{1-a} \le ay + (1-a)z$ for all positive y, z and $0 \le a \le 1$. Determine when equality can hold.

b) Suppose that $x_1, \ldots, x_n$ is a list of n positive real numbers. Prove that $(\Sigma_{i=1}^n x_i)/n \ge (\Pi_{i=1}^n x_i)^{1/n}$, with equality only when $x_1 = \cdots = x_n$. (Hint: part (a) can be applied to give a proof by induction on n.)

c) Suppose $a_1, \ldots, a_n$ are nonnegative real numbers. Find the maximum of $\Pi_{i=1}^n x_i^{a_i}$ subject to $\Sigma x_i = 1$.

d) Use part (c) to give a different proof of part (b).

17.28. Use the unboundedness of $\ln x$ to prove that $\Sigma_{n=1}^\infty 1/n$ diverges.

17.29. (!) Suppose $0 < \varepsilon < 1$, and consider $\int_\varepsilon^1 \ln(x) dx$.

a) Evaluate this integral using Theorem 17.22.

b) Take the limit as $\varepsilon \to 0$ of the answer to part (a) to obtain a value for the improper integral $\int_0^1 \ln(x) dx$.

c) Use upper sums to justify the statement that $\lim_{n \to \infty} \frac{1}{n} \Sigma_{k=1}^n \ln(k/n)$ equals the answer to part (b).

d) Rewrite the expression in part (c) to demonstrate that

$$\lim_{n \to \infty} \frac{(n!)^{1/n}}{n} = \frac{1}{e}.$$

(Comment: This is a weak form of Stirling's Formula, which is used to approximate $n!$. Stirling's Formula states that $n!$ is approximately $n^n e^{-n} \sqrt{2\pi n}$.)

17.30. (+) Suppose f is continous on $[a, b]$. Compute $\lim_{n \to \infty} (\int_a^b |f|^n)^{1/n}$. (Comment: compare with Exercise 15.14.)

17.31. *Root test.* Suppose $\langle a \rangle$ is a sequence such that $L = \limsup |a_n|^{1/n}$.
   a) Prove that if $L < 1$, then $\Sigma_{k=1}^{\infty} a_k$ converges.
   b) Prove that if $L > 1$, then $\Sigma_{k=1}^{\infty} a_k$ diverges.

17.32. (!) Suppose f is continuous and nonnegative on $[0, \infty)$.
   a) Prove that $\int_0^{\infty} f(x)dx$ exists if $\lim_{x \to \infty} \dfrac{f(x+1)}{f(x)}$ exists and is less than 1.
   b) Prove that $\int_0^{\infty} f(x)dx$ exists if $\lim_{x \to \infty} (f(x))^{1/x}$ exists and is less than 1.
   c) In parts (a) and (b), prove that the integrals do not exist if the specified limits exist but exceed 1.

17.33. (!) Suppose that x, y, t are positive real numbers.
   a) Prove that

$$t^2 + t(x + y) + (\tfrac{x+y}{2})^2 \geq t^2 + t(x + y) + xy \geq t^2 + 2t\sqrt{xy} + xy.$$

   b) After taking reciprocals of the expressions in part (a), integrate from 0 to ∞ with respect to t to prove that

$$\frac{x+y}{2} \geq \frac{x-y}{\ln(x) - \ln(y)} \geq \sqrt{xy}$$

   c) Suppose $u \in \mathbb{R}$. Use part (b) to show that

$$\frac{e^u + e^{-u}}{2} \geq \frac{e^u - e^{-u}}{2u} \geq 1$$

   d) Prove part (c) directly using power series.

17.34. Suppose $n \in \mathbb{N}$. Use integration by parts to prove that $n! = \int_0^{\infty} e^{-x}x^n dx$.

17.35. The function Γ defined by $\Gamma(y) = \int_0^{\infty} e^{-x}x^{y-1}dx$ for $y > 0$ extends the notion of factorial to real arguments, with $\Gamma(n + 1) = n!$.
   a) Prove that the improper integral defining $\Gamma(y)$ is convergent when $y \geq 1$. (Hint: use Exercise 17.32a.)
   b) (+) When $0 < y < 1$, the integral defining $\Gamma(y)$ is also improper at the endpoint 0. Prove that this improper integral also converges.
   c) Prove that $\Gamma(y + 1) = y\Gamma(y)$.
   d) Given that $\Gamma(\tfrac{1}{2}) = \sqrt{\pi}$, evaluate $\int_0^{\infty} e^{-x^2} dx$.
   e) (++) Prove that $\Gamma(\tfrac{1}{2}) = \int_0^{\infty} e^{-x}x^{-1/2}dx = \sqrt{\pi}$.

Chapter 18

Complex Numbers

The complex number system extends the real number system by allowing solutions to the equation $t^2 = -1$. The resulting number system has many unexpected and useful properties essential to modern science as well as to pure mathematics. One particularly beautiful application is the Fundamental Theorem of Algebra, which states that every nonconstant polynomial with complex coefficients has a zero. We will prove this result by extending to the complex numbers the ideas we have developed about convergence.

PROPERTIES OF THE COMPLEX NUMBERS

The real Cartesian plane $\mathbb{R}^2$ becomes the Euclidean plane when we define the distance between points by the usual Euclidean distance formula. It is useful in geometry and physics to conceive of the points as vectors, but we will not do this here. Instead we will define operations of arithmetic on $\mathbb{R}^2$ to make it into a field called $\mathbb{C}$. We then call the elements of $\mathbb{C}$ "complex numbers", and the Euclidean distance from the origin to a complex number is its "magnitude". We now give the definitions of addition and multiplication that make $\mathbb{R}^2$ into a field with an element i satisfying $i^2 = -1$.

18.1. Definition. A *complex number* z is an ordered pair of real numbers. We write $z = (x, y)$ or $z = x + iy$, treating i as a formal symbol. The *sum* and *product* of complex numbers $z = (x, y)$ and $w = (a, b)$ are $z + w = (x + a, y + b)$ and $zw = (xa - yb, xb + ya)$. We write $\mathbb{C}$ for the set of complex numbers with these operations.

This definition for multiplication, written in terms of ordered pairs, is what results when we write $z = x + iy$ and $w = a + ib$, formally expand the product using the distributive law, and finally set $i^2 = -1$.

318

18.2. Example. $(1+i)^2 = (1+i)(1+i) = 1 + 2i + i^2 = 2i.$ ∎

18.3. Proposition. Under these operations of sum and product, $\mathbb{C}$ is a field. The identity element for addition is $0 + 0i$, and the identity element for multiplication is $1 + 0i$. The multiplicative inverse of $z \neq 0 + 0i$ is $(x - iy)/(x^2 + y^2)$.

Proof. (Exercise 18.1-3). ∎

The expressions $x - iy$ and $x^2 + y^2$ in the formula for z^{-1} play prominent roles in complex analysis.

18.4. Definition. Given $z = x + iy$, the *conjugate* of z is the complex number $\bar{z} = x - iy$. The *magnitude* or *absolute value* of $z = x + iy$ is $|z| = \sqrt{x^2 + y^2} = \sqrt{z\bar{z}}$, which is the distance from (x, y) to the origin. We call x the *real part* and y the *imaginary part* of the complex number $z = x + iy$, writing $x = \mathrm{Re}(z)$ and $y = \mathrm{Im}(z)$.

18.5. Remark. When $\mathrm{Im}z = y = 0$, addition and multiplication reduce to addition and multiplication of real numbers. We therefore identify $x + i0$ with $x \in \mathbb{R}$. In this sense, the field $\mathbb{R}$ is contained in the field $\mathbb{C}$.

Observe furthermore that $|x + i0|$ equals $|x|$, the ordinary absolute value of the real number x. Hence magnitude extends the notion of absolute value from real numbers to complex numbers, and this justifies calling it the absolute value. We also have $|\bar{z}| = |z|$. Using the conjugate, we can write the multiplicative inverse of z as $z^{-1} = \bar{z}/|z|^2$. ∎

The triangle inequality also holds for complex numbers.

18.6. Proposition. (Triangle Inequality). For $z, w \in \mathbb{C}$,
$$|z + w| \leq |z| + |w|.$$
Furthermore, equality holds if and only if one of these numbers is a real multiple of the other.

Proof. If $w = 0$, the inequality is trivial. Otherwise, we will prove that $|z + w|^2 \leq (|z| + |w|)^2$, which yields the desired inequality by taking positive square roots. Expanding the squares and simplifying shows that the inequality is equivalent to $\mathrm{Re}(z\bar{w}) \leq |z|\,|w|$ (Exercise 18.4).

For each real number t, we compute
$$0 \leq |z + tw|^2 = |z|^2 + 2t\,\mathrm{Re}(z\bar{w}) + t^2|w|^2.$$
We now choose $t = -\mathrm{Re}(z\bar{w})/|w|^2$. This yields
$$0 \leq |z|^2 - \frac{2(\mathrm{Re}(z\bar{w}))^2}{|w|^2} + \frac{(\mathrm{Re}(z\bar{w}))^2}{|w|^4}|w|^2 = |z|^2 - \frac{(\mathrm{Re}(z\bar{w}))^2}{|w|^2}.$$
Multiplying by $|w|^2$ yields $0 \leq |z|^2|w|^2 - (\mathrm{Re}(z\bar{w}))^2$, as desired. ∎

The triangle inequality will also extend to infinite sums, after we have defined convergence for sequences and series of complex numbers. The definitions of limit and convergence in terms of absolute value are the same in $\mathbb{C}$ as in $\mathbb{R}$.

18.7. Definition. Suppose $\langle z \rangle$ is a sequence of complex numbers. We say that $\langle z \rangle$ *converges to* L or *has limit* L (written as $z_n \to L$ or $\lim(z_n) = L$) if for every positive real number ε, there exists $N \in \mathbb{N}$ such that $n > N$ implies $|z_n - L| < \varepsilon$. A *Cauchy sequence* of complex numbers is a sequence such that for every positive real number ε, there exists $N \in \mathbb{N}$ such that $n, m > N$ implies $|z_n - z_m| < \varepsilon$.

The Cauchy Convergence Criterion (a sequence converges if and only if it is a Cauchy sequence) holds in $\mathbb{C}$ as well as in $\mathbb{R}$. This extension follows from the corresponding result in $\mathbb{R}$ and the observation that $z_n \to L$ if and only if $\operatorname{Re} z_n \to \operatorname{Re} L$ and $\operatorname{Im} z_n \to \operatorname{Im} L$ (Exercise 18.9). We apply this to study the convergence of series.

18.8. Definition. A series $\Sigma_{n=0}^{\infty} w_n$ of complex numbers *converges* if its sequence of partial sums converges. A series $\Sigma_{n=0}^{\infty} w_n$ *converges absolutely* if $\Sigma_{n=0}^{\infty} |w_n|$ converges.

As in the real case, absolute convergence implies convergence (Corollary 14.31).

18.9. Proposition. If $\langle z \rangle$ is a sequence of complex numbers such that $\Sigma_{n=0}^{\infty} |z_n|$ converges, then $\Sigma_{n=0}^{\infty} z_n$ converges, and $|\Sigma_{n=0}^{\infty} z_n| \leq \Sigma_{n=0}^{\infty} |z_n|$.

Proof. Since $|\Sigma_{n=0}^{N} z_n| \leq \Sigma_{n=0}^{N} |z_n|$ for each N, the inequality follows if $\Sigma_{n=0}^{\infty} z_n$ converges. When $\Sigma_{n=0}^{\infty} |z_n|$ converges, its sequence of partial sums is a Cauchy sequence. Hence there exists N' such that $N > M > N'$ implies $\Sigma_{n=M+1}^{N} |z_n| < \varepsilon$. This implies that

$$|\Sigma_{n=0}^{N} z_n - \Sigma_{n=0}^{M} z_n| = |\Sigma_{n=M+1}^{N} z_n| \leq \Sigma_{n=M+1}^{N} |z_n| < \varepsilon.$$

Hence the partial sums of $\Sigma_{n=0}^{\infty} z_n$ form a Cauchy sequence, and the series converges. ∎

For complex power series, one useful test of convergence is the "ratio test". The root test also applies (Exercise 18.24).

18.10. Proposition. (Ratio Test) Suppose $f(z) = \Sigma_{n=0}^{\infty} a_n z^n$ is a complex power series and that $|a_{n+1}/a_n|$ converges to L. Then $f(z)$ converges absolutely for all z with $L|z| < 1$. In particular, when $L = 0$ it converges for all z.

Proof. The ratio test for the real series $\Sigma_{n=0}^{\infty} |a_n z^n|$ considers $\frac{|a_{n+1} z^{n+1}|}{|a_n z^n|} = |\frac{a_{n+1}}{a_n}| \, |z|$. By hypothesis, the limit is $L|z|$. When this is less than 1, $f(z)$ converges absolutely. ∎

The ratio test can also be used to show divergence. The series $g(z)$ diverges when $|z| > 1/L$, but the test is inconclusive when $|z| = 1/L$ (compare Σz^n with $\Sigma(z^n/n^2)$).

For real numbers, we defined the exponential function to be the inverse of the logarithm function. We also derived a formula for the exponential function as a convergent power series. We use the power series to extend the exponential function to $\mathbb{C}$.

18.11. Definition. For every $z \in \mathbb{C}$, the value of the exponential function is defined to be $e^z = \Sigma_{n=0}^{\infty} z^n/n!$.

To show that the exponential function is well-defined, we apply the ratio test. Since $|a_{n+1}/a_n| = 1/(n+1) \to 0$, the series converges for every z. The familiar property $e^z e^w = e^{z+w}$ also extends to $\mathbb{C}$ (Exercise 18.10). These results about the exponential allow us to define the sine and cosine functions. For $\theta \in \mathbb{R}$, the complex numbers of the form $e^{i\theta}$ have magnitude 1 and form the unit circle centered at $(0,0)$ (Exercise 18.12). We define $\cos\theta = \text{Re } e^{i\theta}$ and $\sin\theta = \text{Im } e^{i\theta}$, and thus $e^{i\theta} = \cos\theta + i\sin\theta$. The series expansions for sine and cosine follow from the series for e^z:

$$\sin\theta = \theta - \theta^3/3! + \theta^5/5! - \cdots$$
$$\cos\theta = 1 - \theta^2/2! + \theta^4/4! - \cdots$$

18.12. Example. The Fourier series $\Sigma_{n=0}^{\infty}(a_n \sin(nx) + b_n \cos(nx))$ can be expressed as the imaginary part of a complex power series, using $a_n \sin(nx) + b_n \cos(nx) = \text{Im}((a_n + ib_n)e^{inx})$. ∎

When we view the complex number z as a vector, we can express (x, y) as $|z|$ times a unit vector in the direction of (x, y). This leads to the polar coordinate representation of z, which is $z = |z|e^{i\theta}$ for some real number θ, called an *argument* of z. Because $e^{i(\theta+2n\pi)} = e^{i\theta}$ for all $n \in \mathbb{Z}$, we may assume that $0 \le \theta < 2\pi$. In taking roots of complex numbers, we must consider all possible choices for the argument.

18.13. Lemma. If z is a nonzero complex number and m is a positive integer, then $w^m = z$ has the m solutions $w = |z|^{1/m} e^{i(\theta+2k\pi)/m}$ for $0 \le k \le m-1$. In the geometric view of $\mathbb{C}$, these are equally spaced on a circle centered at the origin.

Proof. (Exercise 18.13). ∎

LIMITS AND CONVERGENCE

Before discussing the Fundamental Theorem of Algebra, we must add some "topological ingredients": the study of open and closed sets.

18.14. Definition. Given a complex number w, we define the *open ball* of radius ε around w to be $\{z \in \mathbb{C}: |z - w| < \varepsilon\}$, written $B_\varepsilon(w)$. A subset S of $\mathbb{C}$ is *open* if for every $w \in S$ there exists $\varepsilon > 0$ such that $B_\varepsilon(w) \subset S$. A subset S of $\mathbb{C}$ is *closed* if $\mathbb{C} - S$ is open.

An open ball is an open set. These definitions apply equally well for real numbers, where open intervals replace open balls in the definition. Closed sets can be characterized by convergent sequences.

18.15. Theorem. A subset S of $\mathbb{C}$ is closed if and only if for every convergent sequence in S, the limit of the sequence also belongs to S.

Proof. Suppose S is closed and $\langle z \rangle$ is a sequence in S converging to L. If $L \notin S$, then by the definition of closed set there is an open ball $B_\varepsilon(L)$ that is entirely outside S. This implies $|z_n - L| > \varepsilon$ for all z_n, which contradicts the definition of convergence to L. Hence $L \in S$.

Conversely, suppose the limit of every convergent sequence of elements in S also belongs to S. If S is not closed, then the complement of S is not open. This implies there is some $L \in \mathbb{C} - S$ such that no open ball around L is contained in $\mathbb{C} - S$. In particular, for every $n \in N$, the ball $B_{1/n}(L)$ contains a point of S. We define a sequence in S by choosing $z_n \in B_{1/n}(L) \cap S$. This sequence converges to L, but $L \notin S$. From this contradiction, we conclude that S must be closed. ∎

18.16. Definition. A subset S of $\mathbb{C}$ is *bounded* if there exists a positive real number M such that $|z| \leq M$ for all $z \in S$. A subset S of $\mathbb{C}$ is *compact* if every sequence $\langle z \rangle$ of elements of S has a subsequence $\{z_{n_k}\}$ converging to a limit that belongs to S. A *closed rectangle* in $\mathbb{C}$ is a set $\{z = x + iy: a \leq x \leq b, c \leq y \leq d\}$ for some $a, b, c, d \in \mathbb{R}$.

18.17. Theorem. Every closed rectangle in $\mathbb{C}$ is compact.

Proof. (Sketch) This proof is very similar to the proof of the Bolzano-Weierstrass Theorem (Theorem 14.17). Given $\langle z \rangle$, we extract a convergent subsequence by dividing the rectangle into four subrectangles at each stage (Exercise 18.19), and choosing a subrectangle that contains z_n for infinitely many n. ∎

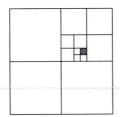

18.18. Theorem. A closed subset of a compact set in $\mathbb{C}$ is also compact.

Proof. Suppose $T \subseteq S$, with T closed and S compact. Let $\langle z \rangle$ be a sequence in T. Since $T \subseteq S$ and S is compact, $\langle z \rangle$ has a convergent subsequence $\{z_{n_k}\}$ whose limit belongs to S. Since T is closed, the limit is also in T. Hence T is compact. ∎

18.19. Theorem. A subset of $\mathbb{C}$ is compact if and only if it is closed and bounded.

Proof. Suppose first that S is closed and bounded. Because S is bounded, it is a subset of a closed rectangle. By the preceding two theorems, S is compact.

We prove the converse by contradiction. Suppose S is compact. If S is not closed, then S contains a convergent sequence $\langle z \rangle$ whose limit L is not in S. Every subsequence of $\langle z \rangle$ also converges to L. Since $L \notin S$, this violates the definition of compactness for S.

If S is not bounded, then we can define $\langle z \rangle$ by letting z_n be an element of S with magnitude greater than n. This has no convergent subsequence, again violating the definition of compactness. ∎

The definitions of limit and continuity for functions of a complex variable are much like those for functions of a real variable.

18.20. Definition. A complex-valued function f defined on an open ball around p is *continuous* at p if for each $\varepsilon > 0$ there exists $\delta > 0$ so that $|z - p| < \delta$ implies $|f(z) - f(p)| < \varepsilon$. In other words, for each $\varepsilon > 0$ there exists $\delta > 0$ so that $f(B_\delta(p)) \subseteq B_\varepsilon(f(p))$.

18.21. Proposition. Suppose $f \colon \mathbb{C} \to \mathbb{C}$. The following four statements are equivalent:
A) f is continuous.
B) for every open set T, $I_f(T)$ is open.
C) for every closed set T, $I_f(T)$ is closed.
D) for each sequence $\langle z \rangle$, $z_n \to w$ implies $f(z_n) \to f(w)$.

Proof. (Exercise 18.21). ∎

Since the interval $[0, r]$ is closed in $\mathbb{R}$, the set $\{z \in \mathbb{C} : |f(z)| \leq r\}$ is closed whenever f is a continuous function from $\mathbb{C}$ to $\mathbb{R}$.

18.22. Theorem. Let f be a real-valued function that is defined and continuous on a compact subset S of $\mathbb{C}$. Then S contains elements at which f achieves its maximum and its minimum on S.

Proof. We first prove that f is bounded. If not, then for each $n \in \mathbb{N}$ we can find z_n such that $|f(z_n)| > n$. Since S is compact, $\langle z \rangle$ has a convergent subsequence $\{z_{n_k}\}$. By statement (D) of Proposition 18.21, $f(z_{n_k})$ converges. This contradicts $|f(z_{n_k})| > n_k$.

We mimic the proof of the Minimum-Maximum Theorem (Theorem 15.25). Since f is bounded and real-valued, the set $f(S)$ has a supremum β. By the elementary properties of the supremum (Exercise 13.3), $f(S)$ contains a sequence $\langle y \rangle$ converging to β. Let $\langle z \rangle$ be a sequence such that $f(z_n) = y_n \to \beta$. By compactness, the sequence $\langle z \rangle$ has a convergent subsequence $\{z_{n_k}\}$. Let $w = \lim z_{n_k}$. Since f is continuous, $f(w) = \beta$. The proof for achieving the minimum is similar (Exercise 18.22). ∎

THE FUNDAMENTAL THEOREM OF ALGEBRA

For the proof of the Fundamental Theorem of Algebra, we need to consider infinite limits.

18.23. Definition. We write $\lim_{z \to w} f(z) = L$ if for every $\varepsilon > 0$ there exists $\delta > 0$ such that $0 < |z - w| < \delta$ implies $|f(z) - L| < \varepsilon$. When $\langle z \rangle$ is a sequence of nonzero complex numbers, we write $z_n \to \infty$ if $|z_n|^{-1} \to 0$. We also write "$f(z) \to \infty$ as $z \to \infty$" if for every $\varepsilon > 0$, there exists $\delta > 0$ such that $|z| > 1/\delta$ implies $|f(z)| > 1/\varepsilon$.

A *nonconstant complex polynomial* is a function $p : \mathbb{C} \to \mathbb{C}$ defined by $p(z) = \Sigma_{j=0}^{k} a_j z^j$, where the coefficients $a_0, \ldots, a_k$ are complex numbers, $k \geq 1$, and $a_k \neq 0$. We call k the *degree* of p. If $p(z) = 0$, then z is a *zero* of p.

18.24. Lemma. If p is a nonconstant complex polynomial, then $p(z) \to \infty$ as $z \to \infty$.

Proof. A polynomial of degree k has at most k zeros, since a polynomial with zero α can be expressed as $(z - \alpha)$ times a polynomial of lower degree. (The proof of Theorem 4.15 remains valid in this setting.) Since the set of zeros is finite, we can choose M such that $p(z) \neq 0$ for $|z| > M$.

Let $w = 1/z$. It suffices to show that $\lim_{w \to 0} 1/[p(w^{-1})] = 0$. For $|w| < 1/M$, the function is well-defined. We compute

$$\frac{1}{p(w^{-1})} = \frac{1}{\Sigma_{j=0}^{k} a_j w^{-j}} = \frac{w^k}{\Sigma_{j=0}^{k} a_j w^{k-j}} = \frac{w^k}{\Sigma_{j=0}^{k} a_{k-j} w^j} = \frac{w^k}{a_k + \Sigma_{j=1}^{k} a_{k-j} w^j}$$

As $w \to 0$, the numerator approaches 0 and the denominator approaches $a_k \neq 0$, so $1/[p(w^{-1})] \to 0$. ∎

To prove the Fundamental Theorem of Algebra, we will find a z with $|p(z)|$ as small as possible. Compactness guarantees a minimum for $|p|$ on any closed ball, and Lemma 18.24 implies that $|p|$ is large outside a closed ball with large radius. These ideas enables us to reduce the problem of finding the zeros of an arbitrary polynomial to taking the rth root of a complex number.

18.25. Theorem. (Fundamental Theorem of Algebra) Every nonconstant complex polynomial has a zero in $\mathbb{C}$.

Proof. Suppose p is a nonconstant complex polynomial. By Lemma 18.24, there is an M such that $|z| > M$ implies $|p(z)| > |p(0)|$. Let $S_M = \{z \in \mathbb{C}: |z| \leq M\}$. Since S_M is closed and bounded, it is compact (Theorem 18.19). Since p and the absolute value function are continuous, their composition $|p|$ is also continuous. By Theorem 18.22, $|p|$ achieves its minimum on S_M. Thus there is a z' with $|p(z')| \leq |p(z)|$ for $|z| \leq M$ and $|p(z')| \leq |p(0)| < |p(z)|$ for $|z| > M$. Thus $|p(z')|$ is the minimum value for $|p|$ on all of $\mathbb{C}$.

We now use proof by contradiction to prove that $p(z') = 0$. If $p(z') \neq 0$, then we define

$$h(w) = \frac{p(z' + w)}{p(z')} .$$

Observe that $|h(w)| \geq 1$ for all $w \in \mathbb{C}$. Also, since z' is a constant, h is a polynomial in w with $h(0) = 1$. Thus $h(w) = 1 + \Sigma_{j=1}^{n} d_j w^k$. Let r be the smallest index for which $d_r \neq 0$.

We can now write $h(w) = 1 + d_r w^r + g(w)$, where $\lim_{w \to 0} \frac{g(w)}{w^r} = 0$. By the definition of limit, there is a $\delta > 0$ such that $0 < |w| < \delta$ implies $\frac{|g(w)|}{|w|^r} < \frac{1}{2}|d_r|$.

Choose a positive number α such that $\alpha < \delta^r |d_r|$. By Lemma 18.13, we can find $\zeta \in \mathbb{C}$ satisfying $d_r \zeta^r = -\alpha$. We have $|\zeta| < \delta$, since $|\zeta|^r = |\frac{\alpha}{d_r}| < \frac{\delta^r |d_r|}{|d_r|} = \delta^r$. From $|\zeta| < \delta$, we have $|g(\zeta)| \leq \frac{1}{2}|d_r||\zeta|^r = \frac{1}{2}\alpha$. From $d_r \zeta^r = -\alpha$, we have $h(\zeta) = 1 - \alpha + g(\zeta)$. By the triangle inequality,

$$|h(\zeta)| \leq 1 - \alpha + |g(\zeta)| < 1 - \alpha + \frac{1}{2}\alpha < 1.$$

This contradicts the observation that $|h(w)| \geq 1$ for all $w \in \mathbb{C}$. We conclude that $p(z') = 0$. ∎

18.26. Corollary. Every nonconstant complex polynomial of degree n can be expressed as a product of linear factors,
$$p(z) = c\Pi_{j=1}^{n}(z - \alpha_i),$$
where c is a nonzero constant and each α_i is a zero of p in $\mathbb{C}$. ∎

One theme throughout this book has been the solution of equations. When we considered an equation within a given number system that had no solution, we introduced a larger number system in which it had solutions. We began with the natural numbers $\mathbb{N}$. We introduced the integers $\mathbb{Z}$ to solve $x + n = 0$, the rational numbers $\mathbb{Q}$ to solve $mx + n = 0$, the real numbers $\mathbb{R}$ to solve $x^2 - 2 = 0$, and finally the complex numbers $\mathbb{C}$ to solve $x^2 + 1 = 0$. In $\mathbb{C}$ we can find the zeros of all polynomials. For this reason, the field of complex numbers is called the *algebraic closure* of $\mathbb{R}$.

EXERCISES

18.1. (−) Prove that $\mathbb{C}$ is a group under addition, with identity $(0, 0)$.

18.2. (−) *Multiplication of complex numbers.*
a) Prove that $(1, 0)$ is an identity for multiplication.
b) Prove that if $a^2 + b^2 \neq 0$, then $(\frac{a}{a^2+b^2}, \frac{-b}{a^2+b^2}) \cdot (a, b) = (1, 0)$. (Comment: This proves that $z^{-1} = \bar{z}/|z|^2$.)
c) Prove that $\mathbb{C} - \{0\}$ is a group under multiplication.

18.3. (−) Prove that addition and multiplication of complex numbers are associative and commutative and satisfy the distributive law.

18.4. For complex numbers z, w, prove that $|zw| = |z|\,|w|$ and that $|z + w|^2 = |z|^2 + |w|^2 + 2\,\mathrm{Re}(z\bar{w})$.

18.5. Suppose w_1 and w_2 are distinct points in $\mathbb{C}$. Give a geometric description of the set $\{z: |z - w_1| = |z - w_2|\}$.

18.6. Prove the following properties of complex conjugation for all $z, w \in \mathbb{C}$:
a) $\overline{zw} = \bar{z}\,\bar{w}$.   b) $\overline{z+w} = \bar{z} + \bar{w}$.   c) $|\bar{z}| = |z|$.

18.7. (−) Suppose $z = (x, y) \in \mathbb{C}$. Prove that $x = (z + \bar{z})/2$ and $y = (z - \bar{z})/2i$.

18.8. (−) Express the complex cube roots of 1 both in the form $re^{i\theta}$ and in the form $x + iy$.

18.9. Prove that $z_n \to A$ if and only if $\mathrm{Re}(z_n) \to \mathrm{Re}(A)$ and $\mathrm{Im}(z_n) \to \mathrm{Im}(A)$. Use this to prove that a sequence of complex numbers converges if and only if it is a Cauchy sequence.

18.10. (!) *Trigonometry and the exponential function.*
a) Prove that $e^z e^w = e^{z+w}$ for all $z, w \in \mathbb{Z}$. (Hint: compare corresponding terms in the series).

b) Use part (a) and the formula $e^{i\theta} = \cos\theta + i\sin\theta$ to prove that $\cos(n\theta)$ and $\sin(n\theta)$ are polynomials in the variables $\cos\theta$ and $\sin\theta$.

c) Prove the trigonometric identity $\cos(3\theta) = 4\cos^3\theta - 3\cos\theta$. Obtain a similar formula for $\sin(3\theta)$.

18.11. Prove that conjugation is a continuous function. Apply the continuity of conjugation to prove that $\exp(\bar{z}) = \overline{\exp(z)}$.

18.12. Use Exercises 18.10 and 18.11 to conclude that $|e^{i\theta}| = 1$ for all $\theta \in \mathbb{R}$ and that $e^{2\pi i} = 1$.

18.13. Suppose z is a nonzero complex number and m is a positive integer. Prove that $w^m = z$ has the m distinct solutions $w = |z|^{1/m} e^{i(\theta + 2k\pi)/m}$ for $0 \le k \le m - 1$. Plot these solutions for the case $m = 8$ and $z = 256i$.

18.14. Define $f: \mathbb{C} \to \mathbb{C}$ by $f(z) = iz$. Describe the functional digraph of f.

18.15. For $z \in \mathbb{C}$ and $z \ne 1$, prove that $\sum_{k=0}^{n-1} z^k = (1 - z^n)/(1 - z)$. Give a geometric interpretation of this result when z is an nth root of 1.

18.16. (!) Suppose $z^n = 1$. Obtain a simple formula for $\prod_{k=0}^{n-1} z^k$.

18.17. Prove that the set of nth roots of 1, under multiplication, form a group "isomorphic" to $\mathbb{Z}_n$.

18.18. Use the characteristic equation method (Theorem 12.24) to solve the recurrence $a_n = -a_{n-2}$ for $a_0 = 2$ and $a_1 = 4$, giving a single formula for a_n.

18.19. Fill in the details of the proof in Theorem 18.17 that every closed rectangle in $\mathbb{C}$ is compact.

18.20. Given $w \in \mathbb{C}$ and $r \in \mathbb{R}$, prove that the set $\{z \in \mathbb{C}: |z - w| \le r\}$ is closed.

18.21. Suppose $f: \mathbb{C} \to \mathbb{C}$. Prove that the following statements are equivalent:
  A) f is continuous.
  B) for every open set T, $I_f(T)$ is open.
  C) for every closed set T, $I_f(T)$ is closed.
  D) for each sequence $\langle z \rangle$, $z_n \to w$ implies $f(z_n) \to f(w)$.

18.22. Suppose f is a real-valued function that is defined and continuous on a compact subset S of $\mathbb{C}$. Prove that S contains an element at which f achieves its minimum on S. (Comment: this completes the proof of Theorem 18.22.)

18.23. Prove that on $\mathbb{C}$, every polynomial is continuous, the absolute value function is continuous, and the composition of continuous functions is continuous. Conclude that $|p|$ is continuous when p is a complex polynomial.

18.24. (Root test) Let $\sum_{n=0}^{\infty} a_n z^n$ be a power series. Let $L = \limsup |a_n|^{1/n}$. Prove that $\sum_{n=0}^{\infty} a_n z^n$ converges absolutely if $|z| < 1/L$ and diverges if $|z| > 1/L$.

18.25. Define $f: \mathbb{R} \to \mathbb{C}$ by $f(t) = (1 + it)^2/(1 + t^2)$. Prove that the image of f is the unit circle minus the point -1, and that $\lim_{t \to \pm\infty} f(t) = -1$. Describe the trigonometric relationship between t and θ if θ satisfies $f(t) = e^{i\theta}$. How does this problem relate to Pythagorean triples?

Appendix A

From $\mathbb{N}$ to $\mathbb{R}$

It is possible to base mathematics on elementary set theory, but it seems more satisfying to begin with the natural numbers. From the natural numbers $\mathbb{N}$ we construct first the integers $\mathbb{Z}$, then the rational numbers $\mathbb{Q}$, and finally the real numbers $\mathbb{R}$. At each step, we must also define the arithmetic operations and verify that the desired arithmetic properties hold. The concepts of sets, functions, and relations (particularly Cartesian products, composition, and equivalence relations) are available to us, as they do not depend on numbers.

Each time we extend a system to a larger system, we build an equivalence relation and define the elements of the larger system to be the equivalence classes of the relation. We also want the smaller system to appear within the larger system in a natural way. The treatment of the arithmetic operations and their properties is different for $\mathbb{N}$. Here we rely on induction; for this reason we include the Well-Ordering Property of $\mathbb{N}$ among our assumptions. When we study $\mathbb{N}$, the only operations we construct are addition and multiplication; we verify associativity, commutativity, the distributive law, and several other elementary properties.

To construct $\mathbb{Z}$, we consider pairs of natural numbers. We define a relation $\sim$ on $\mathbb{N} \times \mathbb{N}$ by $(a, b) \sim (c, d)$ if $a + d = b + c$. Each resulting equivalence class consists of pairs having the same "difference". We define arithmetic operations on these equivalence classes and prove that they behave as we expect integer arithmetic to behave.

We can add, subtract and multiply within $\mathbb{Z}$. We construct the rational numbers to permit division also, except by 0. We consider pairs of integers, specifically the Cartesian product $\mathbb{Z} \times (\mathbb{Z} - \{0\})$. As in Chapter 8, we define $\sim$ on $\mathbb{Z} \times (\mathbb{Z} - \{0\})$ by $(a, b) \sim (c, d)$ if $ad = bc$. Each resulting equivalence class consists of pairs having the same "ratio". In Chapter 8, we defined arithmetic operations on these equivalence classes and discussed how to prove that they behave as we expect rational arithmetic to behave. In particular, having assumed the arithmetic properties of $\mathbb{Z}$, we concluded that $\mathbb{Q}$ is an ordered field. We will not repeat this part of the construction.

We have observed that $\mathbb{Q}$ does not contain quantities such as $\sqrt{2}$, which we believe exist. To remedy this defect we "complete" $\mathbb{Q}$ into $\mathbb{R}$ by introducing limiting processes. To do so, we consider the set S of Cauchy sequences of elements of $\mathbb{Q}$. We define a relation ~ on S by $\langle a \rangle \sim \langle b \rangle$ if $\langle a \rangle - \langle b \rangle$ converges to 0. We prove that this is an equivalence relation and define $\mathbb{R}$ to be the set of equivalence classes. We define arithmetic on elements of $\mathbb{R}$ using arithmetic of sequences. We prove that the resulting structure satisfies all the properties of a complete ordered field.

THE NATURAL NUMBERS

We assume the natural numbers with their familiar order relation "<" and the Well-Ordering Property. Given the relation < , we also have the relations > , ≤ , ≥ arising from it. We assume the following axioms about these concepts.

A.1. Axioms.
1) (Trichotomy) For $m, n \in \mathbb{N}$, exactly one of $\{n = m, n < m, m < n\}$ holds.
2) (Transitivity) For $l, m, n \in \mathbb{N}$, $l < m$ and $m < n$ imply $l < n$.
3) (Well-Ordering) Every non-empty subset of $\mathbb{N}$ has a least element.

Using the Well-Ordering Property, we define the successor function $\sigma: \mathbb{N} \to \mathbb{N}$. The Well-Ordering Property guarantees that $\mathbb{N}$ itself has a least element, which we call 1. For $n \in \mathbb{N}$, we define $\sigma(n)$ to be the least element in $\{k: n < k\}$. We assume as an axiom that the image of σ is $\mathbb{N} - \{1\}$.

A.2. *Properties of the successor function* σ.
1) (Axiom) The image of σ is $\mathbb{N} - \{1\}$.
2) σ is injective.
3) For all $n \in \mathbb{N}$, $n < \sigma(n)$.
Here (1) is an axiom, (2) follows from the definition of σ and the Well-Ordering Property, and (3) follows from the definition of σ. ∎

By statements A.1(1) and A.1(2), σ is a bijection from $\mathbb{N}$ to $\mathbb{N} - \{1\}$. Together with A.1(3), this implies that every natural number occurs when we start with 1 and successively apply σ. In other words, $\mathbb{N} = \{1, \sigma(1), \sigma(\sigma(1)), \cdots\}$. We give the natural numbers the usual names $1, 2, 3, \cdots$ corresponding to this order.

As in Theorem 4.3, the Well-Ordering Property implies the Principle of Mathematical Induction. Induction allows us to define addition and multiplication and to prove their fundamental properties. The successor function σ defines the operation of "adding one". Given any natural number n, we define "$n+1$" to be (another name for) the natural number $\sigma(n)$. We define addition of other natural numbers using iteration of the successor function.

A.3. Definition. "Adding k" is a function $a_k \colon \mathbb{N} \to \mathbb{N}$. For $k = 1$, a_k is defined by $a_k(n) = \sigma(n)$. Given a_k, the function a_{k+1} is defined by $a_{k+1}(n) = \sigma(a_k(n))$. *Addition* is a binary operator on $\mathbb{N}$. For $n, k \in \mathbb{N}$, the *sum*, written $n + k$, is the natural number $a_k(n)$.

Because every natural number is reached by applying σ to 1, this defines a_k for each $k \in \mathbb{N}$. In essence, we have defined "adding $k+1$" to mean adding k and then adding 1.

A.4. Proposition. Addition of natural numbers is associative and commutative.

Proof. For associativity, we prove that $a + (b + c) = (a + b) + c$ by induction on c. For $c = 1$, the statement is $a + (b + 1) = (a + b) + 1$; this is the definition of adding $b + 1$. For the inductive step, suppose the claim holds for $c = k$. Using the induction hypothesis for the third equality and definition of addition for the others, we have $a + [b + (k + 1)] = a + [(b + k) + 1] = [a + (b + k)] + 1 = [(a + b) + k] + 1 = (a + b) + (k + 1)$.

We prove commutativity in two steps. First, we prove by induction on n that $n + 1 = 1 + n$. For $n = 1$, we have the same expression on both sides. For the inductive step, suppose $k + 1 = 1 + k$. Then $(k + 1) + 1 = (1 + k) + 1 = 1 + (k + 1)$, using the induction hypothesis and associativity. Now we consider a fixed $n \in \mathbb{N}$ and prove that $n + m = m + n$ by induction on m. We have already proved the basis step. For the inductive step, suppose $n + k = k + n$. Then $n + (k + 1) = (n + k) + 1 = (k + n) + 1 = k + (n + 1) = k + (1 + n) = (k + 1) + n$. $\blacksquare$

A.5. Definition. "Multiplying by k" is a function $m_k \colon \mathbb{N} \to \mathbb{N}$. For $k = 1$, m_k is defined by $m_k(n) = n$ (the identity function). Given the function m_k, the function m_{k+1} is defined by $m_{k+1}(n) = m_k(n) + n$. *Multiplication* is a binary operator on $\mathbb{N}$. For $n, k \in \mathbb{N}$, the *product*, written $k \cdot n$ or kn, is defined to be the natural number $m_k(n)$.

"Multiplying by k" means adding up k copies of n. The validity of this definition rests on induction (equivalently, on the Well-Ordering Property). This is why we said in Remark 4.8 that Gauss's argument to evaluate $\Sigma_{i=1}^{n} i$ involved induction. By convention, multiplicative operations always take precedence over additive operations when both appear in an expression without parentheses: $ab + c$ means $(ab) + c$.

A.6. Proposition. The *distributive law* $a(b + c) = ab + ac$ holds in $\mathbb{N}$.

Proof. We use induction on a. For $a = 1$, we have $1(y + z) = y + z = 1y + 1z$. Assuming the statement when $a = n$, we use the definition of multiplication by $n + 1$, the induction hypothesis, and the properties of addition to compute $(n + 1)(y + z) = n(y + z) + (y + z) = (ny + nz) + (y + z) = (ny + y) + (nz + z) = (n + 1)y + (n + 1)z$. ∎

A.7. Proposition. Multiplication of natural numbers is associative and commutative.

Proof. First, we prove by induction on n that $n \cdot 1 = 1 \cdot n$. For $n = 1$, we have the same expression on both sides. Assuming the statement when $n = k$, we use also the definition of multiplication (by 1 and by k) to compute $(k + 1) \cdot 1 = k \cdot 1 + 1 = 1 \cdot k + 1 = k + 1 = 1 \cdot (k + 1)$. Next, we prove by induction on m that $nm = mn$. We have proved the statement for $m = 1$. Assume that it also holds for $m = k$. Using these, the distributive law, and the definition of multiplication, we compute $n(k + 1) = nk + n1 = kn + 1n = kn + n = (k + 1)n$.

From commutativity, we obtain the alternative form of the distributive law: $(a + b)c = c(a + b) = ca + cb = ac + bc$.

For associativity, we prove that $a(bc) = (ab)c$ by induction on a. For $a = 1$, we have $1(bc) = bc = (1b)c$. If the claim holds for $a = k$, we have $(k + 1)(bc) = k(bc) + bc = (kb)c + bc = (kb + b)c = ((k + 1)b)c$. ∎

The properties of functions yield the cancellation properties of equalities in natural numbers.

A.8. Proposition. For every $k \in \mathbb{N}$, addition of k and multiplication by k are injective functions from $\mathbb{N}$ to $\mathbb{N}$. Furthermore, if $a, b, c \in \mathbb{N}$, then $a + c = b + c$ implies $a = b$, and $ca = cb$ implies $a = b$.

Proof. Because the composition of injective functions is injective (Proposition 3.43), the first statement follows inductively from the injectivity of σ and the identity function. The second statement follows immediately from the first. ∎

We close our discussion of the natural numbers by proving that the size of a finite set is a well-defined notion. Definition 5.10 states that a nonempty set S is finite if for some natural number n there is a bijection $f: S \to \{1, \ldots, n\}$. Proposition A.9 (stated also as Proposition 5.11), justifies the statement that such a set has precisely n elements.

A.9. Proposition. If there is a bijection $f: [m] \to [n]$, then $m = n$.

Proof. Let $P(n)$ be the statement claimed. We use induction on n to prove $P(n)$ for all $n \in \mathbb{N} \cup \{0\}$. For the basis step $n = 0$, observe that there is no function from a non-empty set into the empty set, because no elements are available as images; hence m must be 0.

For the induction step, suppose $n \in \mathbb{N}$; we prove $P(n-1) \to P(n)$. Since the inverse of a bijection is a bijection, we may assume by the basis step that $m > 0$. Suppose $f: [m] \to [n]$ is a bijection, and let $x = f^{-1}(n)$. By discarding the element x from the domain, we obtain a bijection $f': [m] - \{x\} \to [n-1]$, defined by $f'(i) = f(i)$ for all $i \in [m] - \{a\}$. Define $h: [m-1] \to [m] - \{x\}$ by $h(i) = i$ if $i < x$ and $h(i) = i + 1$ if $i \geq x$. Since h is a bijection, the composition $f' \circ h$ is a bijection from $[m-1]$ to $[n-1]$. By the induction hypothesis, $m - 1 = n - 1$. Hence $m = n$, which completes the proof of the implication $P(n-1) \to P(n)$. ∎

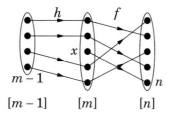

$$[m-1] \qquad [m] \qquad [n]$$

THE INTEGERS

Given the natural numbers, we can define the set $\mathbb{Z}$ of integers in several ways. We could define $\mathbb{Z}$ as a set of symbols. Let 0 be a symbol not in $\mathbb{N}$, and let $-\mathbb{N}$ denote the set of formal symbols $\{-m: m \in \mathbb{N}\}$. We can then define $\mathbb{Z} = \mathbb{N} \cup (-\mathbb{N}) \cup \{0\}$. This defines $\mathbb{Z}$ as a set, but it is difficult to verify that this set satisfies the usual properties of arithmetic.

Instead, we will define $\mathbb{Z}$ using an equivalence relation on $\mathbb{N} \times \mathbb{N}$. We define the relation $\sim$ on $\mathbb{N} \times \mathbb{N}$ by $(a, b) \sim (c, d)$ if $a + d = b + c$.

A.10. Proposition. The relation $\sim$ is an equivalence relation on $\mathbb{N} \times \mathbb{N}$.

Proof. We use arithmetic properties of $\mathbb{N}$. Because $a + d = d + a$, $\sim$ is reflexive. Because $a + d = b + c$ if and only if $c + b = d + a$, $\sim$ is symmetric. If $(a, b) \sim (c, d)$ and $(c, d) \sim (e, f)$, we have $a + d = b + c$ and $c + f = d + e$, which imply $a + f + c + d = b + e + c + d$ (using commutativity of addition on $\mathbb{N}$). By cancellation (Proposition A.8), we have $a + f = b + e$, and hence $\sim$ is transitive. ∎

We write $[(a, b)]$ for the equivalence class containing (a, b). We want negative numbers to represent differences, so we think geometrically of $(4, 0)$ as a "negative" distance from 4 to 0. Thus $\{(4, 0), (5, 1), (6, 2), \cdots\}$ is the class we wish to call "-4". This approach makes it is easy to extend the arithmetic properties of $\mathbb{N}$ to $\mathbb{Z}$. In Exercise A.1 we ask for a bijection between $\mathbb{N} \cup (-\mathbb{N}) \cup \{0\}$ and the set $\mathbb{Z}$ defined here.

A.11. Definition. Addition and multiplication on ℤ are defined by
$$[(a, b)] + [(c, d)] = [(a + c, b + d)].$$
$$[(a, b)] \cdot [(c, d)] = [(ad + bc, ac + bd)].$$

In these expressions, the arithmetic operations on the left are being defined; those on the right involve arithmetic in ℕ.

A.12. Theorem. (Arithmetic Properties of ℤ).
1) Addition and multiplication are well-defined.
2) Addition and multiplication have identity elements $[(n, n)]$ and $[(n, n + 1)]$, respectively.
3) Addition and multiplication are commutative and associative.
4) The distributive law holds.
5) Each element $[(a, b)]$ has an additive inverse $[(b, a)]$.

Proof. We include the proofs for the properties of addition. The proofs for multiplication are similar and are left as Exercise A.3.

(1) Suppose that $(a, b) \sim (a', b')$ and that $(c, d) \sim (c', d')$. To prove that addition is well-defined, we must show that $(a, b) + (c, d) \sim (a', b') + (c', d')$. This becomes $(a + c, b + d) \sim (a' + c', b' + d')$ using the definition of addition in ℤ. By the definition of $\sim$, we must check that $a + c + b' + d' = b + d + a' + c'$. We are given $a + b' = a' + b$ and $c + d' = c' + d$. Adding these equations and using the associativity and commutativity of addition in ℕ gives the result.

(2) To prove that $[(n, n)]$ is the additive identity, note first that it is independent of n. We must verify that $[(a, b)] + [(n, n)] = [(a, b)]$. We compute $[(a, b)] + [(n, n)] = [(a + n, b + n)]$. The associative and commutative properties of addition in ℕ imply that $(a + n) + b = (b + n) + a$, and hence $[(a + n, b + n)] = [(a, b)]$.

(3) For commutativity, we must prove that $[(a, b)] + [(c, d)] = [(c, d)] + [(a, b)]$. When we apply the definition on each side, this becomes $[(a + c, b + d)] = [(c + a, d + b)]$, which holds because addition in ℕ is commutative. The same method yields associativity of addition in ℤ from associativity of addition in ℕ.

(4) The definitions of addition and multiplication on ℤ lead to
$$[(a, b)] \cdot ([(c, d)] + [(e, f)]) = [(a, b)] \cdot [(c + e, d + f)]$$
$$= [(a(d + f) + b(c + e), a(c + e) + b(d + f))].$$
Also,
$$[(a, b)] \cdot [(c, d)] + [(a, b)] \cdot [(e, f)] = [(ad + bc, ac + bd)] + [(af + be, ae + bf)]$$
$$= [(ad + bc + af + be, ac + bd + ae + bf)].$$
By the properties of arithmetic in ℕ, the two results are equal.

(5) We compute $[(a, b)] + [(b, a)] = [(a + b, b + a)] = [(a + b, a + b)]$. Hence the sum of $[(a, b)]$ and $[(b, a)]$ is the additive identity. ∎

We define subtraction by $[(a, b)] - [(c, d)] = [(a, b)] + [(d, c)]$. We can treat $\mathbb{N}$ as a subset of $\mathbb{Z}$ by identifying the number n with the class $[(0, n)]$. Since $[(0, a)] + [(0, b)] = [(0, a + b)]$ and $[(0, a)] \cdot [(0, b)] = [(0, ab)]$. these operations mirror the corresponding operations on $\mathbb{N}$.

Given $n \in \mathbb{N}$, we henceforth write $-n$ for $[(n, 0)]$, 0 for $[(n, n)]$, and 1 for $[(n, n + 1)]$. By A.12(5), this defines $-n$ to be the additive inverse of n. We also use the minus sign for subtraction; it is natural to write $[(a, b)]$ as $b - a$. Now the arithmetic and order operations behave as we wish, and we have introduced subtraction and negative numbers.

A.13. Proposition. Suppose $a, b \in \mathbb{N}$. With $-b$, a, b defined in $\mathbb{Z}$ as above, we have $a - b = a + (-b)$ and $-(-b) = b$.

Proof. First, $a + (-b) = [(0, a)] + [(b, 0)] = [(b, a)] = a - b$. Next, $-(-b)$ is the additive inverse of the additive inverse of $[(0, b)]$, which is $[(0, b)]$. ∎

These remarks imply that $\{[(0, n)]: n \in \mathbb{N}\}$ is a positive set in $\mathbb{Z}$; membership in this set is closed under addition and multiplication, and if $x \neq 0$ exactly one of $x, -x$ belongs to this set.

THE RATIONAL NUMBERS

The construction of $\mathbb{Q}$ from $\mathbb{Z}$ is discussed in Chapter 8. Here we remark only on the similarity between the equivalence relation defined to construct $\mathbb{Z}$ from $\mathbb{N}$ and the equivalence relation defined to construct $\mathbb{Q}$ from $\mathbb{Z}$. In the first case, $(a, b) \sim (c, d)$ if $a + d = b + c$; in the second, $(a, b) \sim (c, d)$ if $ad = bc$. The reason for the similarity is that both relations are designed to introduce inverses.

As discussed in Chapter 8, the second relation above is an equivalence relation on $\mathbb{Z} \times (\mathbb{Z} - \{0\})$, we define $\mathbb{Q}$ to be the set of equivalence classes under this relation, and we define addition and multiplication operators on these classes that make $\mathbb{Q}$ into an ordered field.

THE REAL NUMBERS

We construct $\mathbb{R}$ from $\mathbb{Q}$ by what mathematicians call "completion". The idea is simply that we want the limits of Cauchy sequences to exist; if a sequence of rational numbers is Cauchy, but does not converge in $\mathbb{Q}$, then we decree that it has a limit in $\mathbb{R}$. Different sequences may approach the same value, so we will need to consider such pairs of sequences to be equivalent. Thus a real number will be an equivalence class of Cauchy sequences of rational numbers. We prove that the set of these equivalence classes forms a complete ordered field.

We begin by listing the things we must do. We must define objects $\alpha, \beta, \gamma, \ldots$ that will be the elements of the set $\mathbb{R}$. We must designate elements $\mathbf{0}$ and $\mathbf{1}$ and define operations of addition and multiplication on $\mathbb{R}$. We must prove that these operations satisfy the algebraic laws of Definition 1.15 (statements (5-7) below). We must specify the subset of $\mathbb{R}$ that is positive and prove that this satisfies the axioms for a positive set (Definition 1.16, statements (4,8) below). Finally, we must prove that our system satisfies the Completeness Property.

A.14. What must be done.
Step 1) Define $\mathbb{R}$.
Step 2) Define $\mathbf{0}$ and $\mathbf{1}$.
Step 3) Define addition and multiplication. Define for each α an additive inverse $-\alpha$ and for each $\beta \neq \mathbf{0}$ a multiplicative inverse β^{-1}.
Step 4) Define a subset P called the *positive set* and verify trichotomy.
Step 5) Verify the following laws for addition.
$$\alpha + \mathbf{0} = \alpha$$
$$\alpha + \beta = \beta + \alpha$$
$$(\alpha + \beta) + \gamma = \alpha + (\beta + \gamma)$$
$$\alpha + -\alpha = \mathbf{0}$$
Step 6) Verify the following laws for multiplication.
$$\alpha \mathbf{1} = \alpha$$
$$\alpha \beta = \beta \alpha$$
$$(\alpha \beta)\gamma = \alpha(\beta \gamma)$$
$$\alpha \alpha^{-1} = \mathbf{1} \text{ if } \alpha \neq \mathbf{0}$$
Step 7) Verify the distributive law $(\alpha + \beta)\gamma = \alpha\gamma + \beta\gamma$.
Step 8) Verify that addition and multiplication preserve order.
$$\alpha, \beta > \mathbf{0} \text{ implies both } \alpha + \beta > \mathbf{0} \text{ and } \alpha\beta > \mathbf{0}$$
Step 9) Prove that every non-empty subset $S \subset \mathbb{R}$ that is bounded above has a least upper bound. ∎

We assume that the rational numbers and their properties are known. Recall that $|x| = \max\{x, -x\}$. Note that reciprocals of large natural numbers are small positive rational numbers; these take the place of the real number ε in the definition of convergence.

We recall that a sequence of rational numbers is a function $a: \mathbb{N} \to \mathbb{Q}$; we write $\langle a \rangle$ to denote a sequence.

A.15. Definition. The sequence $\langle a \rangle$ is a *Cauchy sequence* if for every $k \in \mathbb{N}$ there is an N such that $n, m > N$ implies $|a_n - a_m| < 1/k$. The sequence $\langle a \rangle$ *converges* to $L \in \mathbb{Q}$ if for every $k \in \mathbb{N}$ there exists $N \in \mathbb{N}$ such that $n > N$ implies $|a_n - L| < 1/k$.

A.16. Lemma. If $a_n \to L$, then $\langle a \rangle$ is a Cauchy sequence.

Proof. In the proof of Proposition 14.4, replace ε with $1/k$. ∎

Let S denote the set of Cauchy sequences of rational numbers. A Cauchy sequence of rational numbers need not converge to a rational number (Exercise A.6). Nevertheless, S has many useful properties.

A.17. Proposition. The set S of Cauchy sequences of rational numbers is closed under addition, multiplication, and scalar multiplication:
  1) If $\langle a \rangle \in S$ and $\langle b \rangle \in S$, then $\langle a+b \rangle \in S$.
  2) If $\langle a \rangle \in S$ and $\langle b \rangle \in S$, then $\langle ab \rangle \in S$.
  3) If $\langle a \rangle \in S$ and $c \in \mathbb{Q}$, then $\langle ca \rangle \in S$.

Proof. (Exercise A.7). ∎

A.18. Lemma. If a Cauchy sequence $\langle a \rangle \in S$ has a convergent subsequence, then $\langle a \rangle$ also converges and has the same limit.

Proof. (Exercise A.8). ∎

We have defined convergence of sequences of rational numbers to a rational limit, in particular to zero. We will partition S into equivalence classes using a relation $\sim$. We write $\langle a \rangle \sim \langle b \rangle$ to mean that $\langle a-b \rangle$ converges to the rational number 0. We write **0** for the subset of S consisting of sequences converging to 0, and we record some of its properties:

A.19. Lemma. Suppose that $\langle a \rangle$, $\langle b \rangle$, $\langle c \rangle$ are Cauchy sequences.
  1) If $\langle a \rangle$ and $\langle b \rangle$ converge to 0, then $\langle a+b \rangle$ converges to 0.
  2) If $\langle a \rangle$ converges to 0, then $\langle ca \rangle$ converges to 0.

Proof. These proofs can be obtained from the proofs of Lemmas 14.6 and 14.8 by replacing ε with $1/k$, using also Lemma 14.5. ∎

Using the notation for **0**, these become the statements "$\langle a \rangle, \langle b \rangle \in \mathbf{0}$ imply $\langle a+b \rangle \in \mathbf{0}$" and "$\langle a \rangle \in \mathbf{0}$, $\langle c \rangle \in S$ imply $\langle ca \rangle \in \mathbf{0}$". In general in algebra, a *ring* is a set equipped with both addition and multiplication satisfying appropriate axioms. A subset of a ring that is closed under addition and under multiplication by elements of the full ring is an *ideal*. Thus **0** is an ideal in the ring S of Cauchy sequences of rational numbers.

A.20. Corollary. The relation $\sim$ is an equivalence relation on S.

Proof. The reflexive property is trivial; $\langle a \rangle \sim \langle a \rangle$ if and only if $\langle a-a \rangle$ converges to 0, but $\langle a-a \rangle$ is the zero sequence. The symmetric property follows because $\langle a-b \rangle$ converges to 0 if and only if $\langle b-a \rangle$ converges to 0. The transitive property follows from the lemma: $\langle a-c \rangle = \langle a-b \rangle + \langle b-c \rangle$, so $\langle a \rangle \sim \langle b \rangle$ and $\langle b \rangle \sim \langle c \rangle$; now the first statement of Proposition A.17 implies that $\langle a-c \rangle$ converges to 0. ∎

A.21. Remark. Our approach here echoes our approach to modular arithmetic. We defined the integers modulo p by considering equivalence classes. Two integers are equivalent (modulo p) when their difference is a multiple of p. The set of multiples of p is an ideal in $\mathbb{Z}$, just as **0** is an ideal in S. In both cases we consider the equivalence classes modulo the ideal. Just as two integers are congruent modulo p when their difference is a multiple of p, so two Cauchy sequences represent the same real number when their difference converges to 0. ∎

We define the real numbers $\mathbb{R}$ to be the set of equivalence classes of S under the relation ~. This completes Step 1.

The class **0** consists of the elements of S converging to zero; this will be the additive identity **0** in $\mathbb{R}$. We let **1** denote the set of sequences converging to the rational number 1; this will be the multiplicative identity. This completes Step 2.

We next define the positive real numbers.

A.22. Definition. The real number α is *positive* if, for each sequence $\langle a \rangle \in \alpha$, there exist $k, N \in \mathbb{N}$ such that $n > N$ implies $a_n > 1/k$. The real number α is *negative* if $-\alpha$ is positive, where $-\alpha = \{\langle -a \rangle : a \in \alpha\}$.

In other words, for every sequence belonging to a positive equivalence class, the numbers in that sequence are eventually bounded away from zero and positive. In order to prove that the trichotomy property holds, we must prove the following lemma.

A.23. Lemma. For any Cauchy sequence of rational numbers, precisely one of the following conditions holds:
  a) The terms are eventually positive and bounded away from zero.
  b) The terms are eventually negative and bounded away from zero.
  c) The sequence converges to zero.

Proof. Suppose $\langle a \rangle \in S$. Observe that no pair of the conditions can both hold for $\langle a \rangle$. We prove that if neither (a) nor (b) holds, then (c) must hold. If a sequence $\langle a \rangle$ is eventually bounded away from zero, meaning there exist $k, N \in \mathbb{N}$ such that $n > N$ implies $|a_n| > 1/k$, but for every $M \in \mathbb{N}$ there exist terms of both signs after a_M, then $\langle a \rangle$ cannot be a Cauchy sequence. Hence we may assume that $\langle a \rangle$ is not bounded away from zero. This means there is a subsequence of $\langle a \rangle$ that converges to zero. By Lemma A.18, if a Cauchy sequence $\langle a \rangle$ has a convergent subsequence, then $\langle a \rangle$ also converges and has the same limit. Hence (c) holds for $\langle a \rangle$. ∎

Suppose now that α is a real number; we emphasize that α is a set of sequences. Each sequence in α satisfies one of these three properties.

We claim that all sequences in α satisfy the same one. If $\langle a \rangle$ and $\langle b \rangle$ are elements of α, then $\langle a - b \rangle$ converges to 0. Therefore, if A.23a holds for $\langle a \rangle$, then it also holds for $\langle b \rangle$, because the terms of $\langle b \rangle$ are eventually arbitrarily close to those of $\langle a \rangle$. Similarly, if A.23b holds for $\langle a \rangle$, it also holds for $\langle b \rangle$. Finally, if A.23c holds for $\langle a \rangle$, then by the definition of the equivalence relation this also holds for $\langle b \rangle$.

These remarks prove that, for each real number α, all the sequences $\langle a \rangle \in \alpha$ satisfy the same property among {A.23a,A.23b,A.23c}. We define $\alpha > 0$, that is, α is *positive*, if A.23a holds for each $\langle a \rangle \in \alpha$. Similarly, $\alpha < 0$ if A.23b holds for each $\langle a \rangle \in \alpha$. If neither holds, then $\alpha = 0$. This proves the trichotomy property in Step 4.

Next we must give careful definitions of the algebraic operations. For the moment we write α, β for real numbers. We think of α as a set of sequences $\{\langle a \rangle\}$, and we write $\langle a \rangle$ for a representative of this equivalence class. Suppose that $\langle a \rangle$ is an element of S. We write $[\langle a \rangle]$ for the set of all elements equivalent to $\langle a \rangle$ (the class containing $\langle a \rangle$).

A.24. Definition. Suppose $\langle a \rangle, \langle b \rangle$ are sequences contained in the real numbers α, β, respectively. The *sum* and *product* of α and β are defined by

$$\alpha + \beta = [\langle a + b \rangle].$$
$$\alpha \cdot \beta = [\langle ab \rangle].$$

For the definitions to be valid, we must show that the results do not depend upon which elements we choose from the classes.

A.25. Lemma. Addition and multiplication of real numbers are well-defined.

Proof. Select two sequences $\langle a \rangle, \langle a' \rangle$ in α and two sequences $\langle b \rangle, \langle b' \rangle$ in β. By definition, we know that $\langle a - a' \rangle$ and $\langle b - b' \rangle$ converge to zero. By Proposition A.17, $\langle a + b \rangle$ and $\langle a' + b' \rangle$ are Cauchy sequences. We must prove that $\langle a + b \rangle \sim \langle a' + b' \rangle$, which requires $\langle a + b - (a' + b') \rangle$ to converge to zero. We can write $\langle a + b - (a' + b') \rangle$ as $\langle (a - a') + (b - b') \rangle$ and apply Lemma A.19(1) (the sum of sequences converging to zero also converges to zero). Thus addition is well-defined.

By Proposition A.17, $\langle ab \rangle$ and $\langle a'b' \rangle$ are Cauchy sequences. To prove that the definition is independent of the representatives, we must show that $\langle ab - a'b' \rangle \sim 0$. We add and subtract $a'b$ to write $\langle ab - a'b' \rangle$ as $\langle (a - a')b + a'(b - b') \rangle$. By Lemma A.19, each term converges to zero, and then so does their sum. Thus multiplication is well-defined. ∎

We have now defined zero, one, positive, negative, sum, and product. To define the additive inverse, we put $-\beta = [\langle -b \rangle]$, where $\langle b \rangle$ is any element of β. This definition is valid: if $\langle b \rangle, \langle b' \rangle \in \beta$, then $[\langle - b' \rangle] = [\langle - b \rangle]$, because $-\langle b \rangle - \langle - b' \rangle = \langle b' - b \rangle$ converges to 0.

To define the reciprocal of a nonzero real number, we need a preliminary observation. Let β be a nonzero real number. By Lemma A.23, we know that each sequence in β is eventually bounded away from zero. Therefore, omitting finitely many terms from some member of β yields a representative $\langle b \rangle$ all of whose terms are nonzero. For such a sequence $\langle b \rangle$, we define $\langle b^{-1} \rangle$ to be the sequence whose nth term is b_n^{-1}. Using such a representative $\langle b \rangle$ of β, we define $\beta^{-1} = [\langle b^{-1} \rangle] = \{c : \langle c \rangle \sim \langle b^{-1} \rangle\}$. Again the definition is valid.

We have now accomplished everything through Step 4. The laws in Steps 5,6,7,8 have similar proofs. It is possible to prove a general lemma about sequential continuity of functions of several variables and then derive these laws from it. Instead we prove two of them, leaving the rest to the exercises.

A.26. Lemma. Addition of real numbers is commutative.

Proof. Given real numbers α and β, we wish to prove that $\alpha + \beta = \beta + \alpha$. We select arbitrary representatives of each real number; we select distinct representatives for each occurrence of the number in the formula we wish to prove. Choose $\langle a \rangle, \langle a' \rangle \in \alpha$ and $\langle b \rangle, \langle b' \rangle \in \beta$. We must prove that $[\langle a \rangle + \langle b \rangle] = [\langle a' \rangle + \langle b' \rangle]$. This is equivalent to proving that $\langle a + b - (a' + b') \rangle$ converges to 0. From the distributive and commutative laws for $\mathbb{Q}$, we have $\langle a + b - (a' + b') \rangle = \langle (a - a') + (b - b') \rangle$. By definition, $\langle (a - a') \rangle$ and $\langle (b - b') \rangle$ converge to 0. By Lemma A.19, their sum also converges to 0. Hence $[\langle (a - a') + (b - b') \rangle] = \mathbf{0}$, and the result has been proved. ∎

A.27. Lemma. The distributive law holds for real numbers.

Proof. We wish to prove $(\alpha + \beta)\gamma = \alpha\gamma + \beta\gamma$, where $\alpha, \beta, \gamma \in \mathbb{R}$. We choose representatives $\langle a \rangle$, $\langle b \rangle$, $\langle c \rangle$ for the left side of the law, and representatives $\langle a' \rangle$, $\langle b' \rangle$, $\langle c' \rangle$, and $\langle c'' \rangle$ for the right side. We need to prove that $(\langle a + b \rangle)\langle c \rangle \sim \langle a'c' + b'c'' \rangle$. This is equivalent to proving that $\langle (a + b)c - a'c' - b'c'' \rangle$ converges to 0. As usual, we add and subtract until we get a sum of sequences that all tend to zero. Using the field axioms for $\mathbb{Q}$, we obtain

$$\langle (a + b)c - a'c' - b'c'' \rangle = \langle ac + bc - a'c' - b'c'' \rangle$$

$$= \langle (a - a')c + a'(c - c') + (b - b')c + b'(c - c'') \rangle.$$

The last expression is a sum of four products. Each product has a factor converging to 0 and hence itself converges to 0 by Lemma A.19(2). Now the sum of these products converges to 0 by Lemma A.19(1). This completes the proof. ∎

Steps 5-8 are completed in Exercises A.9-12.

We treat a rational number A as a real number α in the following way. Let $\alpha = [\langle a \rangle]$, where $\langle a \rangle$ is the constant sequence such that $a_n = A$ for all n. Thus it makes sense to write an inequality between a rational number and a real number. Furthermore the additive identities and multiplicative identities of $\mathbb{Q}$ and $\mathbb{R}$ correspond.

Finally, we prove the Completeness Property. Recall that a non-empty subset $S \subset \mathbb{R}$ is bounded above if there exists a real number β such that $\alpha \in S$ implies $\alpha \leq \beta$. Adding a positive number to an upper bound yields another upper bound; hence every set having an upper bound has a rational upper bound. Similarly, for every non-empty set there are rational numbers that are not upper bounds. We remark that the limit of a convergent sequence of upper bounds for S must be an upper bound for S (Exercise A.13).

A.28. Theorem. The set $\mathbb{R}$ of real numbers satisfies the Completeness Property.

Proof. Suppose S is a non-empty set of real numbers having an upper bound. Let a_1 be a rational number that is not an upper bound for S, and let b_1 be a rational number that is an upper bound for S. Note that $a_1 < b_1$. Put $c_1 = \frac{a_1 + b_1}{2}$, and observe that c_1 is the average of the two numbers. Hence $a_1 < c_1 < b_1$, and c_1 is rational.

We define $\langle a \rangle$, $\langle b \rangle$, $\langle c \rangle$ inductively. Given a_n, b_n, we define c_n to be the average of a_n and b_n. If c_n is not an upper bound for S, then we put $a_{n+1} = c_n$ and $b_{n+1} = b_n$. If c_n is an upper bound for S, then we put $b_{n+1} = c_n$ and $a_{n+1} = a_n$. This defines three sequences of rational numbers. By induction on n, we observe that $a_n < c_n < b_n$ holds for each n. We also check that $|b_n - a_n| = |b_1 - a_1|/2^{n-1}$. We claim that the sequences $\langle a \rangle$, $\langle b \rangle$, $\langle c \rangle$ are equivalent Cauchy sequences. Thus $[\langle a \rangle] = [\langle c \rangle] = [\langle b \rangle]$, and this equivalence class represents a real number.

To prove this claim, observe first that $a_n \leq a_m < b_m \leq b_n$ when $m > n$. Thus $|b_n - b_m| \leq |b_n - a_n| = |b_1 - a_1|/2^{n-1}$. Since $1/2^{n-1} \to 0$, we can guarantee that $|b_n - b_m| < \varepsilon$ by making n sufficiently large. Hence $\langle b \rangle$ is a Cauchy sequence. By similar reasoning, $\langle a \rangle$ is a Cauchy sequence. Since the values of $\langle c \rangle$ are even closer together than corresponding values of $\langle a \rangle$ or $\langle b \rangle$, also $\langle c \rangle$ is a Cauchy sequence. The sequences are equivalent because their differences converge to 0.

We claim that $[\langle c \rangle]$ is the least upper bound for S. By the remark in the previous paragraph, it is an upper bound. There cannot be a smaller upper bound; to prove this we proceed by contradiction. If x is a smaller upper bound, then $x - [\langle a \rangle]$ is negative. Choose the reciprocal of an integer smaller than $|x - [\langle a \rangle]|$. Definition A.22 guarantees that we can find an a_n such that $|a_n - [\langle a \rangle]| < |x - [\langle a \rangle]|$. This implies $a_n > x$, which contradicts the statement that a_n is not an upper bound. ∎

We have now constructed the real numbers and proved that they form a complete ordered field. There is only one complete ordered field, in the sense that we could label the elements of any complete ordered field **F** by the real numbers in such a way that **F** behaves just like ℝ. We sketch the proof, meaning that we describe the bijection and leave as exercises the details of verifying the behavior of the operations.

A.29. Theorem. ℝ is the only complete ordered field.

Proof. (sketch). Suppose **F** is an ordered field; to indicate the possibility that the elements of **F** differ from those of ℝ, we write the elements of **F** in bold type. We write **0** and **1** for the additive and multiplicative identity elements in **F**. We define a bijection $f: ℝ \to \mathbf{F}$ that preserves arithmetic and order. We define f in stages, first defining it on 0 and ℕ, then extending it to ℤ and ℚ before using the completeness axiom to extend it to ℝ.

Define $f(0) = \mathbf{0}$, $f(1) = \mathbf{1}$, and $f(n) = \mathbf{1} + \mathbf{1} + \cdots + \mathbf{1}$, meaning that $f(n)$ is the sum in **F** of n copies of **1**. Using the existence of additive and multiplicative inverses in **F**, we extend f to negative integers by defining $f(-n) = -f(n)$ and then to rational numbers by defining $f(m/n) = f(m)/f(n)$. The operations in boldface are operations in **F**. We leave to Exercise A.15 the verification that f preserves the order relation on ℚ.

Next we define f on the irrational numbers. Given $x \in ℝ$, let S denote the set of rational numbers less than x, and let $S' = \{f(y): y \in S\}$. Since f preserves the order relation on ℚ, the set S' is bounded above in **F** by the image of some rational number bigger than x. Since S' has an upper bound and **F** is complete, S' has a supremum in **F**. Let **x** be the supremum of S' in **F**; we define $f(x)$ to be **x**. We leave to Exercise A.16 the proof that f is a bijection and that f preserves addition, multiplication, and positivity on ℝ. Once these claims are verified, we see that **F** behaves exactly like ℝ, with the role of $x \in ℝ$ played by its boldface counterpart $f(x) = \mathbf{x}$ in **F**. ∎

EXERCISES

A.1. Establish a bijection between ℕ∪(−ℕ)∪{0} and the set of equivalence classes of ℕ×ℕ under the relation ∼ defined by putting $(a, b) \sim (c, d)$ if $a + d = b + c$.

A.2. Write an inductive definition of exponentiation by a natural number and prove that $x^{m+n} = x^m x^n$ when $x, m, n \in ℕ$.

A.3. Complete the proof of Theorem A.12 by verifying that multiplication in ℤ is well-defined, has identity element $[(n, n + 1)]$, and is commutative and associative.

A.4. Use induction and the definition of multiplication to prove that the product of two nonzero integers is nonzero. Use this and the distributive law to prove that multiplication by a nonzero integer is an injective function from $\mathbb{Z}$ to $\mathbb{Z}$.

A.5. Prove that multiplication by a natural number is an order-preserving function from $\mathbb{Z}$ to $\mathbb{Z}$ ($x > y$ implies $f(x) > f(y)$), and use this to prove that multiplication by a nonzero integer is an injective function from $\mathbb{Z}$ to $\mathbb{Z}$.

A.6. Define $\langle a \rangle$ by $a_1 = 2$ and $a_{n+1} = \frac{1}{2}(a_n + \frac{2}{a_n})$ for $n \in \mathbb{N}$. Prove that $\langle a \rangle$ is a Cauchy sequence of rational numbers. Prove that $\langle a \rangle$ has no limit in $\mathbb{Q}$. What does this say about Lemma A.16?

A.7. Prove that the set S of Cauchy sequences of rational numbers is closed under addition, multiplication, and scalar multiplication.

A.8. Prove that if a Cauchy sequence of rational numbers has a convergent subsequence, then $\langle a \rangle$ also converges and has the same limit.

A.9. Prove that multiplication of real numbers is commutative.

A.10. Prove that addition and multiplication of real numbers are associative.

A.11. Prove that 0 is an identity element for addition and that 1 is an identity element for multiplication of real numbers. Given $\alpha \in \mathbb{R}$ with $\alpha \neq 0$, prove that $\alpha + (-\alpha) = 0$ and that $\alpha \cdot \alpha^{-1} = 1$.

A.12. Prove that addition and multiplication of real numbers preserve the positive set.

A.13. Prove that the limit of any convergent sequence of upper bounds for S is an upper bound for S.

A.14. Suppose that $|a_{n+1} - a_n| \leq M/2^n$ for some constant $M > 0$. Prove that $\langle a \rangle$ is a Cauchy sequence. (Hint: estimate $|a_m - a_n|$ by using a telescoping sum, and use the convergence of $\Sigma_{k=0}^{\infty} 1/2^k$.)

A.15. Prove that the function f constructed in Theorem A.29 preserves the order relation on $\mathbb{Q}$.

A.16. Prove that the function f constructed in Theorem A.29 is a bijection and preserves addition, multiplication and positivity on $\mathbb{R}$.

Appendix B

Hints to Selected Exercises

In this appendix we provide some general guidelines and more specific suggestions for many of the exercises. This should help students who have trouble getting started in finding and writing proofs.

GENERAL DISCUSSION

The first step is making sure that one understands exactly what the problem is asking. Some problems request a verification of a mathematical statement. Definitions may provide a road map for what needs to be verified. Sometimes, the desired statement follows from a theorem already proved, and then one needs to verify that its hypotheses hold.

Other problems may involve some experimentation to discover the mathematical statement that needs to be proved. In the Handshake Problem (Solution 4.20), for example, one examines small cases to discover a general pattern and then proves it by induction. Keep in mind, though, that examples do not generally provide a proof.

When seeking a direct proof of a conditional statement, one can work from both ends. List statements that follow from the hypothesis. List statements that suffice to imply the conclusion. When some statement appears in both lists, the problem is solved.

When unsuccessful with the direct method, list what would be implied if the conclusion were not true. If something in this list contradicts something in the list of statements that follow from the hypothesis (or other known true statements), then again the problem is solved, using the method of contradiction.

Contradiction is particularly appropriate for statements of impossibility; Theorem 2.3 provides a typical example of this kind of proof. To prove that something exists, often one can construct an example and prove that it has the desired properties; this is the direct method. The proof of a universally quantified statement must be valid for every value of the variable in the given universe.

Induction often works to prove statements that involve a natural number as a parameter. Many proofs in algebra and discrete mathematics use extremality or the pigeonhole principle (consider the proofs of Theorem 7.26 and Examples 10.2-10.6). Another technique is to consider a smallest counterexample to a desired statement and then use its existence to obtain a smaller counterexample. This can be viewed as induction or contradiction or extremality (see Theorems 6.7 and 8.13).

When using the epsilon/delta definition of limit, it is useful to think of $|x - a|$ as the distance between x and a. Combining this with the right picture often makes the result obvious. The problem is then to convert this intuition into a precise explanation of why the hypothesis implies the conclusion. This may involve carefully using the definitions or citing appropriate theorems.

In mathematics, one often wants to prove that two things are the same. A common technique for showing that two numbers are equal is to show that their difference is zero. This appears in Corollary 4.19 and in several exercises in Part IV. In analysis, this technique often involves showing that their difference can be made arbitrarily small. This leads to the definition of a limit. To show that a limit exists or that it equals a specific number, one needs to show that some positive quantity is small. It suffices to show that something larger (but simpler) is small (Lemma 14.9 is a typical example).

We can also prove that $a = b$ by proving $a \leq b$ and $b \leq a$. This is closely related to proving equality of sets by proving set inclusion in both directions. This in turn is analogous to a bijection being both injective and surjective. The "two inequality" approach to equality arises also in analysis; we prove that $\sup A = \alpha$ by proving that α is an upper bound and that there is no smaller upper bound. We can also prove equalities by "counting two ways"; this technique amounts to geometric or combinatorial understanding of algebraic identities.

It may not be obvious what technique works in a particular problem. Several techniques may work; we gave many different proofs of Fermat's Little Theorem. The techniques from one area of mathematics may apply unexpectedly in another; this is part of the beauty of mathematics. Mathematicians find proofs by working hard; stubbornness is a virtue. One tries all imaginable techniques to solve a problem. Practice increases understanding and speed in finding proofs.

The final step is to produce a careful and complete exposition of the solution. Writing out a proof can reveal hidden subtleties or cases that have been overlooked. It can also expose thoughts that turn out to be irrelevant. Producing a well-written solution often involves repeated revision. The process of writing solutions helps develop a useful skill: the ability to express oneself concisely, clearly, and accurately.

SPECIFIC HINTS

1.9. Explain why there is no missing dollar!

1.10. Consider all the different ways to factor 36 into a product of three positive integers. The given story eliminates all but one possiblity.

1.11. Apply similar reasoning as in Exercise 1.10 for various possible ages of the mailman, until obtaining a case that fits the scenario.

1.12. Relate this to the Babylonian Problem.

1.19. When is the inequality satisfied?

1.20. When x, y have the same sign, multiplying by xy gives an equivalent inequality.

1.21. If $a \neq 0$, then the graph is a parabola. Place the parabola at different heights, facing both up and down, to see the possibilities.

1.24. How many of the factors must be negative?

1.28. What does the order axiom say about sums of positive elements?

2.2. Let x, y, z be the three ages. Rewrite each piece of information in terms of x, y, z and solve the resulting equations. Check your answer.

2.5. There is more than one interpretation.

2.9. Vowel implies Odd is the same as Not Odd implies Not Vowel.

2.10. In one case the same δ must work for every a; in the other case δ is allowed to depend on a.

2.13. The front child must use the information that the other children were unable to decide instantly. Therefore consider what the other children would see if the front child had two red hats, two black hats, and one of each.

2.16. In both cases count the number of squares of each color in the checkerboard with deleted squares. With T-shapes, an additional argument is needed.

2.23. "x is odd" means we can write $x = 2n + 1$ for some n; "$x^2 - 1$ is divisible by 8" means we can write $x^2 - 1 = 8m$ for some m. Note that $n(n + 1)$ is always even. For part (b), use the contrapositive.

2.24. Pay attention to the scope of the quantifiers.

2.26-31. Convert intuition obtained from Venn diagrams into careful language.

2.32. Is it possible to reach a configuration with an odd number of white tokens inside every circle?

3.2. Think about leap years.

3.3. Read the discussion about "well-defined".

3.4. Square both sides. See exercise 1.13.

3.8c. Think of the point $(p, q) \in \mathbb{R}^2$ as an arrow from $(0, 0)$ to (p, q). Then locate the tail of the arrow $f(x, y)$ at the point (x, y).

3.9. Find $f(\text{Sunday})$ and $f(\text{Monday})$.

3.12. Start with $f(x) = f(y)$ and use the inequality.

3.14. The constant term is irrelevant. Replace x with $Ax + B$ for an appropriate choice of A and B to eliminate the quadratic term.

3.16. Consider the sum and the difference of $f(x)$ and $f(-x)$.

3.19. Pay careful attention to negation of quantifiers.

3.20-21. The definitions of injection and surjection say what must be verified.

3.22. Mimic example 3.41.

3.23a. What happens if A has more elements than B?

3.24-26. The definitions of injection and surjection say what must be verified.

3.25. Suppose that $f(a) = f(b)$, and apply f again.

3.27. Consider the pictures above Remark 3.30 and after Definition 3.40.

3.29. Explicitly list all permutations of [4] and count those without fixed points. Devise a more efficient approach for [5].

3.31. Compute $f(f(f(x)))$ and set it equal to x. Be careful not to divide by zero, and be sure not to confuse fixed points with points in a 3-cycle.

4.1. Relate the expression to $4(1 + 2 + \cdots n)$.

4.2. Relate the expression to $4(0 + 1 + \cdots n)$.

4.5-7. In the algebraic computations for the induction step, factor out the desired factors as early as possible.

4.8. Use induction and be precise.

4.10. Use partial fractions to rewrite the fraction.

4.11. Try small values to guess the formula. Be sure to replace $2i - 1$ by $2(i + 1) - 1 = 2i + 1$ when replacing n by $n + 1$.

4.13-14. Try small values to guess the formula.

4.15. Relate this to a familiar sum.

4.16. Try weighing a different number of balls from each box.

4.18. Modify the geometric sum to obtain the desired formula.

4.19. Apply the geometric sum.

4.20. More generally, determine who wins when the goal is a multiple of 4.

4.21. Use induction; apply the known inequality for the first n factors to obtain the desired inequality for $n + 1$ factors.

4.23. Induction on k is quicker than induction on n.

4.26. Use induction on n.

4.29. Use induction on n and the inequality $1 < 5$.

4.35. Find r and s, and then use the Method of Undetermined Coefficients.

4.36. Use induction on k.

4.37. Try small cases until the pattern becomes clear. Two proof are needed, showing that good configurations can be completely removed and that bad configurations cannot.

4.38. A person who can say "Nov. 30" will win. Work backwards from the end, determining the winning dates. (Strong induction permits a formal proof.)

4.41. One way is to consider a system of linear equations for the coefficients.

4.42. Assuming calculus, take logarithmic derivatives of $\Pi(x - \alpha_i)$. Without calculus, try the cases $n = 1, 2, 3$ to guess the pattern, and prove it by induction.

5.2. Think about "carrying".

5.7. When can two elements be interchanged?

5.9. Use induction.

5.14. Imitate the proof that $N \times N$ is countable.

5.19. Assume without loss of generality that $n \le m < k$; cancel common factors.

5.20. Find the first n which can be used as a basis step of an inductive proof; consider smaller values separately.

5.21. Consider the difference of successive factorials.

5.22. Use induction or factorization. (Later we learn a better method).

5.23. What determines a rectangle within the grid?

5.25. Use the rule of product. In parts (a) and (b), pick the ranks and then the cards within the ranks.

5.26. Think of x as the captain of one of two teams; count two ways.

5.28. Consider lists of zeros and ones where not all terms are zero.

5.29. Consider pairs of integer points in the interval $[1, n]$.

5.32a. Focus on one element. Delete it from subsets that contain it and add it to subsets that don't.

5.33. Describe a procedure by which the entries in an element of B_n specify how to build a permutation in A_n by successively inserting the elements $1, 2, \ldots, n$.

5.34. For part (b), interchange rows and columns to define a bijection.

6.1-2. Let x be the number of coins of each type. Determine the amount of money as a function of x in each case. Use the notion of "relatively prime".

6.3-4. Mimic the procedure in 6.1.

6.10. Consider the prime factorization.

6.12. Consider the numbers modulo 3.

6.13. Use induction or case analysis.

6.15. Combine induction and case analysis.

6.16. Use induction or cancel common factors.

6.17. If S is the set of all primes, build a number that is not divisible by any member of S.

6.21. Let $x - 4$ be the original number of coconuts and follow the scenario.

6.23. Consider that $\Sigma_{i=1}^{2n} i = n(2n + 1)$.

6.24. Provide an explicit factorization, using the factorization of $x^{2r+1} - y^{2r+1}$.

6.30. Compare with example 6.12.

6.32. Clear fractions.

6.39. Try small examples to guess the pattern; then prove it by induction. The proof is easy when the pattern is expressed using divisibility considerations.

7.1-5. Rely on the definitions.

7.6. Consider leap years separately. One approach is to work mod 7, where Friday is congruent to 6 mod 7. Starting with the value of January 13 mod 7, consider the value of the 13th of each month. Explain how to avoid doing seven computations.

7.7. Perform computations modulo 8.

7.9. List the squares modulo 5.

7.10. Start with k congruent to 1 mod (k-1). The proof is then one line!

7.13. Write each number as a sum of powers of 10. What is the congruence class of 10^k modulo 9?

7.15. What is the congruence class of 10^k modulo 11?

7.16. Use $x^2 - y^2 = (x + y)(x - y)$. When can we divide by $(x + y)$?

7.20-22. Use modular arithmetic.

7.24-25. Use the Chinese remainder theorem.

7.28. Use Fermat's Little Theorem.

7.30. What is the congruence class of 2^{12} mod 13?

7.35. Multiply together all of $\{a, 2a, \ldots, (p-1)a\}$.

7.36. First observe that (-1) is a square modulo $m^2 + 1$.

8.1. Follow the definitions and prove that $an + bm$ and mn have a common factor if and only if m and n have a common factor.

8.3. Clear fractions, simplify, and consider the discussion before Solution 3.37. Alternatively, use inequalities: compare $1/(x + y)$ and $1/x + 1/y$ to $1/x$ according to the sign of y.

8.7. Consider parity.

8.8. Use a bijection between $\mathbb{N}$ and $\mathbb{N} \times \mathbb{N}$.

8.10. Consider the equation $x^k - n = 0$ and imitate Example 8.15.

8.13. The answer is not 1/2.

8.14. See the discussion in Example 9.15.

8.18. Make a substitution so that Theorem 8.21 applies.

9.5. Consider the dots on the diagonal.

9.6. Generalize the committee-with-chair argument in Lemma 9.8.

9.7c. What happens when each variable x_i is set to 1?

9.10. Consider selecting k people from m men and n women.

9.11. Consider $m + n + 1$ locations in a row. The right side represents the number of ways to fill $r + s + 1$ of them. Show that the terms in the sum count the same set, grouped by the position of the $r + 1$th selected location.

9.12. Combine selections-with-repetition with the idea in Exercise 9.10.

9.14. Consider a rectangular arrangement of points.

9.15. Use induction or expand a cleverly chosen product of n factors.

9.16-17. Count each class the same number of times.

9.18. Generalize from the case $k = 2$ done in the text.

9.19. This can be done by evaluating summations or by a more direct method.

9.21. Describe the portions of the sample space that correspond to the event that the length of the chord exceeds $\sqrt{3}$.

9.22. The switching argument of the Ballot Problem can be used.

9.23. For part (a), use induction on n. For part (b), establish a bijection between these arrangements (when $m = n + 1$) and ballot sequences.

9.24. Consider the height of the path at each horizontal step.

9.27. Use Bayes' Formula.

9.28. Given that A does not win on the first flip, what is the probability that A wins later?

9.29-30. The expected payoff of switching can be computed by using conditional probability or by using a more direct ad hoc argument.

9.31. Show that each sample point contributes the same amount to each side of the equation.

9.33. The total number of meals purchase is the sum, for i from 1 to n, of the number of meals purchase after receiving the $i - 1$st type of coupon until receiving the ith type of coupon.

9.34b. Use the linearity of expectation.

9.35. There are several ways to express the desired random variable as a sum of random variables that take only the values 0 and 1.

9.37. Since the monomials are equally likely, this does not involve multinomial coefficients. Monomials correspond to selections with repetition.

9.40. Build a generating function with one factor for each type of coin, where the desired value is the coefficient of x^{12}.

9.41. Expand $(x + 1)^{2n}$ and $(x + 1)^n$ by the binomial theorem. Square the second, and then compare the coefficients of x^n in the two cases.

9.43. Consider multiplication of formal power series.

9.44,46. Build a generating function using a factor for the contribution involving integers of each size.

10.2. How many committees can be formed from n professors?

10.5. Work modulo 10. If no two are congruent, then divide into classes by considering sums.

10.6. Consider the average sum of 3 consecutive numbers.

10.8. Partition the square into regions such that two points in the same region have the desired property.

10.9. Prove the contrapositive.

10.12. Use partial sums.

10.14. Consider the acquaintances of one student.

10.24-41. Generally speaking, define a universe and appropriate subsets $A_1, \ldots, A_m$ so that the desired set is the set of elements in the universe outside all of $A_1, \ldots, A_m$, and then apply the Inclusion-Exclusion formula.

11.2. Consider the vertex degrees in $G - e$.

11.10. A complete graph with n vertices has $\binom{n}{2}$ edges.

11.12. Define a graph to model the possible moves. What is the condition for reaching the desired configuration?

11.14. If P and Q have no common vertex, obtain a longer path.

11.17. Use induction on $n - k$, or use the properties of trees.

11.18. Use induction or apply Theorem 11.25 to appropriate subgraphs.

11.24. Use the degree-sum formula, proving the stronger result that there is a bipartite subgraph using at least half of the edges at each vertex.

11.27. Use induction or contradiction.

11.28. Describe the maximal matchings in C_n.

11.29. Count edges.

11.30. Prove that Hall's condition holds.

11.31. When there is no perfect matching, use Hall's Theorem to obtain an independent set with more than half the vertices.

11.38. The Petersen graph can be described as the graph whose vertices are the 2-element subsets of $\{1, 2, 3, 4, 5\}$, with two vertices adjacent if and only if the pairs are disjoint.

11.42. Prove the contrapositive.

11.43. Consider coloring the vertices in order, always using the least-indexed color not yet appearing on a neighbor.

11.44. Color the vertices in a wise order.

11.46. For the second part, use Theorem 11.50.

11.49. The minimum vertex degree is at most the average vertex degree.

11.51. Show that the boundary of the unbounded face has length at least n.

12.1. The answer is obvious. Prove it by induction.

12.4. Use the characteristic equation method.

12.6. The recurrence is not linear! Don't try to solve it.

12.7. Where does the last object go?

12.8. How many regions does the last circle add?

12.11. Partition the pairings according to the mate of one particular person.

12.16. How can the paths with n reflections be obtained from paths with fewer reflections?

12.20-21. Use the model of 1,2-lists summing to n.

12.27. Suppose α is the name of the largest of the k cards that appear at the top. Show that the number of flips until α appears at the top is the same as the number of flips in some pile in which at most $k - 2$ cards appear at the top.

12.33. To solve the recurrence when n is a power of 2, let $b_k = a_{2^k}$, for $k \geq 0$.

12.35. Where does the last object go?

12.40. Travel around the circle from a fixed point, using the pairing to define a list of 1's and 0's that will be a ballot list.

12.46. Treat each partition as an array of dots, with the number of dots in the ith row being the size of the ith largest part. Split these arrangements into two sets with the desired sizes.

12.48. For part (a), partition the spanning trees into appropriate subsets.

13.1. $\frac{1}{10} = 0\frac{1}{3} + 0\frac{1}{9} + 2\frac{1}{27} + \cdots$.

13.2. For a set S, consider the set $\{x : -x \in S\}$.

13.3. Both directions are required. For one implication, find for each $n \in \mathbb{N}$ an element of S within $1/n$ of α. For the other implication, use contradiction. Assume there is a smaller upper bound m and use $\alpha - m$ as ε in the definition of convergence. A picture may help clarify the discussion.

13.4. Caution: $\sup A$ need not be in A, and $\sup B$ need not be in B.

13.6. For the first part, verify the decreasing property directly. Boundedness below is easy.

13.7. Apply the Monotone Convergence Theorem.

13.12. Can this method ever list an irrational number?

13.14. Relate the binary expansion of a number in T to a subset of $\mathbb{N}$.

13.15. Use decimal or binary expansions.

13.16. Generalize the technique of Theorem 5.18.

14.1. $\frac{2}{7} = 0\frac{1}{2} + 1\frac{1}{4} + \cdots$ for the first part. For the second part, try squaring binary numbers directly and comparing with 2. For example, in binary $(1.0)^2 = 1_{(2)} < 2_{(10)}$, while $(1.1)^2 = 10.01_{(2)} > 2_{(10)}$.

14.2. This follows immediately by combining the definition of convergence with the definition of subsequence. Write it out!

14.3. To show that $\frac{1}{b_n}$ converges to $\frac{1}{M}$, write $|\frac{1}{b_n} - \frac{1}{M}| = |M - b_n|\frac{1}{|M||b_n|}$, and bound the denominator by an appropriate constant for sufficiently large n.

14.4. For (a) and (b), draw a picture and then use proof by contradiction.

14.7. Use induction.

14.8. Similar to example 15.10.

14.10. Graph both $x^2 - 4x + 6$ and x; translate x by 2 to simplify the situation.

14.11. Translate to reduce to the case $f(x) = x^2 + c$. Graph this parabola and the line $y = x$ on the same graph. Iterate the function.

14.14. Consider how long the fly travels! (Avoid summing a series.)

14.15. Use an $\varepsilon/2$ argument and the notion of partial sums.

14.19. Use the technique in the proof of Solution 14.24.

14.23. What would happen to Σx_j if the limit were not zero?

14.27. For (a), write $\frac{1}{n(n+1)}$ as $\frac{1}{n} - \frac{1}{n+1}$ to obtain a telescoping series. For (b), use $\frac{1}{(n+1)^2} < \frac{1}{n(n+1)} < \frac{1}{n^2}$.

14.32. Compare to $2\Sigma\frac{1}{n}$.

14.35. First add up enough positive terms to exceed L, then enough negative ones to be less than L, and continue this process.

14.41. Factor b_1 from a partial sum. Write $\frac{b_{k+1}}{b_1}$ as a product of $\frac{b_{j+1}}{b_j}$. Use the hypothesis to obtain bounds in terms of corresponding expressions for $\{a_j\}$.

14.43. First reduce to the case of a small positive rational number x. Then study what happens to the numerator after subtracting the largest reciprocal of a integer that is less than x.

15.2. Draw a picture like that in Example 15.5 with $a = .5$ and $f(x) = 1/x$.

15.4. Let $\varepsilon = 1/n$ in the definition of convergent sequence.

15.9. Apply the Intermediate Value Theorem to the function $f - g$.

15.10. Study the behavior for x near both positive and negative infinity, and use the Intermediate Value Theorem.

15.11. Start with $0 \le (ax - by)^2$ and choose a, b appropriately.

15.14. Consider separately the cases when $x \le 1$ and $x > 1$. For the second part, reduce to the first part.

15.20. First draw a picture of such a function.

16.2. Use an $\varepsilon/2$ argument and the definition of derivative.

16.4. "Derationalize" the difference of two cube roots.

16.8. Use the difference quotient definition; prove that the derivative is 1.

16.9. Use the difference quotient definition; prove that the limits must be zero.

16.13. Think of V and r as functions of time and use the chain rule.

16.14. Maximize the function defined by $f(x) = x - x^2$.

16.17. For part (a), apply the induction hypothesis for each possible value of m_k, then choose the best m_k. For part (b), think of a set counted by $\Sigma_{i<j} m_i m_j$; how does a small change in $\{m_i\}$ without changing Σm_i affect the size of this set?

16.19. Use $g(x) = x$.

16.20. Let $g_y(x) = f(x) - yx$. Prove that g_y has a minimum. What happens when $(g_y)'(x) = 0$?

16.21. Assume that f has two fixed points and use the Mean Value Theorem to obtain a contradiction.

16.23. Think about increases and decreases.

16.25. Can f' be positive somewhere without also being negative somewhere?

16.30. Use the chain rule and the product rule to compute higher derivatives.

16.35. If $t \in (a, b)$, then $(t, f(t))$ cannot be above the line segment connecting $(a, f(a))$ and $(b, f(b))$.

16.43. After finding the limit $f(x)$, compute $|f_n(x) - f(x)|$ and simplify. Can this be less than epsilon independently of x?

16.45. Use the second derivative test for convexity.

16.46. Start with $\Sigma x^n = 1/(1 - x)$ and differentiate twice. Then use $n^2 = n(n - 1) + n$.

16.47. Use the method of Exercise 16.46 to sum series representing the expected number of runs.

16.48. Differentiate both sides of the formula for the finite geometric series.

16.52b. Use calculus to find the maximum value of $f_n(x)$ as a function of n.

17.6. Show that the upper sums and lower sums must differ by one.

17.7. For (b), Use the idea in Exercise 17.6.

17.9. The formula for the difference between upper and lower sums simplifies!

17.10a. If $f(t) \neq 0$, then $f(x) \neq 0$ for x near t. From this obtain a partition for which the lower sum must be strictly positive.

17.11. Use the Fundamental Theorem of Calculus and the chain rule: find $g'(x)$.

17.15. Use the Fundamental Theorem of Calculus to get a differential equation.

17.17. Do what the problem says and no more!

17.20. Use the chain rule to differentiate the identity $(f^{-1}(f))(x) = x$.

17.24. For the second part, remember that ln is an increasing function.

17.26. One way is to set $h = f^g$ and take logs before differentiating.

17.28. Consider upper sums for $\int_1^N (1/x)dx$ for large N.

17.30. If f is not identically zero, show that it suffices to assume $\max|f| = 1$. Then consider separately regions where $|f|$ is near one and where it is not.

17.32. The proofs correspond to the proofs for series.

17.33. For part (a), use the Arithmetic Mean / Geometric Mean Inequality. For part (b), use the techniques of calculus to do the integrals. For part (c), substitute into (b).

17.35c. If calculus of two variables is available, square the integral and change variables appropriately.

18.4. Square both sides to make the first part easier.

18.5. Think in terms of distance.

18.7. Write $\bar{z} = x - iy$. Compute $z + \bar{z}$ and $z - \bar{z}$.

18.14. Iterate f a few times.

18.15. Use induction for the proof. Think of a complex number as a vector or force to do the second part.

18.16. Take the product by summing the exponents. Consider the odd and even cases separately.

18.20. Show that the complement is open by using the definition of open.

18.25. Compare the real and imaginary parts of $f(t)$ with the formulas in Theorem 8.18.

Appendix C

Suggestions for Further Reading

1. *The Second Scientific American Book of Mathematical Puzzles and Diversions*, by Martin Gardner, Simon and Schuster, New York, 1961.

2. *Number Theory*, by Andre Weil (with the collaboration of Maxwell Rosenlicht), Springer-Verlag, 1979.

3. *Number Theory: An Approach through History, from Hammurapi to Legendre*, by Andre Weil, Birkhauser, 1983.

4. *Galois Theory*, by Harold Edwards, Springer-Verlag, 1984.

5. *Introduction to Probability Theory*, by Paul Hoel, Sidney Port, and Charles Stone, Houghton-Mifflin, 1971.

6. *Aspects of Combinatorics*, by Victor Bryant, Cambridge, 1993.

7. *Applied Combinatorics* (third edition), by Alan Tucker, Wiley, 1995.

8. *Combinatorics: Topics, Techniques, and Algorithms*, by Peter Cameron, Cambridge, 1994.

9. *Introduction to Graph Theory*, by Douglas West, Prentice-Hall, 1996.

10. *Calculus*, by Michael Spivak, Publish or Perish Inc., 1980.

11. *Introduction to Analysis*, by Michael Schramm, Prentice-Hall, 1996

12. *Analysis: An Introduction to Proof*, by Steven Lay, Prentice-Hall, 1986.

13. *Introduction to Analysis*, by M. Rosenlicht, Scott-Foresman, 1968.

14. *Complex Variables*, N. Levinson and R. Redheffer, McGraw-Hill, 1970.

15. *Complex Variables*, by Stephen Fisher, Wadsworth-Brooks/Cole, 1990.

16. *The Emperor's New Mind*, by Roger Penrose, Penguin Books, 1989.

Many of our motivating problems appear in some form in books of recreational mathematics like [1]. Our next recommended book [2] is less than 70 pages of impeccable lecture notes. It provides an elegant development of elementary number theory, congruences, and groups. Weil's historical approach to number theory [3] considers many more topics at a higher level. Edwards's scholarly book [4] discusses the mathematics used in solving polynomial equations. It includes historical discussion and some of the original literature.

For readers who wish to study probability, [5] is a good choice. It includes both discrete density functions (combinatorial probability) and continuous density functions. It includes further developments of many topics from our book and provides a point of departure for studying statistics.

Many books in discrete mathematics treat both enumerative combinatorics and graph theory: [6] and [7] are examples that are particularly readable. At a somewhat higher level, [8] introduces a broader selection of discrete topics, including combinatorial designs, and [9] explores graph theory more extensively.

Many books cover elementary real analysis and the theory of calculus, at various levels. Perhaps [10] is the best calculus book ever written. It includes nearly everything in analysis that we have covered, plus all the computational techniques of calculus. It also contains a superb collection of exercises. Two books in elementary real analysis worth noting are [11] and [12]. Some analysis books, such as [13], also discuss the theory of functions of several variables. We make no specific recommendation among the many books on multivariable calculus; what the student should read depends upon previous courses, interest in applications, and other factors.

The theory of complex variables has found many applications in the sciences as well as in other branches of mathematics. The book [14] gives a beautiful and complete treatment of the theory of functions of one complex variable. It begins by defining complex numbers and can be understood by someone who has read our book. Another book that starts at the beginning is [15]; it contains many applications of complex variables to engineering and science.

Mathematicians and physicists write few books that catch the attention of the general public. Penrose's book [16] is a fascinating description of such issues as whether computers can think and feel emotions. It contains a wealth of mathematics and physics, but an educated person can read it easily. Penrose makes a convincing argument that complex numbers provide the best language for modern physics.

Appendix D

List of Notation

We list here the most common notation and conventions used in this book. Many mathematical symbols have different meanings in different contexts, as seen in some items here.

Relations, operators, and
   positional notation

$+, -, /, \cdot$ - arithmetic operations
$\sqrt{}$ - square root function
x^y - exponentiation
$<, \leq, >, \geq$ - numerical order relations
$=, \neq$ - equality, inequality
$\equiv (\bmod\ n)$ - congruence (modulo n)
$\neg, \wedge, \vee$ - logical connectives (not, and, or)
$\exists, \forall$ - existential and universal quantifiers
$\Leftrightarrow$ - logical equivalence
$\Rightarrow$ - implication
$\rightarrow$ - limit
∞ - infinity
$\in, \notin$ - membership, non-membership
$\varnothing$ - empty set
$\subseteq, \subset, \supseteq, \supset$ - relations of set containment
$\cup, \cap$ - set operations (union, intersection)
$\lceil\ \rceil$ - ceiling function
$\lfloor\ \rfloor$ - floor function
$|S|$ - size of a finite set S
$|x|$ - absolute value of x
S^c - complement of set S
$\bar{G}$ - complement of graph G
$\sim$ - typical equivalence relation
$\bar{a}$ - equivalence class of a
$\bar{z}$ - conjugate of complex number z
$\langle a \rangle$ - sequence
$\{x: P(x)\}$ - set description
$[k]$ - $\{1, \ldots, k\}$
k-set - set of size k

$[a, b]$ - closed interval $\{x \in \mathbb{R}: a \leq x \leq b\}$
(a, b) - open interval $\{x \in \mathbb{R}: a < x < b\}$
(a, b) - ordered pair
A^n - set of n-tuples with entries in A
a_n - nth term of a list a or sequence $\langle a \rangle$
$a | b$ - a divides b
$n!$ - n factorial
$f: A \rightarrow B$ - f is a function from A to B
$f \circ g$ - composition of functions
$\int_a^b$ - integral from a to b
$X \times Y$ - Cartesian product of sets
$A - B$ - difference of sets
$G - v, G - e$ - deletion of vertex or edge
   of graph
$\binom{n}{k}$ - binomial coefficient
$\binom{n}{k_1 \cdots k_t}$ - multinomial coefficient

Usage of Roman alphabet

$\langle a \rangle, \langle b \rangle$ - typical sequences
A, B - typical sets
$\mathbb{C}$ - the set of complex numbers
C_n - cycle with n vertices
C_n - Catalan number
cos - cosine function
df/dx - derivative of f
   with respect to x
$d(v)$ - degree of vertex x
D_n - number of derangements
e - exp(1), base of natural logarithms

e - typical edge in a graph
e - typical error function
$E(G)$ - edge set of graph G
$E(X)$ - expectation of
    random variable X
exp - exponential function
f, g, h - typical functions
F, G, H - typical graphs
F_n - Fibonacci number
f', f'' - first and second derivatives of f
gcd - greatest common divisor
I, J - typical intervals on real line
I_f - inverse image under f
i, j, k, l, m, n - typical integers
i - the complex number $\sqrt{-1}$
inf - infimum
K_n - complete graph with n vertices
$K_{m,n}$ - complete bipartite graph
lcm - least common multiple
ln - logarithm function
$L(f, P)$ - lower sum
L, M - typical limits or bounds
max - maximum
min - minimum
$\mathbb{N}$ - the set of natural numbers
$N(x), N(S)$ - neighborhood in a graph
$O(f)$ - functions dominated by
    a multiple of f
p, q - typical polynomials
P, Q, R - typical logical statements
P, Q, R - typical partitions of intervals
P_n - path with n vertices
$\mathbb{Q}$ - the set of rational numbers
Q_n - hypercube of dimension n (a graph)

$\mathbb{R}$ - the set of real numbers
R - typical relation
sup - supremum
sin - sine function
S, T - typical sets
T - typical tree
u, v, w - typical vertices
uv, xy - typical edges
U - typical universal set
$U(f, P)$ - upper sum
$V(G)$ - vertex set of graph G
$\mathbf{v}, \mathbf{e}, \mathbf{f}$ - count of vertices, edges, faces
    in plane graph
X, Y - typical sets
X, Y - typical random variables
x, y, z - typical real numbers
x, y, z - typical vertices
z, w - typical complex numbers
$\mathbb{Z}$ - the set of integers
$\mathbb{Z}_n$ - the set of congruence classes
    modulo n

Usage of Greek alphabet

α, β, γ - typical real numbers
$\Gamma(y)$ - gamma function at y
ε, δ - (small) positive numbers
θ - angle
$\chi(G)$ - chromatic number
$\chi(G; k)$ - chromatic polynomial
$\phi(n)$ - Euler totient
π - area of the unit circle
Π - product
σ, τ - typical permutations
Σ - summation

Index

A page number in italics designates a definition. A page number in bold designates material such as a major result or the main treatment of the concept; it may also include a definition.